Wearable Systems and Antennas Technologies for 5G, IOT and Medical Systems

T0136142

Wearable Systems and Antennas Technologies for 5G, IOT and Medical Systems

Edited By Albert Sabban

CRC Press
Taylor & Francis Group
Boca Raton London New York

CRC Press is an imprint of the
Taylor & Francis Group, an **informa** business

First edition published 2021
by CRC Press
6000 Broken Sound Parkway NW, Suite 300, Boca Raton, FL 33487-2742

and by CRC Press
2 Park Square, Milton Park, Abingdon, Oxon, OX14 4RN

Library of Congress Cataloging-in-Publication Data
Names: Saban, Avraham, 1951- editor.
Title: Wearable systems and antennas technologies for 5G, IOT and medical systems / edited by Albert Sabban.
Description: First edition. | Boca Raton : CRC Press, 2021. | Includes bibliographical references and index. | Summary: "This book covers wearable antennas, RF measurements techniques and measured results in the vicinity of the human body, setups, and design considerations. New topics and design methods are presented for the first time in the area of wearable antennas, metamaterial antennas and fractal antennas"-- Provided by publisher.
Identifiers: LCCN 2020026637 (print) | LCCN 2020026638 (ebook) | ISBN 9780367409135 (hardback) | ISBN 9780367409142 (ebook)
Subjects: LCSH: Wearable technology--Antennas. | Detectors.
Classification: LCC TK7871.6 .W38 2021 (print) | LCC TK7871.6 (ebook) | DDC 621.382/4--dc23
LC record available at https://lccn.loc.gov/2020026637
LC ebook record available at https://lccn.loc.gov/2020026638

ISBN: 978-0-367-40913-5 (hbk)
ISBN: 978-0-367-40914-2 (ebk)

Typeset in Times
by SPi Global, India

This book is dedicated to the memory of my father, mother and sister:
David Sabban, Dolly Sabban and Aliza Sabban.

Contents

Preface

Due to the progress in the development of communication systems, it is now possible to develop low cost wearable communication systems. Wearable antennas are antennas incorporated into clothing or worn close to the body; they are used for communication and medical purposes, which include tracking and navigation, mobile computing, and public health and safety. Examples include smartwatches (with integrated Bluetooth antennas), glasses (such as Google Glass with Wi-Fi and GPS antennas), GoPro action cameras (with Wi-Fi and Bluetooth antennas) and sensors to monitor patient health. They are increasingly common in consumer electronics and for healthcare and medical applications. However, the development of compact efficient wearable antennas is one of the major challenges in the development of wearable communication and medical systems. Technologies, such as printed compact antennas and miniaturization techniques, have been developed to create efficient small wearable antennas which are the main theme of this book. Each chapter in this book covers enough mathematical detail and provides sufficient explanations to enable electrical, electromagnetic and biomedical engineers, and students and scientists from all areas, to follow and understand the topics presented. New topics and design methods are presented for the first time in the area of wearable antennas, metamaterial antennas, fractal antennas and communication systems. The book covers wearable antennas development, textile antennas, reconfigurable antennas, active antennas, RF measurement techniques, measured results in the vicinity of the human body, setups and design considerations. The wearable antennas and devices presented in this book were analyzed by using 3D full-wave electromagnetic commercial software. New active and textile wearable antennas will be presented in this book.

The main objective of this book is to present wearable communication systems and compact wearable antennas for 5G communication, IOT and medical systems. The main goal of wearable wireless body area networks is to provide continuous medical data to physicians.

Key features presented in this new book

- Wearable antennas for 5G, IOT and medical applications.
- IOT antennas and systems for commercial and medical applications not presented in previous books.
- Textile antennas not presented in previous books.
- Wearable reconfigurable antennas for communication and medical applications not presented in previous books.
- New types of systems and antennas for medical applications and IOT systems.
- New types of metamaterial antennas and artificial magnetic conductors.
- New types of helical antenna (UHF) in a wrist device.
- Quad helix GPS antenna mounted on a helmet.

– Several new examples of wearable systems and antennas are presented in this book.
– Dual polarized and circular polarized antennas are presented in this book.
– New results and effects of new wearable systems and antennas on the human body are presented in this book.

Each chapter in this book provides significant information to enable electrical and biomedical engineers, students and scientists from all areas to follow and understand the topics presented. This book begins with elementary communication, electromagnetics, microwave and antenna topics for students and engineers with no background in communication, electromagnetic theory and antenna theory to enable them to study and understand the basic design principles and features of communication systems, antennas, wearable antennas, printed antennas and compact antennas for communication, IOT and medical applications.

Several topics and designs are presented in this book for the first time. This book may serve students and design engineers as a reference book. This book presents new designs in the areas of wearable systems and antennas, metamaterial antennas, fractal antennas and active receiving and transmitting antennas. The text contains significant mathematical detail and explanations to enable electrical engineering and physics students to understand all topics presented in this book.

Several new wearable antennas and communication systems are presented in this book. Design considerations and the computed and measured results of the new wearable systems and antennas are presented in the book.

Acknowledgments

My wife – Mazal Sabban
My daughters – Dolly and Lilach
My son – David Sabban
Grandchildren – Nooa, Avigail, Ido, Shira, Efrat, Yael Hodaia, Tamara

I acknowledge my engineering colleagues who helped me throughout my 40-years of engineering and research career.

I acknowledge all authors who contributed to this book.

Editor

Dr Albert Sabban holds a PhD in Electrical Engineering from the Faculty of Electrical and Computer Engineering, University of Colorado at Boulder, USA (1991), and an MBA from the Faculty of Management, University of Haifa, Israel (2005). He is currently Senior Lecturer and Researcher at the Department of Electrical and Electronic Engineering at Kinneret College and ORT Braude College. In 1976 he joined RAFAEL in Israel where he worked as a Senior Researcher, Group Leader and Project Leader in the Electromagnetic Department until 2007. From 2007 to 2010, Albert Sabban was a RF and Antenna Specialist at biomedical hi-tech companies where he designed wearable compact systems and antennas for medical systems. From 2008 to 2010 he worked as a RF Specialist and Project Leader at hi-tech biomedical companies. He has published over 100 research papers and holds a patent in the antenna area. He has written four books on wearable compact systems and antennas for communication and medical systems. He has written a book on electromagnetics and microwave theory for graduate students, and a book on wideband microwave technologies for communication and medical applications. He also wrote a chapter of a book on microstrip antennas and a chapter of a book on wearable printed antennas for medical applications. He also edited and wrote two chapters of a book on green technologies. He also edited and wrote two chapters of a book on RF novel antennas.

Dr Albert Sabban – Author and Editor

Affiliation: Kinneret Academic College, Israel

Retired: RAFAEL, 2008; ORT Braude College, 2018

Email: sabban@netvision.net.il

Interests: wearable systems and antennas, medical devices and applications, system engineering, IOT, 5G communication systems

LinkedIn: https://www.linkedin.com/in/albert-sabban-38206239/

Contributors

Dr Sema Dumanli
Professional affiliation and full address: Boğaziçi University, Department of Electrical
    and Electronics Engineering, 34342 Bebek, Istanbul
Phone: +90 212 359 6895
Email: sema.dumanli@boun.edu.tr
Professional URL:

1. https://academics.boun.edu.tr/sema.dumanli/
2. http://www.lifesci.boun.edu.tr

Dr Sema Dumanli received her PhD degree from the University of Bristol, UK, in 2010. From 2010 to 2017, she was a Research Engineer and a Senior Research Engineer with Toshiba Research Europe, Bristol, UK. She is currently Assistant Professor at Boğaziçi University, Istanbul, Turkey, where she serves on the Management Board of the University Center for Life Sciences and Technologies. Her current research interests include antenna design for body area networks, implantable and wearable devices, eHealth and multi-in multi-out communications.

Dr Herwansyah Lago
Professional affiliation and full address: Universiti Malaysia Sabah (UMS), Faculty of
    Engineering, Kota Kinabalu, Sabah, Malaysia
Phone: +6010 955 9007
Email: herwansyahlago@gmail.com

Dr Herwansyah Lago received his PhD degree in Communication Engineering from Universiti Malaysia Perlis (UniMAP) in 2017. He was then a Post-Doctoral Research Fellow at the Universiti Teknikal Malaysia Melaka (UTeM). He is currently a Senior Lecturer at the UMS. His research focuses on telecommunication electronics, especially antenna structure, metamaterials, dielectric materials, RF and microwave design.

Prof Ely Levine
Professional affiliation and full address: Afeka Academic College of Engineering, Tel
    Aviv, Israel
Phone: +97237668722
Email: levineel@zahav.net.il

Prof Ely Levine is a well-known expert in antennas and radio engineering. He holds a BSEE degree and a MSEE degree in Electrical Engineering from the Technion, Haifa, Israel, and a PhD in Applied Physics from the Weizmann Institute of Science, Rehovot, Israel. Prof. Levine held senior development positions in leading electronics companies (ELTA, Elop and others). He joined Afeka Academic College of Engineering, Tel Aviv, in 2006 where he teaches communications, antennas, microwave systems and components, wireless radio and radar systems. He has published more than 65 papers and conference proceedings and co-authored two books.

Dr Mohammad Nazri Md Noor
Professional affiliation and full address: Centre of Diploma Studies (CDS), Universiti
    Malaysia Perlis (UniMAP)
Phone: +6097673000
Email: nazrry@um.edu.my
Professional URL: http://www.researcherid.com/rid/S-8623-2016

Dr Mohammad Nazri Md Noor received a bachelor's degree (with honors) in Electrical Engineering majoring in Communications from Universiti Teknologi Malaysia (UTM), Skudai, Malaysia, in 2000. He is currently a Senior Engineering Instructor under the CDS, UniMAP. Dr Nazri is a Senior Lecturer in Management and the Coordinator for Bachelor of Business Administration (BBA) at the Faculty of Business and Accountancy, Universiti Malaya. He holds a doctorate in Business Administration from the University Teknologi MARA. His industrial experience with RHB Investment Bank spanned over 10 years. His research interests include management, organization behavior and knowledge management. His major areas of teaching include human capital, leadership, business ethics and Islamic perspectives in business.

Prof Mohd Ridzuan Mohd Nor
Professional affiliation and full address: Head of Network Operations, WeBe Digital,
    Shah Alam, Selangor, Malaysia
Phone: +60 3 5514 6000
Email: zuanzack@gmail.com
Professional URL:
LinkedIn: https://www.linkedin.com/in/mohd-ridzuan-mohd-nor-12028626/

Mohd Ridzuan Mohd Nor received his bachelor's degree (with honors) in Computer Engineering from the Universiti Malaysia Perlis (UniMAP). He first joined UniMAP under the School of Computer and Communications Engineering (SCCE) as an Engineering Instructor. He is currently a Senior Engineering Instructor in the Centre of Diploma Studies (CDS), UniMAP. He teaches courses such as embedded system design, object-oriented programming, programming languages, electrical technologies, digital electronics, computer networking and electromagnetic theory.

He has 20 years of working experience in the telecommunications industry with a mixed background from technical engineering, project management, operations management and leadership.

He has technical and management experience in various voice and data technologies, such as 1G, 2G, 3G and 4G with different transmission technologies, such as PDH, SPDH, SDH, IPRan, DWDM, MPLS, IPCore, IX and IP and fiber infrastructures.

Prof Kashif Nisar Paracha, PhD
Assistant Professor
GC University Faisalabad
Department of Electrical Engineering,
Iqbal Block, GCUF New Campus,
Jhang Road, 38000, Faisalabad, Pakistan
Email: kashifnisar@gcuf.edu.pk

Kashif Nisar Paracha was born in Faisalabad, Pakistan. He received his B.S. degree with honors in Electrical Engineering (EE) from University of Engineering and Technology (UET), Taxila, Pakistan, and M.S. in Electrical Engineering from King Fahd University of Petroleum and Mineral (KFUPM), Dhahran, Saudia Arabia, in 2004 and 2008, respectively. He has received his PhD degree in Electrical Engineering from Universiti Teknologi Malaysia (UTM), Malaysia in 2019. He is the author and co-author of about eight research journal papers and four conference articles. He is a senior member of IEEE and Life time Professional Member of Pakistan Engineering Council (PEC). His research interests include communication, antenna design: wearable antennas, arrays, metasurfaces; on-body communication Metamaterials, inkjet Printing Methods and Algorithms. Currently, he is supervising two Masters Students as well.

K.N. Paracha is currently serving as assistant professor in EE Department, Government College University Faisalabad (GCUF), since 2019 and worked as Lecturer in the same department from 2011 to 2019. He was Research Assistant in EE department at KFUPM, from 2006 to 2008, and taught in EE department, The University of Faisalabad (TUF), Pakistan, from 2008 to 2011.

Prof Ping Jack Soh
Professional affiliation and full address: Associate Professor, The School of Computer and Communications Engineering (SCCE), Universiti Malaysia Perlis (UniMAP), Pauh Putra Campus, 02600 Arau, Perlis, Malaysia
Phone: +6010 955 9007
Email: pjsoh@unimap.edu.my or pjsoh@ieee.org
Professional URL: https://www.esat.kuleuven.be/telemic/People-of-telemic/ping-jack-soh
LinkedIn: https://www.linkedin.com/in/pjsoh/

Prof Ping Jack Soh received his PhD degree in Electrical Engineering from KU Leuven, Belgium, in 2013. He is currently Associate Professor at the SCCE, UniMAP, and a Research Affiliate at KU Leuven, Belgium. He researches actively in his areas of interest: wearable antennas, arrays, metasurfaces; on-body communication; electromagnetic safety and absorption; and wireless and radar techniques for healthcare applications. To date, he has/is leading six internationally and nationally funded projects, besides being involved collaboratively in other projects. He currently (co)supervises 15 postgraduate projects, and has successfully (co)supervised 7 PhD and 16 MSc students to completion.

Prof Guy A. E. Vandenbosch
Professional affiliation and full address: Universiteit Leuven, Leuven, Belgium
TELEMIC, Kasteelpark Arenberg 10 – box 2444 3001, Leuven, Belgium
Phone: +32 16 32 11 10 or +32 16 37 40 85
Email: guy.vandenbosch@esat.kuleuven.be

Prof Guy A. E. Vandenbosch received his PhD degree in Electrical Engineering from the Katholieke Universiteit Leuven, Leuven, Belgium, in 1991. He was a Research and Teaching Assistant from 1985 to 1991 with the Telecommunications and Microwaves section of the Katholieke Universiteit Leuven, where he worked on the modeling of microstrip antennas with the integral equation technique. From 1991 to 1993, he held a postdoctoral research position at the Katholieke Universiteit Leuven. Since 1993, he has been a Lecturer, and since 2005, a Full Professor at the same university. Guy Vandenbosch has taught or teaches courses on: electromagnetic waves; antennas; electromagnetic compatibility; fundamentals of communication and information theory; and electrical engineering, electronics and electrical energy.

1 Wearable Communication and IOT Systems Basics

Albert Sabban

CONTENTS

INTRODUCTION

Wireless communication systems play a major role in shaping people's lifestyles and every minute of their lives. Almost every simple procedure in our life is accompanied and monitored by cellular phones and smartphones. Since 1980, cellular phone technology has progressed every year. Communication engineers developed five generations of cellular phones and smartphones.

There was a huge improvement from the first generation of cellular phones up to the fifth generation of cellular phones and smartphones. Communication theory and design were presented in several books and scientific papers, see [1–14]. Theory and design of wearable communication systems were presented in several books and scientific papers, see [4–14].

1.1 GENERATIONS OF MOBILE NETWORKS

1.1.1 First Generation (1G)

1G is the first generation of wireless cellular technology. Cell phones began with 1G technology in the 1980s. The first generation of mobile networks were reliant upon analog radio systems. 1G supports voice-only calls. Mobile network users could only make phone calls, they could not send or receive text messages. The maximum speed of 1G technology is 2.4 Kbps.

1.1.1.1 1G Basic Features
- Speed 2.4 Kbps.
- Allows voice calls in one country.
- Uses analog signal.
- Poor voice quality.
- Poor battery life.
- Large phone size.
- Limited capacity.
- Poor handoff reliability.
- Poor security.
- Offers a very low level of spectrum efficiency.

1.1.1.2 Bits Per Second
The data rate of a communication system is normally measured in units of bits per second. Kbps is 1000 bits per second. Mbps is 1000 Kbps. Gbps is 1000 Mbps or 1,000,000 Kbps.

1.1.1.3 Global System for Mobile Communications (GSM)

GSM is a cell phone standard called Global System for Mobile Communications. GSM provides standard features like phone call encryption, data networking, caller ID, call forwarding, call waiting, short message service (SMS) and conferencing. GSM cell phone technology works in the 1900 MHz band in the United States and the 900 MHz band in Europe and Asia. GSM phones use a subscriber identification module (SIM) card to store the subscriber's information, such as phone number and other data, that proves that a user is in fact a subscriber to that carrier. Several cellular phones need a SIM card in order to identify the owner and communicate with the mobile network.

1.1.2 SECOND GENERATION (2G)

The 2G cellular phones employ the GSM standard. GSM took cellular phones from analog technology to digital communication technology. The 2G telephone technology introduced call and text encryption, along with data services such as SMS and multimedia messaging service (MMS). The maximum speed of 2G with General Packet Radio Service (GPRS) is 50 Kbps.

1.1.2.1 SMS

SMS stands for short message service, which is the formal name for the technology used for text messaging. It is a way to send short text-only messages from one phone to another. These messages are usually sent over a cellular data network.

1.1.2.2 MMS

MMS, which is a multimedia messaging service, allows cellphone and smartphone users to send each other messages with images, videos and more. The service is based on SMS but adds those features to it. Using MMS, the cell phone can send audio files, contact details, ringtones, photos, videos and other data to any other phone with a text messaging plan.

1.1.2.3 Enhanced Data Rates for GSM Evolution (EDGE)

EDGE is a faster version of GSM. EDGE, which is an acronym for Enhanced Data Rates for GSM Evolution, is a speed and latency advancement in GSM technology. EDGE networks were designed to deliver multimedia applications, such as streaming television, audio and video, to cellular phones at speeds up to 384 Kbps.

1.1.2.4 2G Basic Features

- Data speed up to 64 Kbps.
- Uses digital signals.
- Enables services such as text messages, picture messages and MMS (multimedia message).
- Provides better quality and capacity.
- Unable to handle complex data such as videos.
- Requires strong digital signals to help mobile phones work.

1.1.2.5 2.5G and 2.75G

2.5G introduced a new packet-switching technique that was more efficient than 2G technology. 2.75G provided a theoretical threefold speed increase.

1.1.2.6 2.5G Basic Features

- Provides phone calls.
- Users can send/receive email messages.
- Web browsing.
- Speed rate 64–144 Kbps.
- Camera phones.
- Takes 6–9 minutes to download a 3-minute MP3 song.

1.1.3 Third Generation (3G)

3G enables high-speed access to data and voice services. A 3G network is a high-speed mobile broadband network, offering a data speed rate of at least 144 Kbps. 3G networks can offer speeds of 3.1 Mbps.

1.1.3.1 3G Basic Features

- Speed rate 2 Mbps.
- Increased bandwidth and data transfer rates to accommodate web-based applications and audio and video files.
- Provides faster communication.
- Users can send and receive large email messages.
- High-speed web/more security/video conferencing.
- Three-dimensional gaming.
- Large capacities and broadband capabilities.
- TV streaming/mobile TV/phone calls.
- Users can download a 3-minute MP3 song.
- Expensive fees for 3G licenses services.
- High bandwidth requirement.
- Expensive 3G phones.

1.1.4 Fourth Generation (4G)

4G supports mobile web access, gaming services, high-definition (HD) mobile TV, video conferencing, 3D TV and other features that demand high speeds. The maximum speed of a 4G network for moving devices is 100 Mbps.

1.1.4.1 4G Basic Features

- May provide 10 Mbps to 1 Gbps speed rate.
- High quality streaming video.
- Combination of Wi-Fi and Wi-Max.
- High security.
- Provides any kind of service at any time as per user requirements anywhere.
- Expanded multimedia services.
- Low cost per bit.

1.1.5 Fifth Generation (5G)

5G will provide significantly faster data rates, higher connection density, much lower latency, energy savings and other improvements. The maximum speed of a 5G network will be up to 20 Gbps.

1.1.5.1 5G Basic Features

- It is highly supportable to Wireless World Wide Web.
- High speed, high capacity.
- Provides large broadcasting of data in Gbps.
- Multimedia newspapers, watch TV programs with HD clarity.
- Faster data transmission than that of the previous generation.
- Large phone memory, swift dialing speed, clarity in audio and video.
- 5G technology offers high resolution for cell phone user and bi-directional large bandwidth sharing.

1.2 RECEIVERS: DEFINITIONS AND FEATURES

Figure 1.1 presents a basic receiver block diagram.

1.2.1 RECEIVERS: DEFINITIONS

Radio frequency (RF), intermediate frequency (IF) and local oscillator (LO): When a receiver uses a mixer, we refer to the input frequency as the RF frequency. The system must provide a signal to mix down the RF, this is called the LO signal. The resulting lower frequency is called the IF, because it is somewhere between the RF frequency and the base band frequency.

Base band frequency: The base band is the frequency at which the information you want to process is.

Pre-selector filter: A pre-selector filter is used to keep undesired radiation from saturating a receiver. For example, we do not want our cell phone to pick up air traffic control radar.

Amplitude and phase matching versus tracking: In a multi-channel receiver (more than one receiver), it is important for the channels to match and track each other over frequency. Amplitude and phase *matching* means that the relative magnitude and phase of signals that pass through the two paths must be almost equal.

Tunable bandwidth versus instantaneous bandwidth: Instantaneous bandwidth is what we get with a receiver when we keep the LO at a fixed frequency and sweep the input frequency to measure the response. The resulting bandwidth is a function of the frequency responses of everything in the chain. Their instantaneous bandwidth has a direct effect on the minimum detectable signal. Tunable bandwidth implies that we change the frequency of the LO to track the RF frequency. The bandwidth in this case is only a function of the pre-selector filter, the

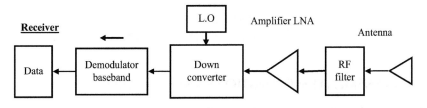

FIGURE 1.1 Basic receiver block diagram for wearable communication system.

low noise amplifier (LNA) and the mixer. Tunable bandwidth is often many times greater than instantaneous bandwidth.

Gain: The gain of a receiver is the ratio of input signal power to output signal power.

Noise figure: The noise figure of a receiver is a measure of how much the receiver degrades the ratio of signal to noise of the incoming signal. It is related to the minimum detectable signal. If the LO signal has a high amplitude modulation (AM) and/or frequency modulation (FM) noise, it could degrade the receiver noise figure because, far from the carrier, the AM and FM noise originate from thermal noise. Remember that the effect of LO AM noise is reduced by the balance of the balanced mixer.

1 dB compression point: 1 dB compression point is the power level where the gain of the receiver is reduced by 1 dB due to compression.

Linearity: A receiver operates linearly if a 1 dB increase in input signal power results in a 1 dB increase in IF output signal strength.

Dynamic range: The dynamic range of a receiver is a measurement of the minimum detectable signal, to the maximum signal that will start to compress the receiver.

Signal to noise ratio (S/N) SNR: SNR is a measure of how far a signal is above the noise floor.

Noise factor, noise figure and noise temperature

- Noise factor is a measure of how the SNR is degraded by a device:

  F = noise factor = $(S_{in}/N_{in})/(S_{out}/N_{out})$,
  S_{in} is the signal level at the input,
  N_{in} is the noise level at the input,
  S_{out} is the signal level at the output,
  N_{out} is the noise level at the output.

- The noise factor of a device is specified with noise from a noise source at room temperature ($N_{in} = KT$), where K is Boltzmann's constant and T is approximately room temperature in Kelvin. KT is approximately −174 dBm/Hz. Noise figure is the noise factor, expressed in decibels:

  NF (decibels) = noise figure = $10 \times \log (F)$.
  T = noise temperature = $290 \times (F−1)$.
  1 dB NF is about 75°, and 3 dB is 288°.
  The noise factor contributions of each stage in a four-stage system are given in Equation 1.1.

$$F = F1 + \frac{F2-1}{G1} + \frac{F3-1}{G1G2} + \frac{F4-1}{G1G2G3} \qquad (1.1)$$

1.3 TRANSMITTERS: DEFINITIONS AND FEATURES

Figure 1.2 presents a basic transmitter block diagram.

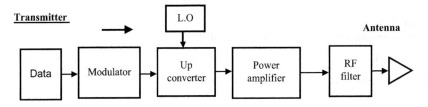

FIGURE 1.2 Basic transmitter block diagram for wearable communication system.

1.3.1 AMPLIFIERS

Class A: The amplifier is biased at close to half of its saturated current. The output conducts during all 360° of phase of the input signal sine wave. Class A does not give maximum efficiency but provides the best linearity. Drain efficiencies of 50% are possible in Class A.

Class B: The power amplifier is biased at a point where it draws nearly zero direct current (DC); for a Field-Effect Transistor (FET), this means that it is biased at pinch-off. During one half of the input signal sine wave it conducts, but not the other half. Class B amplifiers can be very efficient, with theoretical efficiency of 80%–85%. However, we are giving up 6 dB of gain when we move from Class A to Class B.

Class C: Class C occurs when the device is biased so that the output conducts for even less than 180° of the input signal. The output power and gain decrease.

Power density: This is a measure of power divided by transistors' size. In the case of FETs, it is expressed in watts/mm. GaN transistors have a power density of more than 10 W/mm.

Saturated output power (PSAT): PSAT is the output power where the Pin/Pout curve slope goes to zero.

Load pull: Load pull is the process of varying the impedance seen by the *output* of an active device to other than 50 Ω in order to measure performance parameters, in the simplest case, gain. In the case of a power device, a load pull power bench is used to evaluate large signal parameters such as compression characteristics, saturated power, efficiency and linearity as the output load is varied across the Smith chart.

Harmonic load pull: This is the process of varying the impedance at the output of a device, with separate control of the impedances at F0, 2F0, 3F0 and so on.

Source pull: This is the process of varying the impedance seen by the input of an active device to other than 50 Ω in order to measure performance parameters. In the case of a low noise device, source pull is used in a noise parameter extraction setup to evaluate how SNR (noise figure) varies with source impedance.

Amplifiers temperature considerations: In the case of a FET amplifier, the gain drops and the noise figure increases. The gain drop is around −0.006 dB/stage/°C. The noise figure of an LNA increases by +0.006 dB/°C. In an LNA, the first stage will dominate the temperature effect.

Power amplifiers: Power amplifiers are used to boost a small signal to a large signal. Solid state amplifiers and tube-amplifiers are usually employed as power amplifiers. Power amplifiers' output power capabilities are listed in Table 1.1.

TABLE 1.1
Power amplifiers' output power capabilities

Frequency band	Solid state	Tube type
L-band through C-band	200 W (LDMOS) GaN	
X-band	50 W (GaN HEMT device)	3000 W (TWT)
Ka-band	6 W (GaAs PHEMT device)	1000 W (klystron)
Q-band	4 W (GaAs PHEMT device)	

1.4 BASIC ELECTROMAGNETIC WAVE DEFINITIONS

Field: A field is a physical quantity that has a value for each point in space and time.

Wavelength: The wavelength is the distance between two sequential equivalent points; see Equation 1.2. Wavelength, λ, is measured in m.

Wave period: The period is the time, T, for one complete cycle of an oscillation of a wave, see Equation 1.2. The period is measured in seconds.

Frequency: The frequency, f, is the number of periods per unit time (second) and is measured in hertz.

Phase velocity: The phase velocity, v, of a wave is the rate at which the phase of the wave propagates in space. Phase velocity and light velocity, c, are measured in m/s.

$$\lambda = v * T$$
$$T = 1/f \tag{1.2}$$
$$\lambda = v/f$$

Electromagnetic waves propagate in free space in phase velocity of light, see Equation 1.3.

$$v = c = 3 \times 10^8 \tag{1.3}$$

Wavenumber: A wavenumber, k, is the spatial frequency of the wave in radians per unit distance (per m). $k = 2\pi/\lambda$.

Angular frequency: The angular frequency, ω, represents the frequency in radians per second. $\omega = v \times k = 2\pi f$.

Polarization: A wave is polarized if it oscillates in one direction or plane. The polarization of a transverse wave describes the direction of oscillation in the plane perpendicular to the direction of propagation.

Antenna: An antenna is used to efficiently radiate electromagnetic energy in desired directions. Antennas match radio frequency systems to space. All antennas may be used to receive or radiate energy. Antennas transmit or receive electromagnetic waves. Antennas convert electromagnetic radiation into electric current, or vice versa. Antennas transmit and receive electromagnetic radiation at radio frequencies. Antennas are a necessary part of all communication links and radio equipment. Antennas are used in systems such as radio and television broadcasting, point-to-point radio communication, wireless Local Area Network (LAN), cell phones, radar, medical systems and spacecraft communication. Antennas are most commonly employed in air or outer space, but can also be operated under water, on and inside the human body, or even through soil and rock at low frequencies for short distances.

1.4.1 Free Space Propagation

Fundamentals of wireless communication systems are presented in several books and journals [1–10]. Flux density at distance R of an isotropic source radiating P_t Watts uniformly into free space is given by Equation 1.4. At distance R, the area of the spherical shell with center at the source is $4\pi R^2$.

$$F = \frac{P_t}{4\pi R^2} \text{ W/m}^2 \qquad (1.4)$$

$$G(\theta) = \frac{P(\theta)}{P_0/4\pi} \qquad (1.5)$$

$P(\theta)$ is variation of power with angle.
$G(\theta)$ is gain at the direction θ, Equation 1.5.
P_0 is total power transmitted.
Sphere $= 4\pi$ solid radians.

Gain is a usually expressed in decibels (dB). G (dB) $= 10 \log 10\ G$. Gain is realized by focusing power. An isotropic radiator is an antenna which radiates in all directions equally. Effective isotropic radiated power (EIRP) is the amount of power the transmitter would have to produce if it was radiating to all directions equally. The EIRP may vary as a function of direction because of changes in the antenna gain versus angle. We now want to find the power density at the receiver. We know that power is conserved in a lossless medium. The power radiated from a transmitter must pass through a spherical shell on the surface of which is the receiver.

The area of this spherical shell is $4\pi R^2$.

Therefore, spherical spreading loss is $1/4\pi R^2$.

We can rewrite the power flux density, as given in Equation 1.6, now considering the transmit antenna gain:

$$F = \frac{EIRP}{4\pi R^2} = \frac{P_t G_t}{4\pi R^2} \text{ W/m}^2 \qquad (1.6)$$

The power available to a receive antenna of area A_r is given in Equation 1.7:

$$P_r = F \times A_r = \frac{P_t G_t A_r}{4\pi R^2} \qquad (1.7)$$

Real antennas have effective flux collecting areas which are less than the physical aperture area. A_e is defined as the antenna effective aperture area.

Where $A_e = A_{phy} \times \eta$; where $\eta =$ aperture efficiency.

Antennas have maximum gain G related to the effective aperture area as written in Equation 1.8. Where A_e is effective aperture area.

$$G = \text{Gain} = \frac{4\pi A_e}{\lambda^2} \tag{1.8}$$

Aperture antennas (horns and reflectors) have a physical collecting area that can be easily calculated from their dimensions:

$$A_{phy} = \pi r^2 = \pi \frac{D^2}{4} \tag{1.9}$$

Therefore, using Equation 1.7 and Equation 1.8 we can obtain the formula for aperture antenna gain as given in Equation 1.9. Antenna gain equations are listed in Equations 1.10–1.13.

$$\text{Gain} = \frac{4\pi A_e}{\lambda^2} = \frac{4\pi A_{phy}}{\lambda^2} \times \eta \tag{1.10}$$

$$\text{Gain} = \left(\frac{\pi D}{\lambda}\right)^2 \times \eta \tag{1.11}$$

$$\text{Gain} \cong \eta \left(\frac{75\pi}{\theta_{3dB}}\right)^2 = \eta \frac{(75\pi)^2}{\theta_{3dBH}\theta_{3dBE}} \tag{1.12}$$

Where $\theta_{3dB} \cong \dfrac{75\lambda}{D}$

θ_{3dB} is the antenna half power beam width. Assuming for instance a typical aperture efficiency of 0.55, gives:

$$\text{Gain} \cong \frac{30,000}{(\theta_{3dB})^2} = \frac{30,000}{\theta_{3dBH}\theta_{3dBE}} \tag{1.13}$$

1.5 FRIIS TRANSMISSION FORMULA

The Friis transmission formula is presented in Equation 1.14.

$$P_r = P_t G_t G_r \left(\frac{\lambda}{4\pi R}\right)^2 \tag{1.14}$$

Free space loss (Lp) represents propagation loss in free space. Losses due to attenuation in atmosphere, La, should also be accounted for in the transmission equation.

Where, $L_p = \left(\dfrac{4\pi R}{\lambda}\right)^2$. The received power may be given as: $P_r = \dfrac{P_t G_t G_r}{L_p}$

Losses due to polarization mismatch, L_{pol}, should also be accounted for. Losses associated with receiving antenna, L_{ra}, and with the receiver, L_r, cannot be neglected in computation of transmission budget, see Equations 1.15 and 1.16. Losses associated with the transmitting antenna are written as, L_{ta}.

$$P_r = \frac{P_t G_t G_r}{L_p L_a L_{ta} L_{ra} L_{pol} L_o L_r}$$ (1.15)

$P_t = Pout/L_t$
EIRP $= P_t G_t$

Where:
 P_t is the transmitting antenna power.
 L_t is the loss between power source and antenna.
 EIRP is the effective isotropic radiated power.

$$P_r = \frac{P_t G_t G_r}{L_p L_a L_{ta} L_{ra} L_{pol} L_{other} L_r}$$
$$= \frac{EIRP \times G_r}{L_p L_a L_{ta} L_{ra} L_{pol} L_{other} L_r}$$ (1.16)
$$= \frac{P_{out} G_t G_r}{L_t L_p L_a L_{ta} L_{ra} L_{pol} L_{other} L_r}$$

Where,

$$G = 10\log\left(\frac{P_{out}}{P_{in}}\right) \text{dB Gain in dB.}$$

$$L = 10\log\left(\frac{P_{in}}{P_{out}}\right) \text{dB Loss in dB.}$$

Gain may be derived as given in Equation 1.17.

$$P_{in} = \frac{V_{in}^2}{R_{in}} \quad P_{out} = \frac{V_{out}^2}{R_{out}}$$

$$G = 10\log\left(\frac{P_{out}}{P_{in}}\right) = 10\log\left(\frac{\frac{V_{out}^2}{R_{out}}}{\frac{V_{in}^2}{R_{in}}}\right)$$ (1.17)

$$G = 10\log\left(\frac{V_{out}^2}{V_{in}^2}\right) + 10\log\left(\frac{R_{in}}{R_{out}}\right) = 20\log\left(\frac{V_{out}}{V_{in}}\right) + 10\log\left(\frac{R_{in}}{R_{out}}\right)$$

The surface area of a sphere of radius d is $4\pi d^2$, so that the power flow per unit area W (power flux in W/m²) at distance d from a transmitter antenna with input power P_T and antenna gain G_T is given in Equation 1.18.

$$W = \frac{P_t G_r}{4\pi d^2} \tag{1.18}$$

The received signal strength depends on the "size" or aperture of the receiving antenna. If the antenna has an effective area A, then the received signal strength is given in Equation 1.19.

$$P_R = P_T G_T \left(A/\left(4\pi d^2\right)\right) \tag{1.19}$$

Define the receiver antenna gain $G_R = 4\pi A/\lambda^2$.
Where, $\lambda = c/f$.

1.6 COMMUNICATION SYSTEMS LINK BUDGET

A link budget determines if the received signal is larger than the receiver sensitivity. A link budget analysis determines if there is enough power at the receiver to recover the information. A link budget must account for effective transmission power. A transmitter block diagram for a wireless wearable communication system is shown in Figure 1.3. A receiver block diagram for a wireless wearable communication system is shown in Figure 1.4. A link budget takes into account the following parameters.

Transmitter
- Transmission power
- Antenna gain
- Losses in cable and connectors

Path losses
- Attenuation
- Ground reflection
- Fading (self-interference)

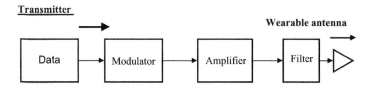

FIGURE 1.3 Transmitter block diagram for wireless wearable communication system.

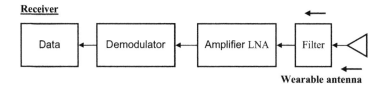

FIGURE 1.4 Receiver block diagram for wireless wearable communication system.

TABLE 1.2

Transmitting channel power budget for wireless wearable communication system

Component	Gain (dB)/Loss (dB)	Power (dBm)	Remarks
Input power		−20	
Transmitter gain	50		
Power amplifier output power		30	
Filter loss	1	29	
Line loss	1	28	
Matching loss	1	27	
Radiated power		27	

TABLE 1.3

Receiving channel power budget for wireless wearable communication system

Component	Gain (dB)/Loss (dB)	Power (dBm)	Remarks
Input power		−30	
Receiver gain	33	3	
Line losses	1	2	
Filter loss	1	1	
Matching loss	1	0	
LNA amplifier output power		0	

Receiver
- Receiver sensitivity
- Losses in cable and connectors

Transmitting channel power budget is presented in Table 1.2.
Receiving channel power budget is listed in Table 1.3.

1.7 PATH LOSS

Path loss is a reduction in a signal's power, which is a direct result of the distance between the transmitter and the receiver in the communication path.

There are many models used in industry today to estimate the path loss and the most common are the Free Space and Hata models. Each model has its own requirements that need to be met in order to be utilized correctly. The Free Space path loss is the reference point used by other models.

1.7.1 FREE SPACE PATH LOSS

Free space path loss(dB) = $20 \log_{10} f + 20 \log_{10} d - 147.56$
   Where
      f is frequency in Hz and
      d is the distance in m.

The Free Space model typically underestimates the path loss experienced for mobile communications. The Free Space model predicts point-to-point fixed path loss.

1.7.2 HATA MODEL

The Hata model is used extensively in cellular communications. The basic model is for urban areas, with extensions for suburbs and rural areas.

The Hata model is valid only for these ranges:

- Distance 1–20 km
- Base height 30–200 m
- Mobile height 1–10 m
- 150–1500 MHz.

The Hata formula for urban areas is:

$$L_H = 69.55 + 26.16\log_{10} f_c - 13.82\log_{10} h_b - env(h_m) + (44.9 - 6.55\log_{10} h_b)\log_{10} R$$

- h_b is the base station antenna height in m.
- h_m is the mobile antenna height also measured in m.
- R is the distance from the cell site to the mobile in km.
- f_c is the transmit frequency in MHz.
- $env(h_m)$ is an adjustment factor for the type of environment and the height of the mobile. $env(h_m) = 0$ for urban environments with a mobile height of 1.5 m.

1.8 RECEIVER SENSITIVITY

Sensitivity describes the weakest signal power level that the receiver is able to detect and decode. Sensitivity is determined by the lowest SNR at which the signal can be recovered. Different modulation and coding schemes have different minimum SNRs. Sensitivity is determined by adding the required SNR to the noise present at the receiver.

1.8.1 NOISE SOURCES

- Thermal noise
- Noise introduced by the receiver's amplifier.

$$\text{Thermal noise} = N = kTB(\text{Watts})$$

- $k = 1.3803 \times 10^{-23}$ J/K
- T = temperature in Kelvin
- B = receiver bandwidth

$$N(\text{dBm}) = 10\log_{10}(kTB) + 30$$

Thermal noise is usually very small for reasonable bandwidths.

1.9.2 Basic Receiver Sensitivity Calculation

$$\text{Sensitivity}\left(W\right) = kTB \times NF\left(\text{linear}\right) \times \text{minimum SNR required}\left(\text{linear}\right)$$

$$\text{Sensitivity}\left(dBm\right) = 10\log_{10}\left(kTB \times 1000\right) + NF\left(dB\right) + \text{minimum SNR required}\left(dB\right)$$

$$\text{Sensitivity}\left(dBm\right) = 10\log_{10}\left(kTB\right) + 30 + NF\left(dB\right) + \text{minimum SNR required}\left(dB\right)$$

Sensitivity decreases in communication systems when:

- Bandwidth increases.
- Temperature increases.
- Amplifier introduces more noise.
- Losses in space, rain and snow.

1.9 INTERNET OF THINGS (IOT) BASICS

The Internet of Things, which is known as IOT, is a new twenty-first century emerging technology. The IOT is a system of interrelated computing devices, mechanical and digital machines, personal devices, objects, animals, people, healthcare devices and wireless communication devices that are provided with unique identifiers (UIDs) and the ability to transfer data over a network without requiring human-to-human or human-to-computer interaction.

The IOT helps people every day to live, work smarter and get complete control over daily services and procedures. IOT provides smart devices to automate homes, companies and healthcare centers. IOT is essential to business. IOT provides businesses with real-time observation and inspection of how their companies' systems really work. IOT enables companies to automate processes and reduce labor costs. It also cuts down on waste and improves service delivery, making it less expensive to manufacture and deliver products as well as offering transparency in customer transactions.

IOT touches every industry, including healthcare, finance, retail and manufacturing. Smart cities help citizens reduce waste and energy consumption and connected sensors are even used in farming to help monitor crop and cattle yields and predict growth patterns.

IOT has evolved from the convergence of wireless technologies, microelectromechanical systems and the internet. The convergence has helped tear down the silos between operational technology and information technology (IT), enabling unstructured machine-generated data to be analyzed for insights to drive improvements.

IOT evolved from machine-to-machine (M2M) communication: machines connecting to each other via a network without human interaction. M2M refers to connecting a device to the cloud, managing it and collecting data. Taking M2M to the next level, IOT is a sensor network of billions of smart devices that connect people, systems and other applications to collect and share data. At its foundation, M2M offers the connectivity that enables IOT.

The IOT is also a natural extension of supervisory control and data acquisition (SCADA), a category of software application program for process control: the gathering of data in real time from remote locations to control equipment and conditions. SCADA systems include hardware and software components. The hardware gathers and feeds data into a computer that has SCADA software installed, where it is then processed and presented in a timely manner. The evolution of SCADA is such that late-generation SCADA systems developed into first-generation IOT systems.

An IOT ecosystem consists of web-enabled smart devices that use embedded processors, sensors and communication hardware to collect, send and act on data they acquire from their environments. IOT devices share the sensor data they collect by connecting to an IOT gateway or other edge device where data are analyzed locally or sent to the cloud to be analyzed. Sometimes, these devices communicate with other related devices and act on the information they get from one another. The devices do most of the work without human intervention, although people can interact with the devices, to set them up, give them instructions or access the data. The connectivity, networking and communication protocols used with these web-enabled devices largely depend on the specific IOT applications deployed. An IOT system block diagram is shown in Figure 1.5.

1.9.1 IOT Benefits to Companies and Organizations

- Monitor their overall business processes.
- Improve customer experience.
- Save time and money.
- Enhance employee productivity.
- Integrate and adapt business models.

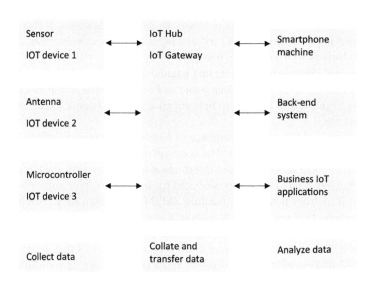

FIGURE 1.5 Internet of Things system.

- Make better business decisions.
- Generate more revenue.

IOT gives companies and organizations the tools to improve their business strategies.

1.9.2 IOT ADVANTAGES

- Ability to access information from anywhere at any time on any device.
- Improved communication between connected electronic devices.
- Transferring data packets over a connected network saves time and money.
- Automating tasks helps improve the quality of a business's services and reduces the need for human intervention.

1.9.3 IOT DISADVANTAGES

- As the number of connected devices increases and more information is shared between devices, the potential that a hacker could steal confidential information also increases.
- Enterprises may eventually have to deal with massive numbers, maybe even millions, of IOT devices and collecting and managing the data from all those devices will be challenging.
- If there is a bug in the system, it is likely that every connected device will become corrupted.
- Since there is no international standard of compatibility for IOT, it is difficult for devices from different manufacturers to communicate with each other.

1.10 LOGARITHMIC RELATIONS

Logarithms are very useful in communication and microwave engineering. Important logarithmic operations are listed in Equations 1.20–1.23.

$$
\begin{aligned}
10\log_{10}\left(A \times B\right) &\\
= 10\log_{10}\left(A\right) + 10\log_{10}\left(B\right) &\\
= A\text{dB} + B\text{dB} &\\
= \left(A + B\right)\text{dB} &
\end{aligned}
\tag{1.20}
$$

$$
\begin{aligned}
10\log_{10}\left(A / B\right) &\\
= 10\log_{10}\left(A\right) - 10\log_{10}\left(B\right) &\\
= A\text{dB} - B\text{dB} &\\
= \left(A - B\right)\text{dB} &
\end{aligned}
\tag{1.21}
$$

$$
\begin{aligned}
10\log_{10}\left(A^{2}\right) &\\
= 2 \times 10\log_{10}\left(A\right) &\\
= 20\log_{10}\left(A\right) &\\
= 2 \times \left(A \text{ in dB}\right) &
\end{aligned}
\tag{1.22}
$$

$$10 \log_{10} \left(\sqrt{A} \right)$$

$$= \frac{10}{2} \log_{10} \left(A \right) \tag{1.23}$$

$$= \frac{1}{2} \times \left(A \text{ in dB} \right)$$

In Table 1.4 are listed linear ratios versus logarithmic ratios.

The received power P_r in dBm is given in Equation 1.24. The received power P_r is commonly referred to as "carrier power," C.

$$P_r = EIRP - L_{ta} - L_p - L_a - L_{pol} - L_{ra} - L_{other} + G_r - L_r \tag{1.24}$$

The power flow per unit area W (power flux in W/m²) at distance d from a transmitting antenna with input power P_t and antenna gain G_t is given in Equation 1.25.

$$W = \frac{PtGt}{4\pi d^2} \tag{1.25}$$

The received signal strength depends on the "size" or aperture of the receiving antenna. If the antenna has an effective area A_e, then the received signal strength is given in Equation 1.26.

$$P_R = P_T \, G_T \left(A_e / \left(4\pi d^2 \right) \right) \tag{1.26}$$

The receiver antenna gain is $G_R = 4\pi A/\lambda^2$. Where, $\lambda = c/f$.

TABLE 1.4
Linear ratio versus logarithmic ratio

Linear ratio	dB	Linear ratio	dB
0.001	−30.0	2.000	3.0
0.010	−20.0	3.000	4.8
0.100	−10.0	4.000	6.0
0.200	−7.0	5.000	7.0
0.300	−5.2	6.000	7.8
0.400	−4.0	7.000	8.5
0.500	−3.0	8.000	9.0
0.600	−2.2	9.000	9.5
0.700	−1.5	10.000	10.0
0.800	−1.0	100.000	20.0
0.900	−0.5	1000.000	30.0
1.000	0.0	18.000	12.6

1.11 WIRELESS COMMUNICATION SYSTEM LINK BUDGET, AN EXAMPLE

$F = 2.4$ GHz => $\lambda = 3e8$ m/s/2.4e9/s $= 12.5$ cm.

At 933 MHz => $\lambda = 32$ cm.

Receiver signal strength: $P_R = P_T \, G_T \, G_R \, (\lambda/4\pi d)^2$

P_R (dBm) $= P_T$ (dBm) $+ G_T$ (dBi) $+ G_R$ (dBi) $+ 10 \log_{10} ((\lambda/4\pi)^2) - 10\log_{10}(d^2)$.

For $F = 2.4$ GHz => $10 \log_{10} ((\lambda/4\pi)^2) = -40$ dB.

For $F = 933$ MHz => $10 \log_{10} ((\lambda/4\pi)^2) = -32$ dB.

1.11.1 MOBILE PHONE DOWNLINK

$\lambda = 12.5$ cm

$f = 2.4$ GHz

P_R (dBm) $= (P_T \, G_T \, G_R L)$ (dBm) -40 dB $+ 10 \log_{10} (1/d^2)$

$P_R - (P_T + G_T + G_R + L) -40$ dB $= 10 \log_{10}(1/d^2)$

Or $155 - 40 = 10 \log_{10} (1/d^2) =$

Or $(155 - 40)/20 = \log_{10} (1/d)$

$d = 10\verb|^| ((155 - 40)/20) = 562$ km

1.11.2 MOBILE PHONE UPLINK

$d = 10\verb|^| ((153 - 40)/20) = 446$ km

For standard 802.11

- $P_R - P_T = -113.2$ dBm.
- 6 Mbps
  - $d = 10^{(113.2-40)/20} = 4500$ m
  - $d = 10^{(113.2-40 - 3)/20} = 3235$ m with 3 dB gain margin
  - $d = 10^{(113.2-40 - 3 - 9)/20} = 1148$ m with 3 dB gain margin and neglecting antenna gains.

- 54 Mbps needs -85 dBm
  - $d = 10^{(99.2-40)/20} = 912$ m
  - $d = 10^{(99.2-40 - 3)/20} = 646$ m with 3 dB gain margin
  - $d = 10^{(99.2-40 - 3 - 9)/20} = 230$ m with 3 dB gain margin and neglecting antenna gains.

Signal strength

- Measure signal strength in
  dBW $= 10 \times \log$ (power in Watts)
  dBm $= 10 \times \log$ (power in mW).
- 802.11 can legally transmit at 30 dBm.
- Most 802.11 PCMCIA cards transmit at 10–20 dBm.
- Mobile phone base station: 20 W, but 60 users, so 0.3 W/user, but antenna has gain $= 18$ dBi.
- Mobile phone handset: 21 dBm.

REFERENCES

[1] W.C.Y. Lee, *Mobile Communication Engineering*, McGraw Hill, New York, 1982.

[2] L.A. Belov, S.M. Smolskiy, V.N. Kochemasov, *Handbook of RF, Microwave, and Millimeter-Wave Components*, Artech House, Boston, USA, 27–28, 2012. ISBN 978-1-60807-209-5.

[3] M.L. Skolnik, *Introduction to Radar Systems*, McGraw Hill, New York, 1980.

[4] A. Sabban, *Wideband RF Technologies and Antenna in Microwave Frequencies*, Wiley Sons, New York, July 2016.

[5] A. Sabban, *Wearable Communication Systems and Antennas for Commercial, Sport, and Medical Applications*, IET Publication, London, UK, December 2018.

[6] A. Sabban, *Novel Wearable Antennas for Communication and Medical Systems*, Taylor & Francis Group, FL, USA, October 2017.

[7] A. Sabban, *Low Visibility Antennas for Communication Systems*, Taylor & Francis Group, New York, 2015.

[8] A. Sabban, "Ultra-Wideband RF Modules for Communication Systems," *PARIPEX, Indian Journal of Research* 5(1), 91–95, January 2016.

[9] A. Sabban, "Wideband RF Modules and Antennas at Microwave and MM Wave Frequencies for Communication Applications," *Journal of Modern Communication Technologies & Research*, 3, 89–97, March 2015.

[10] A. Sabban, "New Wideband printed Antennas for Medical Applications," *IEEE Journal, Transactions on Antennas and Propagation*, 61(1), 84–91, January 2013.

[11] S.C. Mukhopadhyay ed., *Wearable Electronics Sensors*, Springer, Switzerland, 2015.

[12] A. Bonfiglio, D. De Rossi eds, *Wearable Monitoring Systems*, Springer, New York, 2011

[13] T. Gao, D. Greenspan, M. Welsh, R.R. Juang, A. Alm, "Vital Signs Monitoring and Patient Tracking over a Wireless Network," Proceedings of IEEE-EMBS 27th Annual International Conference of the Engineering in Medicine and Biology, Shanghai, China, 102–105, September 1–5, 2005.

[14] K. Fujimoto, J.R. James, eds, *Mobile Antenna Systems Handbook*, Artech House, Boston, MA, 1994.

2 Electromagnetics and Transmission Lines for Wearable Communication Systems

Albert Sabban

CONTENTS

INTRODUCTION

The frequency of voltage supplied by electric companies is 50 Hz in several countries around the world. The wavelength at 50 Hz is 6000 km. At low frequencies, 50 Hz up to 1 kHz, the variation of voltage and current along electric cables may be neglected. If the length of an electric component is less than a tenth of a wavelength, then the variation of voltage and current may be neglected. If the length of the device is less than a tenth of a wavelength, then the device is called a lumped element. If the length of an electric component is longer than a tenth of a wavelength, then the variation of voltage and current cannot be neglected. In this case, the device is called a distributed element. Electric circuits that consist of a lumped element are analyzed by employing Kirchhoff's laws and Ohm's law. Kirchhoff's laws and Ohm's law assume that voltage and current do not vary along the length of a component. Circuits that consist of a distributed element cannot be analyzed by using Kirchhoff's laws and Ohm's law and should be analyzed by using electromagnetic theory. This chapter provides basic theory and a short introduction to electromagnetic, microwave and communication engineering. The transmitting and receiving of information in microwave frequencies is based on the propagation of electromagnetic waves.

2.1 ELECTROMAGNETIC SPECTRUM

The electromagnetic spectrum corresponds to electromagnetic waves as presented in references [1–17]. Infrared light was discovered by Sir William Herschel in 1800. Johann Wilhelm Ritter discovered ultraviolet light in 1801. In 1867, James Clerk Maxwell predicted that there should be light with longer wavelengths than infrared light. In 1887, Heinrich Hertz demonstrated the existence of the waves predicted by Maxwell by producing radio waves in his laboratory. The frequency unit is named after Heinrich Hertz and is measured in hertz. Hertz X-rays were first observed and documented in 1895 by Wilhelm Conrad Röntgen in Germany. Gamma rays were first observed in 1900 by Paul Villard when he was investigating radiation from radium.

At very low frequencies, 1 Hz–1 MHz, the wavelength is of much higher order than the size of the circuit components used in electronic circuits. At very low frequencies, voltage, current and impedance do not vary as a function of the device length. At very low frequencies, relations between voltage, current and impedance are evaluated by using Kirchhoff's laws and Ohm's law. Components with dimensions lower than a tenth of the wavelength are called lumped elements. At high frequencies, the wavelength is of the same order of magnitude as the circuit devices used. At high frequencies, conventional circuit analysis based on Kirchhoff's laws and Ohm's law could not analyze and describe the variation of fields, impedance, voltages and currents along the length of the components. Components with dimensions higher than a tenth of the wavelength are called distributed elements. Kirchhoff's laws and Ohm's law may be applied to lumped elements, but they cannot be applied to distributed elements. To prevent interference and to provide efficient use of the radio spectrum, similar services are allocated in bands, see [11–14]. Bands are divided at wavelengths of 10 nm, or frequencies of 3×10 nHz. Each of these bands has a basic band plan, which dictates how it is to be used and shared, to avoid interference and to set protocol for the compatibility of transmitters and receivers. In Table 2.1 the electromagnetic spectrum and applications are listed. In Table 2.2 the IEEE standard for radar frequency bands is listed.

In Table 2.3 the International Telecommunication Union (ITU) bands are given. In Table 2.4 radar frequency bands as defined by North Atlantic Treaty Organization (NATO) for electromagnetic compatibility (EMC) systems are listed, see [13].

TABLE 2.1
Electromagnetic spectrum and applications

Band name	Abbreviation	ITU	Frequency\λ_0	Applications
Tremendously low frequency	TLF		<3 Hz >100,000 km	Natural and artificial EM noise
Extremely low frequency	ELF		3–30 Hz 100,000 km–10,000 km	Communication with submarines
Super low frequency	SLF		30–300 Hz 10,000 km–1000 km	Communication with submarines
Ultra-low frequency	ULF		300–3000 Hz 1000 km–100 km	Submarine communication, communication within mines
Very low frequency	VLF	4	3–30 kHz 100 km–10 km	Navigation, time signals, submarine communication, wireless heart rate monitors, geophysics
Low frequency	LF	5	30–300 kHz 10 km–1 km	Navigation, clock time signals, AM longwave broadcasting (Europe and parts of Asia), RFID, amateur radio
Medium frequency	MF	6	300–3000 kHz 1 km–100 m	AM, broadcasts, amateur radio, avalanche beacons
High frequency	HF	7	3–30 MHz 100 m–10 m	Shortwave broadcasts, radio, amateur radio and aviation, communications, RFID, radar, near-vertical incidence sky wave (NVIS) radio communications, marine and mobile radio telephony
Very high frequency	VHF	8	30–300 MHz 10 m–1 m	FM, television broadcasts and line-of-sight ground-to-aircraft and aircraft-to-aircraft communications, land mobile communications, amateur radio, weather radio
Ultra-high frequency	UHF	9	300–3000 MHz 1 m–100 mm	Television broadcasts, microwave oven, radio astronomy, mobile phones, wireless LAN, Bluetooth, ZigBee, GPS and two-way radios such as land mobile, FRS and GMRS radios
Super high frequency	SHF	10	3–30 GHz 100 mm–10 mm	Radio astronomy, wireless LAN, modern radars, communications satellites, satellite television broadcasting, DBS

(*Continued*)

TABLE 2.1 (*continued*)
Electromagnetic spectrum and applications

Band name	Abbreviation	ITU	Frequency\λ_0	Applications
Extremely high frequency	EHF	11	30–300 GHz 10 mm–1 mm	Radio astronomy, microwave radio relay, microwave remote sensing, directed-energy weapon, scanners
Terahertz or tremendously high frequency	THz or THF	12	300–3000 GHz 1 mm–100 µm	Terahertz imaging, ultrafast molecular dynamics, condensed-matter physics, terahertz time-domain spectroscopy, terahertz computing/communications

TABLE 2.2
IEEE standard radar frequency bands

Microwave frequency bands – IEEE standard and applications

designation	Frequency range (GHz)	Applications
L band	1–2	Television broadcasts, microwave oven, radio astronomy, mobile phones
S band	2–4	Wireless LAN, Bluetooth, ZigBee, GPS
C band	4–8	Radio astronomy, wireless LAN, radars, communications satellites, satellite television
X band	8–12	Radars, radio astronomy, communications satellites, satellite television
K_u band	12–18	Radars, communications satellites
K band	18–26.5	Communication systems
K_a band	26.5–40	Radars, communications satellites
Q band	30–50	MM waves communication systems
U band	40–60	Radars and communication systems
V band	50–75	Microwave remote sensing, directed-energy weapon, scanners
E band	60–90	Scanners
W band	75–110	Scanners
F band	90–140	Scanners, terahertz imaging
D band	110–170	Scanners, terahertz imaging

2.2 ELECTROMAGNETIC FIELDS THEORY FOR MEDICAL AND 5G SYSTEMS

Electric charges are the sources of electromagnetic fields. The strength of an electric field at any point depends on the magnitude, velocity, acceleration and position of the charges involved. If the charges do not move, they generate electrostatic fields. Moving charges generate electromagnetic fields. Electric charges may be positive or negative charges. Coulomb found that between two charges, q_1 and q_2, exists a force that tends to push or pull them apart depending on the polarity of the charge.

TABLE 2.3

International Telecommunication Union bands and medical applications

Band	Symbol	Frequency	Wavelength	Medical and sport applications
4	VLF	3–30 kHz	10–100 km	Wireless heart rate monitors
5	LF	30–300 kHz	1–10 km	RFID tags
6	MF	300–3000 kHz	100–1000 m	Avalanche beacons
7	HF	3–30 MHz	10–100 m	RFID tags
8	VHF	30–300 MHz	1–10 m	Mobile radio telephony
9	UHF	300–3000 MHz	10–100 cm	WBAN and WWBAN
10	SHF	3–30 GHz	1–10 cm	Wireless WBAN
11	EHF	30–300 GHz	1–10 mm	Microwave sensors
12	THF	300–3000 GHz	0.1–1 mm	Terahertz imaging

TABLE 2.4

Radar frequency bands as defined by North Atlantic Treaty Organization (NATO) for electromagnetic compatibility (EMC) systems [13]

Band	Frequency range (GHz)	Commercial and medical applications
A band	0–0.25	RFID commercial and medical tags; communication with submarines
B band	0.25–0.5	Monitoring medical systems, FM, television broadcasts, WBAN, BAN
C band	0.5–1.0	Mobile phones, WBAN
D band	1–2	Mobile phones, GPS
E band	2–3	Wireless LAN, Bluetooth, ZigBee, GPS
F band	3–4	Radio astronomy, radars
G band	4–6	Radars, satellites communication, wireless LAN
H band	6–8	Radars, satellites communication
I band	8–10	Satellite television, radars
J band	10–20	Radars, satellites communication
K band	20–40	Communication systems, seekers
L band	40–60	Radars, communication systems
M band	60–100	Scanners, microwave remote sensing

Coulomb's law is given in Equation 2.1. ϵ is the media dielectric constant. Where, r is the distance between the charges. The charge produces an electric field E. The field strength is measured as the force per unit charge at that point at that direction and is given in Equation 2.2. Electromagnetic and physical symbols are listed in Table 2.5.

$$F = \frac{q_1 q_2}{4\pi\epsilon r^2} \text{ newtons} \tag{2.1}$$

$$E = \frac{q}{4\pi\epsilon r^2} \vec{r} \, \frac{\text{Volt}}{\text{m}} \tag{2.2}$$

$$\oint_S E \cdot ds = \frac{1}{\varepsilon_0} \int_V \rho dv = \frac{1}{\varepsilon_0} q$$

$$D = \varepsilon E \qquad\qquad (2.3)$$

$$\nabla \cdot D = \rho_v$$

2.3 ELECTROMAGNETIC WAVES THEORY FOR MEDICAL AND 5G SYSTEMS

2.3.1 MAXWELL'S EQUATIONS

Maxwell's equations are presented in several books [1–10]. Maxwell's equations describe how electric and magnetic fields are generated and altered by each other. Maxwell's equations are a classical approximation to the more accurate and fundamental theory of quantum electrodynamics. Quantum deviations from Maxwell's equations are usually small. Inaccuracies occur when the particle nature of light is important or when electric fields are strong. Symbols and abbreviations of physical parameters and electromagnetic fields are listed in Tables 2.5 and 2.6.

2.3.2 GAUSS'S LAW FOR ELECTRIC FIELDS

Gauss's law for electric fields states that the electric flux via any closed surface S is equal to the net charge q divided by the free space dielectric constant.

2.3.3 GAUSS'S LAW FOR MAGNETIC FIELDS

Gauss's law for magnetic fields states that the magnetic flux via any closed surface S is equal to zero. There is no magnetic charge in nature.

$$\oint_S B \cdot ds = 0$$

$$B = \mu H$$

$$\psi_m = \int_S B \cdot ds \qquad\qquad (2.4)$$

$$\nabla \cdot B = 0$$

2.3.4 AMPÈRE'SM LAW

The original "Ampère's law" stated that magnetic fields can be generated by electrical current. Ampère's law was corrected by Maxwell who stated that magnetic fields can also be generated by time-variant electric fields. The corrected "Ampère's law" shows that a changing magnetic field induces an electric field and a time-variant electric field induces a magnetic field.

TABLE 2.5
Symbols and abbreviations of physical parameters

Dimensions	Parameter	Symbol
Wb/m	Magnetic potential	A
m/s^2	Acceleration	a
Tesla	Magnetic field	B
F	Capacitance	C
V/m	Electric field displacement	D
V/m	Electric field	E
N	Force	F
A/m	Magnetic field strength	H
A	Current	I
A/m^2	Current density	J
H	Self-inductance	L
m	Length	l
H	Mutual inductance	M
Coulomb	Charge	q
W/m^2	Poynting vector	$\mathbf{P}$
W	Power	P
Ω	Resistance	R
m^3	Volume	V
m/s	Velocity	v
F/m	Dielectric constant	ε
	Relative dielectric constant	ε_r
H/m	Permeability	μ
1/$\Omega \cdot$ m	Conductivity	σ
Wb	Magnetic flux	ψ

TABLE 2.6
Symbols and abbreviations of electromagnetic fields

Parameter	Symbol	Dimensions
Electric field displacement	D	V/m
Electric field	E	V/m
Magnetic field strength	H	A/m
Dielectric constant in space	ε_0	$8.854 \cdot 10^{-12}$ F/m
Permeability in space	μ_0	$\mu_0 = 4\pi \cdot 10^{-7}$ H/m
Volume charge density	ρ_V	C/m^3
Magnetic field	B	Tesla, Wb/m^2
Conductivity	σ	1/$\Omega \cdot$m
Magnetic flux	ψ	Wb
Skin depth	δ_s	m

$$\oint_C \frac{B}{\mu_0} \cdot dl = \int_S J \cdot ds + \frac{d}{dt} \int_s \varepsilon_0 E \cdot ds$$

$$\nabla X H = J + \frac{\partial D}{\partial t}$$

$$J = \sigma E \tag{2.5}$$

$$i = \int_S J \cdot ds$$

2.3.5 FARADAY'S LAW

Faraday's law describes how a propagating time-varying magnetic field through a surface S creates an electric field, as shown in Figure 2.1.

$$\oint_C E \cdot dl = -\frac{d}{dt} \int_s B \cdot ds$$

$$\nabla X E = -\frac{\partial B}{\partial t} \tag{2.6}$$

2.3.6 WAVE EQUATIONS

The variation of electromagnetic waves as a function of time may be written as $e^{j\omega t}$. The derivative as a function of time is $j\omega e^{j\omega t}$. Maxwell's equations may be written as:

$$\nabla X E = -j\omega\mu H$$

$$\nabla X H = (\sigma + j\omega\varepsilon) E \tag{2.7}$$

A ∇X (curl) operation on the electric field E results in:

$$\nabla X \nabla X E = -j\omega\mu \nabla X H \tag{2.8}$$

By substituting the expression of $\nabla X H$ in the equation we get

$$\nabla X \nabla X E = -j\omega\mu (\sigma + j\omega\varepsilon) E \tag{2.9}$$

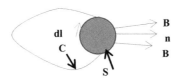

FIGURE 2.1 Faraday's law.

$$\nabla X \nabla X E = -\nabla^2 E + \nabla(\nabla \cdot E) \tag{2.10}$$

In free space there are no charges so $\nabla \cdot E = 0$. Equation 2.11 presents the wave equation for electric fields. Where γ is the complex propagation, constant α represents losses in the medium and β represents the wave phase constant in radian per meter.

$$\nabla^2 E = j\omega\mu(\sigma + j\omega\varepsilon)E = \gamma^2 E$$
$$\gamma = \sqrt{j\omega\mu(\sigma + j\omega\varepsilon)} = \alpha + j\beta \tag{2.11}$$

If we follow the same procedure as for the magnetic field, then we will get the wave equation for magnetic field.

$$\nabla^2 H = j\omega\mu(\sigma + j\omega\varepsilon)H = \gamma^2 H \tag{2.12}$$

The law of conservation of energy imposes boundary conditions on the electric and magnetic fields. When an electromagnetic wave travels from medium 1 to medium 2 the electric and magnetic fields should be continuous as presented in Figure 2.2.

General boundary conditions:

$$n \cdot (D_2 - D_1) = \rho_S \tag{2.13}$$

$$n \cdot (B_2 - B_1) = 0 \tag{2.14}$$

$$(E_2 - E_1) \times n = M_S \tag{2.15}$$

$$n \times (H_2 - H_1) = J_S \tag{2.16}$$

Boundary conditions for dielectric medium:

$$n \cdot (D_2 - D_1) = 0 \tag{2.17}$$

$$n \cdot (B_2 - B_1) = 0 \tag{2.18}$$

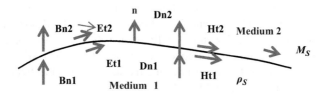

FIGURE 2.2   Fields between two media.

$$\left(E_2 - E_1\right) \times n = 0 \tag{2.19}$$

$$n \times \left(H_2 - H_1\right) = 0 \tag{2.20}$$

Boundary conditions for conductor:

$$n \cdot \left(D_2 - D_1\right) = \rho_S \tag{2.21}$$

$$n \cdot \left(B_2 - B_1\right) = 0 \tag{2.22}$$

$$\left(E_2 - E_1\right) \times n = 0 \tag{2.23}$$

$$n \times \left(H_2 - H_1\right) = J_S \tag{2.24}$$

The condition for a good conductor is: $\sigma \gg \omega\epsilon$.

$$\gamma = \sqrt{j\omega\mu\left(\sigma + j\omega\varepsilon\right)} = \alpha + j\beta \approx \left(1 + j\right)\sqrt{\frac{\omega\mu\sigma}{2}} \tag{2.25}$$

$$\delta_s = \sqrt{\frac{2}{\omega\mu\sigma}} = \frac{1}{\alpha} \tag{2.26}$$

The conductor skin depth is given as δ_s. The wave attenuation is α.

2.4 WAVES PROPAGATION THROUGH HUMAN BODY

The electrical properties of human body tissues should be considered in the design of wearable communication and medical systems. Conductivity and dielectric constants of human body tissues are listed in Table 2.7, see [15, 16]. The complex propagation constant γ, given in Equation 2.11, can be calculated as a function of frequency by using the conductivity and dielectric constant values listed in Table 2.7. The major issue in the design of wearable antennas is the interaction between radio frequency (RF) transmission and the human body. The electrical properties of human body tissues should be considered in the design of wearable antennas. The attenuation, α, of RF transmission through the human body is the real value of the propagation constant γ. Where, $\alpha = Real(\gamma)$. For example, stomach tissue attenuation is around 1.67 dB/cm at 500 MHz and 3.2 dB/cm at 1000 MHz. Blood attenuation is around 3.38 dB/cm at 500 MHz and 4.02 dB/cm at 1000 MHz. Attenuation of stomach tissues, skin and pancreas at frequencies from 150 to 1000 MHz are listed in Table 2.8. Attenuation of fat, small-intestine tissues and blood at frequencies from 150 to 1000 MHz are listed in Table 2.9.

TABLE 2.7
Electrical properties of human body tissues

Tissue	Property	434 MHz	800 MHz	1000 MHz
Skin	σ	0.57	0.6	0.63
	ε	41.6	40.45	40.25
Prostate	σ	0.75	0.90	1.02
	ε	50.53	47.4	46.65
Stomach	σ	0.67	0.79	0.97
	ε	42.9	40.40	39.06
Colon, heart	σ	0.98	1.15	1.28
	ε	63.6	60.74	59.96
Kidney	σ	0.88	0.88	0.88
	ε	117.43	117.43	117.43
Nerve	σ	0.49	0.58	0.63
	ε	35.71	33.68	33.15
Fat	σ	0.045	0.056	0.06
	ε	5.02	4.58	4.52
Lung	σ	0.27	0.27	0.27
	ε	38.4	38.4	38.4

TABLE 2.8
Attenuation of stomach tissues, skin and pancreas

Frequency (MHz)	Attenuation stomach (dB/cm)	Attenuation skin (dB/cm)	Attenuation pancreas (dB/cm)
150	1.10	1.00	1.155
200	1.21	1.1	1.27
250	1.313	1.168	1.375
300	1.414	1.25	1.48
350	1.49	1.30	1.55
400	1.56	1.36	1.62
450	1.62	1.41	1.69
500	1.68	1.45	1.75
550	1.733	1.49	1.81
600	1.79	1.53	1.87
650	1.84	1.58	1.93
700	1.89	1.62	1.99
750	1.94	1.67	2.05
800	1.99	1.715	2.11
850	2.05	1.74	2.16
900	2.11	1.77	2.20
950	2.17	1.78	2.25
1000	2.233	1.825	2.293

TABLE 2.9

Attenuation of fat, small intestine and blood

Frequency (MHz)	Attenuation fat (dB/cm)	Attenuation small intestine (dB/cm)	Attenuation blood (dB/cm)
150	0.163	1.453	2.30
200	0.20	1.863	2.56
250	0.225	2.083	2.77
300	0.25	2.213	2.95
350	0.27	2.295	3.082
400	0.2912	2.35	3.194
450	0.30	2.39	3.29
500	0.315	2.411	3.38
550	0.33	2.43	3.46
600	0.341	2.444	3.54
650	0.354	2.455	3.61
700	0.37	2.464	3.671
750	0.372	2.47	3.73
800	0.376	2.476	3.79
850	0.38	2.48	3.85
900	0.384	2.484	3.91
950	0.389	2.487	3.967
1000	0.394	2.49	4.023

2.5 MATERIALS

In Table 2.10, hard materials are presented. Alumina is the most popular hard substrate in microwave integrated circuits (MICs). Gallium arsenide is the most popular hard substrate in monolithic MICs (MMIC) technology at microwave frequencies.

In Table 2.11, soft materials are presented. Duroid is the most popular soft substrate in MICs and in the printed antennas industry. Dielectric losses in Duroid are significantly lower than dielectric losses in FR-4 substrate. However, the cost of FR-4 substrate is significantly lower than the cost of Duroid. FR-4 is the most common material that is used in fabricated circuit boards. "FR" indicates the material is flame retardant and the "4" indicates woven glass reinforced epoxy resin. Commercial MIC devices usually use FR-4 substrate. Duroid is the most popular soft substrate used in the development of printed antennas with high efficiency at microwave frequencies.

2.6 TRANSMISSION LINES THEORY

Transmission lines are used to transfer electromagnetic energy from one point to another with minimum losses over a wide band of frequencies. There are three major types of transmission lines [1–10]. One type of transmission line has very small cross-sectional dimensions in comparison to wavelength and the dominant mode of propagation is transverse electromagnetic (TEM) mode. The other two types are closed rectangular and cylindrical conducting tubes on which the dominant modes of propagation are transverse electric (TE) mode and transverse magnetic (TM) mode.

TABLE 2.10
Hard materials – ceramics

Material	Symbol or formula	Dielectric constant	Dissipation factor (tan δ)	Coefficient of thermal expansion (ppm/Â°C)	Thermal cond. (W/mÂ°C)	Mass density (gr/cc)
Alumina 99.5%	Al$_2$O$_3$	9.8	0.0001	8.2	35	3.97
Alumina 96%	Al$_2$O$_3$	9.0	0.0002	8.2	24	3.8
Aluminum nitride	AlN	8.9	0.0005	7.6	290	3.26
Beryllium oxide	BeO	6.7	0.003	6.05	250	
Gallium arsenide	GaAs	12.88	0.0004	6.86	46	5.32
Indium phosphide	InP	12.4				
Quartz		3.8	0.0001	0.6	5	
Sapphire		9.3, 11.5				
Silicon (high resistivity)	Si (HRS)	4		2.5	138	2.33
Silicon carbide	SiC	10.8	0.002	4.8	350	3.2

Open boundary structures that have a cross section greater than 0.1λ may support a surface wave mode of propagation.

For TEM modes Ez = Hz = 0. For TE modes Ez = 0. For TM modes Hz = 0.

Voltage and currents in transmission lines may be derived by using the transmission lines equations. Transmission lines equations may be derived by employing Maxwell equations and the boundary conditions on the transmission line section shown in Figure 2.3. Equation 2.27 is the first lossless transmission lines equation. l_e is the self-inductance per length.

$$\frac{\partial V}{\partial Z} = -\frac{\partial}{\partial t} l_e I = -l_e \frac{\partial I}{\partial t} \qquad (2.27)$$

$$\frac{\partial I}{\partial Z} = -c \frac{\partial V}{\partial t} - gV \qquad (2.28)$$

Equation 2.28 is the second lossless transmission lines equation. $c = \dfrac{\Delta C}{\Delta Z} \dfrac{F}{m}$.

$$g = \frac{\Delta G}{\Delta Z} \frac{m}{\Omega}.$$

TABLE 2.11
Soft materials

Manufacturer and material	Symbol or formula	Relative dielectric constant, ε_R	Tolerance on dielectric constant	Tan δ	Mass density (gr/cc)	Thermal conductivity (W/mÅ°C)	Coefficient of thermal expansion (PPM/Å°C) x/y/z
Rogers Duroid 5870	PTFE/Random glass	2.33	0.02	0.0012	2.2	0.26	22/28/173
Rogers Duroid 5880	PTFE/Random glass	2.2	0.02	0.0012	2.2	0.26	31/48/237
Rogers Duroid 6002	PTFE/Random glass	2.94	0.04	0.0012	2.1	0.44	16/16/24
RogersDuroid 6006	PTFE/Random glass	6	0.15	0.0027	2.7	0.48	38/42/24
Rogers Duroid 6010	PTFE/Random glass	10.2-10.8	0.25	0.0023	2.9	0.41	24/24/24
FR-4	Glass/epoxy	4.8		0.022		0.16	
Polyethylene		2.25					
Polyflon CuFlon	PTFE	2.1		0.00045			12.9
Polyflon Polyguide	Polyolefin	2.32		0.0005			108
Polyflon NorCLAD	Thermoplastic	2.55		0.0011			53
Polyflon Clad ULTEM	Thermoplastic	3.05		0.003	1.27		56
PTFE	PTFE	2.1		0.0002	2.1	0.2	
Rogers R/flex 3700	Thermally stable thermoplastic	2.0		0.002			8
Rogers RO3003	PTFE ceramic	3	0.04	0.0013	2.1	0.5	17/17/24
Rogers RO3006	PTFE ceramic	6.15	0.15	0.0025	2.6	0.61	17/17/24

(Continued)

Table 2.11 (*Continued*)
Soft materials

Manufacturer and material	Symbol or formula	Relative dielectric constant, ε_R	Tolerance on dielectric constant	Tan δ	Mass density (gr/cc)	Thermal conductivity (W/mÅ°C)	Coefficient of thermal expansion (PPM/Å°C) x/y/z
Rogers RO3010	PTFE ceramic	10.2	0.3	0.0035	3	0.66	17/17/24
Rogers RO3203	PTFE ceramic	3.02					
Rogers RO3210	PTFE ceramic	10.2					
Rogers RO4003	Thermoset plastic ceramic glass	3.38	0.05	0.0027	1.79	0.64	11/14/46
Rogers RO4350B	Thermoset plastic ceramic glass	3.48	0.05	0.004	1.86	0.62	14/16/50
Rogers TMM 3	Ceramic/thermos	3.27	0.032	0.002	1.78	0.7	15/15/23
Rogers TMM 10	Ceramic/thermoset	9.2	0.23	0.0022	2.77	0.76	21/21/20

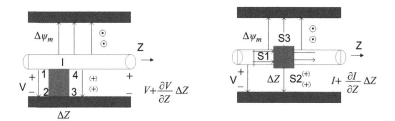

FIGURE 2.3   Transmission line geometry.

Equation 2.29 is the first transmission lines equation with losses.

$$-\frac{\partial v}{\partial z} = Ri + L\frac{\partial i}{\partial t}$$

(2.29)

Equation 2.30 is the second transmission lines equation with losses.

$$-\frac{\partial i}{\partial z} = Gv + C\frac{\partial v}{\partial t}$$

(2.30)

$$\text{where}\quad C = \frac{\Delta C(z)}{\Delta Z}\ \frac{F}{m}\quad L = \frac{\Delta L(z)}{\Delta Z}\ \frac{H}{m}\quad R = \frac{1}{G} = \frac{\Delta R(z)}{\Delta Z}\ \frac{\Omega}{m}$$

By differentiating Equation 2.29 with respect to z and by differentiating Equation 2.30 with respect to t and adding the result we get:

$$-\frac{\partial^2 v}{\partial z^2} = RGv + (RC + LG)\frac{\partial v}{\partial t} + LC\frac{\partial^2 v}{\partial t^2}$$

(2.31)

By differentiating Equation 2.30 with respect to z and by differentiating Equation 2.29 with respect to t and adding the result we get:

$$-\frac{\partial^2 i}{\partial z^2} = RGi + (RC + LG)\frac{\partial i}{\partial t} + LC\frac{\partial^2 i}{\partial t^2}$$

(2.32)

Equations 2.31 and 2.32 are analog to the wave equations. The solution of these equations is a superposition of a forward, $+z$, and backward wave, $-z$.

$$V(z,t) = V_+\left(t - \frac{z}{v}\right) + V_-\left(t + \frac{z}{v}\right)$$

$$I(z,t) = Y_0\left\{V_+\left(t - \frac{z}{v}\right) - V_-\left(t + \frac{z}{v}\right)\right\}$$

(2.33)

Y_0 is the characteristic admittance of the transmission line $Y_0 = \dfrac{1}{Z_0}$.

The variation of electromagnetic waves as a function of time may be written as $e^{j\omega t}$. The derivative as a function of time is $j\omega e^{j\omega t}$. By using these relations we may write phasor transmission lines equations.

$$\frac{dV}{dZ} = -ZI$$

$$\frac{dI}{dZ} = -YV$$

$$\frac{d^2V}{dz^2} = \gamma^2 V \qquad (2.34)$$

$$\frac{d^2I}{dz^2} = \gamma^2 I$$

where:

$$Z = R + j\omega L \quad \frac{\Omega}{m}$$

$$Y = G + j\omega C \quad \frac{m}{\Omega} \qquad (2.35)$$

$$\gamma = \alpha + j\beta = \sqrt{ZY}$$

The solution of the transmission lines equations in harmonic steady state is:

$$v(z,t) = \mathrm{Re}\ V(z)e^{j\omega t}$$
$$i(z,t) = \mathrm{Re}\ I(z)e^{j\omega t} \qquad (2.36)$$

$$V(z) = V_+e^{-\gamma z} + V_-e^{\gamma z}$$
$$I(z) = I_+e^{-\gamma z} + I_-e^{\gamma z} \qquad (2.37)$$

For a lossless transmission line we may write:

$$\frac{dV}{dZ} = -j\omega LI$$

$$\frac{dI}{dZ} = --j\omega CV$$

$$\frac{d^2V}{dz^2} = -\omega^2 LCV \qquad (2.38)$$

$$\frac{d^2I}{dz^2} = -\omega^2 LCI$$

The solution of the lossless transmission lines equations is:

$$V(z) = e^{j\omega t}\left(V^+e^{-j\beta z} + V^-e^{j\beta z}\right)$$
$$I(z) = Y_0\left(e^{j\omega t}\left(V^+e^{-j\beta z} - V^-e^{j\beta z}\right)\right) \qquad (2.39)$$

$$v_p = \frac{\omega}{\beta} = \frac{1}{\sqrt{LC}} = \frac{1}{\sqrt{\mu\varepsilon}}$$

v_p is the phase velocity.
Z_0 is the characteristic impedance of the transmission line.

$$Z_0 = \frac{V_+}{I_+} = \frac{V_-}{I_-} = \sqrt{\frac{(R + j\omega L)}{(G + j\omega C)}}$$

$$\text{for} \quad R = 0 \quad G = 0 \tag{2.40}$$

$$Z_0 = \frac{V_+}{I_+} = \frac{V_-}{I_-} = \sqrt{\frac{L}{C}}$$

2.6.1 WAVES IN TRANSMISSION LINES

A load Z_L is connected, at $z = 0$, to a transmission line with impedance Z_0, (see Figure 2.4). The voltage on the load is V_L. The current on the load is I_L.

$$V(0) = V_L = I(0) \cdot Z_L$$
$$I(0) = I_L \tag{2.41}$$

For $z = 0$ we can write Equation 2.42:

$$V(0) = I(0) \cdot Z_L = V_+ + V_-$$
$$I(0) = Y_0 (V_+ - V_-) \tag{2.42}$$

By substituting $I(0)$ in $V(0)$ we get Equation 2.43:

$$Z_L = Z_0 \frac{V_+ + V_-}{(V_+ - V_-)}$$

$$Z_L = Z_0 \frac{1 + \dfrac{V_-}{V_+}}{1 - \dfrac{V_-}{V_+}} \tag{2.43}$$

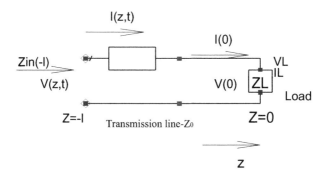

FIGURE 2.4   Transmission line with load.

The ratio $\dfrac{V_-}{V_+}$ is defined as reflection coefficient $\Gamma_L = \dfrac{V_-}{V_+}$.

$$Z_L = Z_0 \frac{1+\Gamma_L}{1-\Gamma_L} \tag{2.44}$$

$$\Gamma_L = \frac{\dfrac{Z_L}{Z_0}-1}{\dfrac{Z_L}{Z_0}+1} \tag{2.45}$$

The reflection coefficient as a function of z may be written as in Equation 2.46:

$$\Gamma(Z) = \frac{V_-}{V_+} = \frac{V_- e^{j\beta z}}{V_+ e^{-j\beta z}} = \Gamma_L e^{2j\beta z} \tag{2.46}$$

The input impedance as a function of z may be written as in Equations 2.47 and 2.48:

$$Z_{in}(z) = \frac{V(z)}{I(z)} = \frac{V_+ e^{-j\beta z} + V_- e^{j\beta z}}{\left(V_+ e^{-j\beta z} - V_- e^{j\beta z}\right)Y_0} =$$

$$= Z_0 \frac{1+\Gamma_L e^{2j\beta z}}{1-\Gamma_L e^{2j\beta z}} \tag{2.47}$$

$$Z_{in}(-l) = Z_L \frac{\cos \beta l + j \dfrac{Z_0}{Z_L} \sin \beta l}{\cos \beta l + j \dfrac{Z_L}{Z_0} \sin \beta l} \tag{2.48}$$

The voltage and current as a function of z may be written as in Equation 2.49:

$$V(z) = V_+ e^{-j\beta z}\left(1+\Gamma(z)\right)$$
$$I(z) = Y_0 V_+ e^{-j\beta z}\left(1-\Gamma(z)\right) \tag{2.49}$$

The ratio between the maximum and minimum voltage along a transmission line is called voltage standing wave ratio (VSWR) S, see Equation 2.50.

$$S = \left|\frac{V(z)\,\mathrm{max}}{V(z)\,\mathrm{min}}\right| = \frac{1+\left|\Gamma(z)\right|}{1-\left|\Gamma(z)\right|} \tag{2.50}$$

2.7 MATCHING TECHNIQUES

Usually, in communication systems, the load impedance is not the same as the impedance of commercial transmission lines. We will get maximum power transfer from source to load if impedances are matched [1–10]. A perfect impedance match corresponds to a VSWR 1:1. A reflection coefficient magnitude of 0 is a perfect match, a value of 1 is perfect reflection. The reflection coefficient (Γ) of a short circuit has a value of -1 (1 at an angle of 180°). The reflection coefficient of an open circuit is 1 at an angle of 0°. The return loss of a load is merely the magnitude of the reflection coefficient expressed in decibels. The correct equation for return loss is: Return loss $= -20 \times \log [\mathrm{mag}\,(\Gamma)]$.

For a maximum voltage V_m in a transmission line the maximum power will be:

$$P_{\mathrm{max}} = \frac{V_m^2}{Z_0} \tag{2.51}$$

For an unmatched transmission line the maximum power will be:

$$P_{\mathrm{max}} = \frac{\left(1-\left|\Gamma\right|^2\right)V_+^2}{Z_0}$$

$$V_{\mathrm{max}} = \left(1+\left|\Gamma\right|\right)V_+ \tag{2.52}$$

$$V_+^2 = \frac{V_{\mathrm{max}}^2}{\left(1+\left|\Gamma\right|\right)^2}$$

$$P_{\mathrm{max}} = \frac{1-\left|\Gamma\right|}{1+\left|\Gamma\right|} \cdot \frac{V_{\mathrm{max}}^2}{Z_0} = \frac{V_{\mathrm{max}}^2}{\mathrm{VSWR}\cdot Z_0} \tag{2.53}$$

A 2:1 VSWR will result in half of maximum power transferred to the load. The reflected power may cause damage to the source.

Equation 2.52 indicates that there is a one-to-one correspondence between the reflection coefficient and input impedance. A movement of distance z along a transmission line corresponds to a change of $e^{-2j\beta z}$ which represents a rotation via an angle of $2\beta z$. In the reflection coefficient plane we may represent any normalized impedance by contours of constant resistance, r, and contours of constant reactance, x. The corresponding impedance moves on a constant radius circle via an angle of $2\beta z$ to a new impedance value. Those relations may be presented by a graphical aid called a Smith chart. Those relations may be presented by the following set of equations:

$$\Gamma_L = \frac{\dfrac{Z_L}{Z_0} - 1}{\dfrac{Z_L}{Z_0} + 1}$$

$$z(l) = \frac{Z_L}{Z_0} = r + jx \tag{2.54}$$

$$\Gamma_L = \frac{r + jx - 1}{r + jx + 1} = p + jq$$

$$\frac{Z(z)}{Z_0} = \frac{1 + \Gamma(z)}{1 - \Gamma(z)} = r + jx$$

$$\Gamma(z) = u + jv \tag{2.55}$$

$$\frac{1 + u + jv}{1 - u - jv} = r + jx$$

$$\left(u - \frac{r}{1+r}\right)^2 + v^2 = \frac{1}{(1+r)^2} \tag{2.56}$$

$$(u - 1)^2 + \left(v - \frac{1}{x}\right)^2 = \frac{1}{x^2}$$

Equation 2.55 presents two families of circles in the reflection coefficient plane. The first family are contours of constant resistance, r, and the second family are contours of constant reactance, x. The center of the Smith chart is $r = 1$. Moving away from the load corresponds to moving around the chart in a clockwise direction. Moving away from the generator toward the load corresponds to moving around the chart in a counter-clockwise direction (see Figure 2.5).

A complete revolution around the chart in a clockwise direction corresponds to a movement of half wavelength away from the load. The Smith chart may be employed to calculate the reflection coefficient and the input impedance for a given transmission line and load impedance. If we are at a matched impedance condition at the center of the Smith chart, any length of transmission line with impedance Z_0 does nothing to the input match. However, if the reflection coefficient of the network (S_{11}) is at some non-ideal impedance, adding transmission line between the network and the reference plane rotates the observed reflection coefficient clockwise about the center of the Smith chart. Further, the rotation occurs at a fixed radius (and VSWR magnitude) if the transmission line has the same characteristic impedance as the source impedance Z_0. Adding a quarter-wavelength, means a 180° phase rotation. Adding one quarter-wavelength from a short circuit moves us 180° to the right side of the chart, to an open circuit.

Smith chart guidelines: The Smith chart contains almost all possible complex impedances within one circle.

- Smith chart horizontal center line represents resistance/conductance.
- Zero resistance is located on the left end of the horizontal center line.
- Infinite resistance is located on the right end of the horizontal center line.

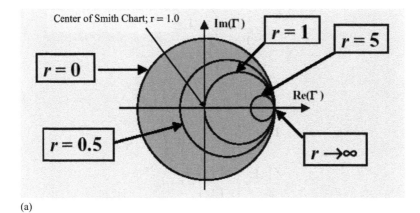

(a)

(b)

(c)

FIGURE 2.5 (a) *r* circles, (b) *x* circles and (c) Smith chart.

- Horizontal centerline – resistive/conductive horizontal scale of the chart.
- Impedances in the Smith chart are normalized to the characteristic impedance of the transmission line and are independent of the characteristic impedance of the transmission.
- The center of the line and of the chart is 1.0 point, where $R = Z_0$ or $G = Y_0$.

At point $r = 1.0$, $Z = Z_0$ and no reflection will occur.

2.7.1 QUARTER-WAVE TRANSFORMERS

A quarter-wave transformer may be used to match a device with impedance Z_L to a system with impedance Z_0, as shown in Figure 2.6. A quarter-wave transformer is a matching network with bandwidth somewhat inversely proportional to the relative mismatch we are trying to match. For a single-stage quarter-wave transformer, the correct transformer impedance is the geometric mean between the impedances of the load and the source. If we substitute to Equation 2.48 $l = \dfrac{\lambda}{4}$, $\beta = \dfrac{2\pi}{\lambda}$ we get:

$$\overline{Z}(-l) = \frac{\dfrac{Z_L}{Z_{02}}\cos\beta\dfrac{\lambda}{4} + j\sin\beta\dfrac{\lambda}{4}}{\cos\beta\dfrac{\lambda}{4} + j\dfrac{Z_L}{Z_{02}}\sin\beta\dfrac{\lambda}{4}} = \frac{Z_{02}}{Z_L} \tag{2.57}$$

$$\overline{Z}(-l) = \frac{Z_{01}}{Z_{02}} = \frac{Z_{02}}{Z_L}$$

We will achieve matching when:

$$Z_{02} = \sqrt{Z_L Z_{01}} \tag{2.58}$$

For complex Z_L values, Z_{02} will also be complex impedance. However, standard transmission lines have real impedance values. To match a complex $Z_L = R + jX$ we transform Z_L to a real impedance $Z_{L1} - jX$ to Z_L. Connecting a capacitor $-jX$ to Z_L is not practical at high frequencies. A capacitor at high frequencies has parasitic inductance and resistance. A practical method to transform Z_L to a real impedance Z_{L1} is to add a transmission line with impedance Z_0 and length l to get a real value Z_{L1}.

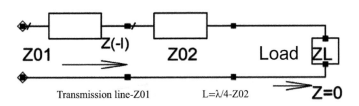

FIGURE 2.6   Quarter-wave transformer.

2.7.2 WIDEBAND MATCHING – MULTI-SECTION TRANSFORMERS

Multi-section quarter-wave transformers are employed for wideband applications. Responses, such as Chebyshev (equi-ripple) and maximally flat, are possible for multi-section transformers. Each section brings us to intermediate impedance. In four-sections transformer from 25 to 50 Ω, intermediate impedances are chosen by using an arithmetic series. For an arithmetic series the steps are equal, $\Delta Z = 6.25\ \Omega$, so the impedances are 31.25 Ω, 37.5 Ω, 43.75 Ω. Solving for the transformers yields $Z_1 = 27.951$, $Z_2 = 34.233$, $Z_3 = 40.505$ and $Z_4 = 46.771\ \Omega$. A second solution to multi-section transformers involves a geometric series from impedance Z_L to impedance Z_S. Here the impedance from one section to the next adjacent section is a constant ratio.

2.7.3 SINGLE STUB MATCHING

A device with admittance Y_L can be matched to a system with admittance Y_0 by using a shunt or series single stub. At a distance l from the load we can get a normalized admittance $\bar{Y}_{in} = 1 + j\bar{B}$. By solving Equation 2.59 we can calculate l.

$$\bar{Y}(l) = \frac{1 + j\dfrac{Z_L}{Z_0}\tan\beta l}{\dfrac{Z_L}{Z_0} + j\tan\beta l} = 1 + j\bar{B} \qquad (2.59)$$

At this location we can add a shunt stub with normalized input susceptance, $-j\bar{B}$ to yield $\bar{Y}_{in} = 1$ as presented in Equation 2.60. $\bar{Y}_{in} = 1$, represents a matched load. The stub can be a short-circuited line or an open-circuited line (Figure 2.7). The susceptance $\bar{B}$ is given in Equation 2.60.

$$\bar{Y}_{in} = 1 + j\bar{B}$$
$$\bar{Y}_{1n} = -j\bar{B} \qquad (2.60)$$
$$\bar{B} = ctg\beta l_1$$

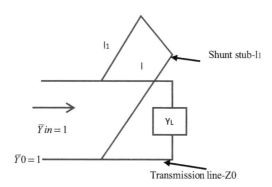

FIGURE 2.7 Single stub matching.

The length l_1 of the short-circuited line may be calculated by solving Equation 2.61.

$$\text{Im}\left\{ \frac{1+j\dfrac{Z_L}{Z_0}\tan\beta l}{\dfrac{Z_L}{Z_0}+j\tan\beta l} \right\} = ctg\,\beta l_1 \tag{2.61}$$

2.8 COAXIAL TRANSMISSION LINE

A coax transmission line consists of two round conductors in which one completely surrounds the other, with the two separated by a continuous solid dielectric [1–9]. The desired propagation mode is TEM. The major advantage of coax over microstrip line is that the transmission line does not radiate. The disadvantages are that coax lines are more expensive. Coax lines are usually employed up to 18 GHz. Coax lines are highly expensive at frequencies higher than 18 GHz. To obtain good performance at higher frequencies, small diameter cables are required to stay below the cutoff frequency. Maxwell laws are employed to compute the electric and magnetic fields in the coax transmission line. A cross section of a coaxial transmission line is shown in Figure 2.8.

$$\oint_S E\cdot ds = \frac{1}{\varepsilon_0}\int_V \rho\,dv = \frac{1}{\varepsilon_0}q \tag{2.62}$$

$$E_r = \frac{\rho_L}{2\pi r\varepsilon}$$

$$V = \int_a^b E\cdot dr = \int_a^b \frac{\rho_L}{2\pi r\varepsilon}\cdot dr = \frac{\rho_L}{2\pi\varepsilon}\ln\frac{b}{a} \tag{2.63}$$

Ampère's law is employed to calculate the magnetic field.

$$\oint H\cdot dl = 2\pi r H_\phi = I$$

$$H_\phi = \frac{I}{2\pi r} \tag{2.64}$$

$$V = \int_a^b E_r\cdot dr = -\int_a^b \eta H_\phi \cdot dr = \frac{I\eta}{2\pi}\ln\frac{b}{a} \tag{2.65}$$

$$Z_0 = \frac{V}{I} = \frac{\eta}{2\pi}\ln\frac{b}{a} \quad \eta = \sqrt{\mu/\varepsilon}$$

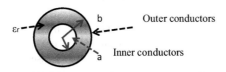

FIGURE 2.8   Coaxial transmission line.

The power flow in the coaxial transmission line may be calculated by calculating the Poynting vector.

$$P = (E \times H) \cdot n = EH$$
$$P = \frac{VI}{2\pi r^2 \ln(b/a)} \tag{2.66}$$
$$W = \int_s P \cdot ds = \int_a^b \frac{VI}{2\pi r^2 \ln(b/a)} 2\pi r \, dr = VI$$

Table 2.12 presents several industry standard coaxial cables. The cables' dimensions, cable impedance and cable cutoff frequency are given in Table 2.12. RG cables are flexible cables. SR cables are semi-rigid cables.

2.8.1 CUTOFF FREQUENCY, F_c, AND WAVELENGTH OF COAX CABLES

The criterion for cutoff frequency is that the circumference at the midpoint inside the dielectric must be less than a wavelength. Therefore, the cutoff wavelength for the TE01 mode is: $\lambda_c = \pi(a+b)\sqrt{\mu_r \varepsilon_r}$.

2.9 MICROSTRIP LINE

Microstrip is a planar printed transmission line. Microstrip is the most popular RF transmission line over the past 20 years [3–10]. Microstrip transmission lines consist of a conductive strip of width "W" and thickness "t" and a wider ground plane, separated by a dielectric layer of thickness, H. In practice, a microstrip line is usually made by etching circuitry on a substrate that has a ground plane on the opposite face. A cross section of the microstrip line is shown in Figure 2.9. The major advantage of microstrip over stripline is that all components can be mounted on top of the board. The disadvantages are that when high isolation is required, such as in a filter or switch, some external shielding is needed. Microstrip circuits

---

TABLE 2.12
Industry standard coaxial cables

Cable type	Outer diameter (in.)	2b (in.)	2a (in.)	Z_0 (Ω)	f_c (GHz)
RG-8A	0.405	0.285	0.089	50	14.0
RG-58A	0.195	0.116	0.031	50	35.3
RG-174	0.100	0.060	0.019	50	65.6
RG-196	0.080	0.034	0.012	50	112
RG-214	0.360	0.285	0.087	50	13.9
RG-223	0.216	0.113	0.037	50	34.6
SR-085	0.085	0.066	.0201	50	60.2
SR-141	0.141	0.1175	0.0359	50	33.8
SR-250	0.250	0.210	0.0641	50	18.9

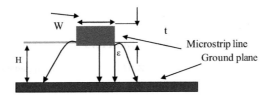

FIGURE 2.9   Microstrip line cross section.

may radiate, causing unintended circuit response. Microstrip is dispersive, signals of different frequencies travel at slightly different speeds. Other microstrip line configurations are offset stripline and suspended air microstrip line. For a microstrip line, not all of the fields are constrained to the same dielectric. At the line edges, the fields pass via air and dielectric substrate. The effective dielectric constant should be calculated.

2.9.1 Effective Dielectric Constant

Part of the fields in the microstrip line structure exists in air and the other part of the fields exists in the dielectric substrate. The effective dielectric constant is somewhat less than the substrate's dielectric constant.

The effective dielectric constant of the microstrip line is calculated by:

$$\text{For} \left(\frac{W}{H}\right) < 1 \tag{2.67a}$$

$$\varepsilon_e = \frac{\varepsilon_r + 1}{2} + \frac{\varepsilon_r - 1}{2}\left[\left(1 + 12\left(\frac{H}{W}\right)\right)^{-0.5} + 0.04\left(1 - \left(\frac{W}{H}\right)\right)^2\right]$$

$$\text{For} \left(\frac{W}{H}\right) \geq 1 \tag{2.67b}$$

$$\varepsilon_e = \frac{\varepsilon_r + 1}{2} + \frac{\varepsilon_r - 1}{2}\left[\left(1 + 12\left(\frac{H}{W}\right)\right)^{-0.5}\right]$$

This calculation ignores strip thickness and frequency dispersion, but their effects are negligible.

2.9.2 Characteristic Impedance

The characteristic impedance Z_0 is a function of the ratio of the width to the height W/H of the transmission line, and has separate solutions depending on the value of W/H. The characteristic impedance Z_0 of microstrip is calculated by:

$$\text{For} \left(\frac{W}{H}\right) < 1 \tag{2.68a}$$

TABLE 2.13

Examples of microstrip line parameters

Substrate	W/H	Impedance (Ω)
Alumina ($\varepsilon_r = 9.8$)	0.95	50
GaAs ($\varepsilon_r = 12.9$)	0.75	50
$\varepsilon_r = 2.2$	3	50

$$Z_0 = \frac{60}{\sqrt{\varepsilon_e}} \ln\left[8\left(\frac{H}{W}\right) + 0.25\left(\frac{H}{W}\right)\right] \Omega$$

$$\text{For}\left(\frac{W}{H}\right) \geq 1 \tag{2.68b}$$

$$Z_0 = \frac{120\pi}{\sqrt{\varepsilon_e}\left[\left(\frac{H}{W}\right) + 1.393 + 0.66 * \ln\left(\frac{H}{W} + 1.444\right)\right]} \Omega$$

We can calculate Z_0 by using Equations 2.68a and 2.68b for a given $\left(\frac{W}{H}\right)$. However, to calculate $\left(\frac{W}{H}\right)$ for a given Z_0 we first should calculate ε_e.

However, to calculate ε_e we should know $\left(\frac{W}{H}\right)$. We first assume that $\varepsilon_e = \varepsilon_r$ and compute $\left(\frac{W}{H}\right)$ for this value of $\left(\frac{W}{H}\right)$ we compute ε_e. Then we compute a new value of $\left(\frac{W}{H}\right)$. Two to three iterations are needed to calculate accurate values of $\left(\frac{W}{H}\right)$ and ε_e. We may calculate $\left(\frac{W}{H}\right)$ with around 10% accuracy by using Equation 2.69. Table 2.13 presents examples of microstrip line parameters.

$$\frac{W}{H} = 8\frac{\sqrt{\left[\frac{Z_0}{e^{42.4}\sqrt{(\varepsilon_r+1)}} - 1\right]\left[\frac{7+\frac{4}{\varepsilon_r}}{11}\right] + \left[\frac{1+\frac{1}{\varepsilon_r}}{0.81}\right]}}{\left[e^{\frac{Z_0}{42.4}\sqrt{(\varepsilon_r+1)}} - 1\right]} \tag{2.69}$$

2.9.3 HIGHER-ORDER TRANSMISSION MODES IN MICROSTRIP LINE

In order to prevent higher-order transmission modes we should limit the thickness of the microstrip substrate to 10% of a wavelength. The cutoff frequency of the higher-order transmission mode is given as $f_c = \frac{c}{4H\sqrt{\varepsilon-1}}$.

EXAMPLES

Higher-order modes will not propagate in microstrip lines printed on alumina substrate 15 mil thick up to 18 GHz. Higher-order modes will not propagate in microstrip lines printed on GaAs substrate 4 mil thick up to 80 GHz. Higher-order modes will not propagate in microstrip lines printed on quartz substrate 5 mil thick up to 120 GHz.

Losses in microstrip line: Losses in microstrip line are due to conductor loss, radiation loss and dielectric loss.

2.9.4 CONDUCTOR LOSS

Conductor loss may be calculated by using Equation 2.70.

$$\alpha_c = 8.686 \log\left(R_S/(2WZ_0)\right) \quad \text{dB/length}$$
$$R_S = \sqrt{\pi f \mu \rho} \quad \text{Skin resistant}$$

(2.70)

Conductor losses may also be calculated by defining an equivalent loss tangent δ_c, given by $\delta_c = \delta_s/h$, where $\delta_s = \sqrt{2/\omega\mu\sigma}$. Where σ is the strip conductivity, h is the substrate height and μ is the free space permeability.

2.9.5 DIELECTRIC LOSS

Dielectric loss may be calculated by using Equation 2.71.

$$\alpha_d = 27.3 \frac{\varepsilon_r}{\sqrt{\varepsilon_{eff}}} \frac{\varepsilon_{eff} - 1}{\varepsilon_r - 1} \frac{tg\delta}{\lambda_0} \, \text{dB/cm}$$
$$tg\delta = \text{dieletric loss coefficent}$$

(2.71)

Losses in microstrip lines are presented in Tables 2.14, 2.15 and 2.16 for several microstrip line structures. For example, total loss of a microstrip line presented in Table 2.15 at 40 GHz is 0.5 dB/cm. For example, total loss of a microstrip line presented in Table 2.16 at 40 GHz is 1.42 dB/cm. We may conclude that losses in microstrip lines limit the applications of microstrip technology at mm wave frequencies.

2.10 WAVEGUIDES

Waveguides are low loss transmission lines.

Waveguides may be rectangular or circular. A rectangular waveguide structure is presented in Figure 2.10. A waveguide's structure is uniform in the z direction. Fields in waveguides are evaluated by solving the Helmholtz equation. A wave equation is given in Equation 2.72. Wave equation in rectangular coordinate system is given in Equation 2.73.

$$\nabla^2 E = \omega^2 \mu \varepsilon E = -k^2 E$$
$$\nabla^2 H = \omega^2 \mu \varepsilon H = -k^2 H \qquad (2.72)$$
$$k = \omega\sqrt{\mu\varepsilon} = \frac{\omega}{v} = \frac{2\pi}{\lambda}$$

TABLE 2.14
Microstrip line losses for alumina substrate 10 mil thick*

Frequency (GHz)	Loss tangent loss (dB/cm)	Metal loss (dB/cm)	Total loss (dB/cm)
10	−0.005	−0.12	−0.124
20	−0.009	−0.175	−0.184
30	−0.014	−0.22	−0.23
40	−0.02	−0.25	−0.27

* Line parameters. Alumina, H = 254 μm (10 mil), W = 247 μm, Er = 9.9, tan δ = 0.0002, 3 μm gold, conductivity 3.5E7 mhos/m.

TABLE 2.15
Microstrip line losses for alumina substrate 5 mil thick*

Frequency (GHz)	Loss tangent loss (dB/cm)	Metal loss (dB/cm)	Total loss (dB/cm)
10	−0.004	−0.23	−0.23
20	−0.009	−0.333	−0.34
30	−0.013	−0.415	−0.43
40	−0.018	−0.483	−0.5

* Alumina, H = 127 μm (5 mil), W = 120 μm, Er = 9.9, tan δ = 0.0002, 3 μm gold, conductivity 3.5E7 mhos/m.

TABLE 2.16
Microstrip line losses for GaAs substrate 2 mil thick*

Frequency (GHz)	Tangent loss (dB/cm)	Metal loss (dB/cm)	Total loss (dB/cm)
10	−0.010	−0.66	−0.67
20	−0.02	−0.96	−0.98
30	−0.03	−1.19	−1.22
40	−0.04	−1.38	−1.42

* GaAs, H = 50 μm (2 mil), W = 34 μm, Er = 12.88, tan δ = 0.0004, 3 μm gold, conductivity 3.5E7 mhos/m.

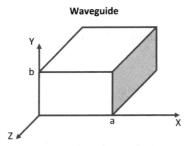

FIGURE 2.10 Rectangular waveguide structure.

$$\nabla^2 E = \frac{\partial^2 E_i}{\partial x^2} + \frac{\partial^2 E_i}{\partial y^2} + \frac{\partial^2 E_i}{\partial z^2} + k^2 E_i = 0 \quad i = x, y, z$$

$$\nabla H = \frac{\partial^2 H_i}{\partial x^2} + \frac{\partial^2 H_i}{\partial y^2} + \frac{\partial^2 H_i}{\partial z^2} + k^2 H_i = 0$$

(2.73)

The wave equation solution may be written as: $E = f(z)g(x, y)$. Field variation in the z direction may be written as $e^{-j\beta z}$. The derivative of this expression in the direction may be written as $-j\beta e^{j\beta z}$. The Maxwell equation may be presented as written in Equations 2.74 and 2.75. A field may be represented as a superposition of waves in the transverse and longitudinal directions.

$$E(x,y,z) = e(x,y)e^{-j\beta z} + e_z(x,y)e^{-j\beta z}$$
$$H(x,y,z) = h(x,y)e^{-j\beta z} + h_z(x,y)e^{-j\beta z}$$
$$\nabla XE = (\nabla_t - j\beta a_z) \times (e + e_z)e^{-j\beta z} = -j\omega\mu(h + h_z)e^{-j\beta z}$$
$$\nabla_t \times e - j\beta a_z \times e + \nabla_t \times e_z - j\beta a_z \times e_z = -j\omega\mu(h + h_z)$$
$$a_z \times e_z = 0$$
$$\nabla_t \times e = -j\omega\mu h_z$$
$$-j\beta a_z \times e + \nabla_t \times e_z = -a_z \times \nabla_t e_z - j\beta a_z \times e = -j\omega\mu h$$

(2.74)

$$\nabla_t \times h = -j\omega\varepsilon e_z$$
$$a_z \times \nabla_t h_z + j\beta a_z \times h = -j\omega\varepsilon e$$

(2.75)

$$\nabla \cdot \mu H = (\nabla_t - j\beta a_z) \cdot (h + h_z)\mu e^{-j\beta z} = 0$$
$$\nabla_t \cdot h - j\beta a_z \cdot h_z = 0$$
$$\nabla \cdot \varepsilon E = 0 \quad \nabla_t \cdot e - j\beta a_z \cdot e_z = 0$$

(2.76)

Waves may be characterized as TEM, TE or TM waves. In TEM waves $e_z = h_z = 0$. In TE waves $e_z = 0$. In TM waves $h_z = 0$ where TEM is transverse electromagnetic, TE is transverse electric and TM is transverse magnetic mode.

2.10.1 TE WAVES

In TE waves $e_z = 0$. h_z is a solution of Equation 2.76. The solution to Equation 2.76 may be written as $h_z = f(x)g(y)$.

$$\nabla_t^2 h_z = \frac{\partial^2 h_z}{\partial x^2} + \frac{\partial^2 h_z}{\partial y^2} + k_c^2 h_z = 0 \tag{2.77}$$

By applying $h_z = f(x)g(y)$ to Equation 2.77 and dividing by fg we get Equation 2.78.

$$\frac{f''}{f} + \frac{g''}{g} + k_c^2 = 0 \tag{2.78}$$

f is a function that varies in the x direction and g is function that varies in the y direction. The sum of f and g may be equal to zero only if they are equal to a constant. These facts are written in Equation 2.79.

$$\frac{f''}{f} = -k_x^2; \quad \frac{g''}{g} = -k_y^2$$
$$k_x^2 + k_y^2 = k_c^2 \tag{2.79}$$

The solutions for f and g are given in Equation 2.80. A_1, $A_2 B_1$, B_2 are derived by applying h_z boundary conditions to Equation 2.80.

$$f = A_1 \cos k_x x + A_2 \sin k_x x$$
$$g = B_1 \cos k_y y + B_2 \sin k_y y \tag{2.80}$$

h_z boundary conditions are written in Equation 2.81.

$$\frac{\partial h_z}{\partial x} = 0 \quad \text{at } x = 0, a$$
$$\frac{\partial h_z}{\partial y} = 0 \quad \text{at } y = 0, b \tag{2.81}$$

By applying h_z boundary conditions to Equation 2.80 we get the relations written in Equation 2.82.

$$-k_x A_1 \sin k_x x + k_x A_2 \cos k_x x = 0$$
$$-k_y B_1 \sin k_y y + k_y B_2 \cos k_y y = 0$$
$$A_2 = 0 \quad k_x a = 0 \quad k_x = \frac{n\pi}{a} \quad n = 0, 1, 2$$
$$B_2 = 0 \quad k_y b = 0 \quad k_y = \frac{m\pi}{b} \quad m = 0, 1, 2 \tag{2.82}$$

The solution for h_z is given in Equation 2.83.

$$h_z = A_{nm} \cos \frac{n\pi x}{a} \cos \frac{m\pi y}{b}$$

$$n = m \neq 0 \quad n = 0,1,2 \quad m = 0,1,2 \tag{2.83}$$

$$k_{c,nm} = \left[\left(\frac{n\pi}{a} \right)^2 + \left(\frac{m\pi}{b} \right)^2 \right]^{1/2}$$

Both n and m cannot be zero. The wave number at cutoff is $k_{c,nm}$ and depends on the waveguide dimensions. The propagation constant γ_{nm} is given in Equation 2.84.

$$\gamma_{nm} = j\beta_{nm} = j\left(k_0^2 - k_c^2 \right)^{1/2} =$$

$$= j\left[\left(\frac{2\pi}{\lambda_0} \right)^2 - \left(\frac{n\pi}{a} \right)^2 - \left(\frac{m\pi}{b} \right)^2 \right]^{1/2} \tag{2.84}$$

For $k_0 > k_{c,nm}$, β is real and wave will propagate. For $k_0 < k_{c,nm}$, β is imaginary and the wave will decay rapidly. Frequencies that define propagating and decaying waves are called cutoff frequencies. We may calculate cutoff frequencies by using Equation 2.85.

$$f_{c,nm} = \frac{c}{2\pi} k_{c,nm} = \frac{c}{2\pi} \left[\left(\frac{n\pi}{a} \right)^2 + \left(\frac{m\pi}{b} \right)^2 \right]^{1/2} \tag{2.85}$$

For $a = 2$ the cutoff wavelength is computed by using Equation 2.86.

$$\lambda_{c,nm} = \frac{2ab}{\left[n^2 b^2 + m^2 a^2 \right]^{1/2}} = \frac{2a}{\left[n^2 + 4m^2 \right]^{1/2}}$$

$$\lambda_{c,10} = 2a \quad \lambda_{c,01} = a \quad \lambda_{c,11} = 2a/\sqrt{5} \tag{2.86}$$

$$\frac{c}{2a} \langle f_{01} \rangle \frac{c}{a}$$

For $\dfrac{c}{2a} \langle f_{01} \rangle \dfrac{c}{a}$ the dominant is TE10.

By using Equations 2.74 to 2.76 we can derive the electromagnetic fields that propagate in the waveguide as given in Equation 2.87.

$$H_z = A_{nm} \cos \frac{n\pi x}{a} \cos \frac{m\pi y}{b} e^{\pm j\beta_{nm}z}$$

$$H_x = \pm j \frac{n\pi \beta_{nm}}{ak_{c,nm}^2} A_{nm} \sin \frac{n\pi x}{a} \cos \frac{m\pi y}{b} e^{\pm j\beta_{nm}z}$$

$$H_y = \pm j \frac{m\pi \beta_{nm}}{bk_{c,nm}^2} A_{nm} \cos \frac{n\pi x}{a} \sin \frac{m\pi y}{b} e^{\pm j\beta_{nm}z} \qquad (2.87)$$

$$E_X = Z_{h,nm} j \frac{m\pi \beta_{nm}}{bk_{c,nm}^2} A_{nm} \cos \frac{n\pi x}{a} \sin \frac{m\pi y}{b} e^{\pm j\beta_{nm}z}$$

$$E_Y = -j Z_{h,nm} \frac{n\pi \beta_{nm}}{ak_{c,nm}^2} A_{nm} \sin \frac{n\pi x}{a} \cos \frac{m\pi y}{b} e^{\pm j\beta_{nm}z}$$

The impedance of the nm modes is given: $Z_{h,nm} = \dfrac{e_x}{h_y} = \dfrac{k_0}{\beta_{nm}} \sqrt{\dfrac{\mu_0}{\varepsilon_0}}$.

The power of the nm mode is computed by using pointing vector calculation as shown in Equation 2.88.

$$P_{nm} = 0.5 \operatorname{Re} \int_0^a \int_0^b E \times H^* \cdot a_z dxdy = 0.5 \operatorname{Re} Z_{h,nm} \int_0^a \int_0^b \left(H_y H_Y^* + H_x H_x^* \right) dxdy$$

$$\int_0^a \int_0^b \cos^2 \frac{n\pi x}{a} \sin^2 \frac{m\pi y}{b} dxdy = \frac{ab}{4} \quad n \neq 0 \ m \neq 0$$

$$\text{or} \quad \frac{ab}{2} \quad n \text{ or } m = 0 \qquad (2.88)$$

$$P_{nm} = \frac{|A_{nm}|^2}{2\varepsilon_{0n}\varepsilon_{0m}} \left(\frac{\beta_{nm}}{k_{c,nm}} \right)^2 Z_{h,nm} \quad \varepsilon_{0n} = 1, \ n = 0 \quad \varepsilon_{0n} = 2, \ n \succ 0$$

The TE mode with the lowest cutoff frequency in rectangular waveguide is TE10. TE10 fields in rectangular waveguide are shown in Figure 2.11.

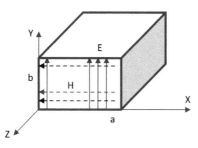

FIGURE 2.11   Transverse electric TE10 mode.

2.10.2 TM WAVES

In TM waves $h_z = 0$. e_z is a solution of Equation 2.89. The solution to Equation 2.89 may be written as $e_z = f(x)g(y)$. e_z should be zero at the metallic walls. e_z boundary conditions are written in Equation 2.90. The solution for e_z is given in Equation 2.91.

$$\nabla_t^2 e_z = \frac{\partial^2 e_z}{\partial x^2} + \frac{\partial^2 e_z}{\partial y^2} + k_c^2 e_z = 0 \tag{2.89}$$

$$\begin{aligned} e_z &= 0 \quad \text{at } x = 0, a \\ e_z &= 0 \quad \text{at } y = 0, b \end{aligned} \tag{2.90}$$

$$e_z = A_{nm} \sin \frac{n\pi x}{a} \sin \frac{m\pi y}{b}$$
$$n = m \neq 0 \quad n = 0,1,2 \quad m = 0,1,2 \tag{2.91}$$
$$k_{c,nm} = \left[\left(\frac{n\pi}{a} \right)^2 + \left(\frac{m\pi}{b} \right)^2 \right]^{1/2}$$

The first propagating TM mode is TM11, $n = m = 1$. By using Equations 2.80 to 2.82 and Equation 2.89 we can derive the electromagnetic fields that propagate in the waveguide as given in Equation 2.92.

$$E_z = \sin \frac{n\pi x}{a} \sin \frac{m\pi y}{b} e^{\pm j\beta_{nm}z}$$
$$E_x = -j \frac{n\pi \beta_{nm}}{ak_{c,nm}^2} \cos \frac{n\pi x}{a} \sin \frac{m\pi y}{b} e^{\pm j\beta_{nm}z}$$
$$E_y = -j \frac{m\pi \beta_{nm}}{bk_{c,nm}^2} A_{nm} \sin \frac{n\pi x}{a} \cos \frac{m\pi y}{b} e^{\pm j\beta_{nm}z} \tag{2.92}$$
$$H_X = \frac{-E_y}{Z_{e,nm}}$$
$$H_Y = \frac{E_x}{Z_{e,nm}}$$

The impedance of the nm modes is $Z_{e,nm} = \dfrac{\beta_{nm}}{k_0} \sqrt{\dfrac{\mu_0}{\varepsilon_0}}$.

The TM mode with the lowest cutoff frequency in rectangular waveguide is TM11. TM11 fields in rectangular waveguide are shown in Figure 2.12.

2.11 CIRCULAR WAVEGUIDE

A circular waveguide is used to transmit an electromagnetic wave in circular polarization. At high frequencies, the attenuation of several modes in circular waveguide

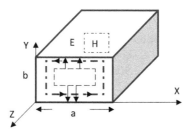

FIGURE 2.12   Transverse magnetic, TM11 mode.

is lower than in rectangular waveguide. The circular waveguide's structure is uniform in the z direction. Fields in waveguides are evaluated by solving Helmholtz equation in cylindrical coordinate system. A circular waveguide in cylindrical coordinate system is presented in Figure 2.13. A wave equation is given in Equation 2.93. A wave equation in cylindrical coordinate system is given in Equation 2.94.

$$\nabla^2 E = \omega^2 \mu \varepsilon E = -k^2 E$$
$$\nabla^2 H = \omega^2 \mu \varepsilon H = -k^2 H \qquad (2.93)$$
$$k = \omega\sqrt{\mu\varepsilon} = \frac{\omega}{v} = \frac{2\pi}{\lambda}$$

$$\nabla^2 E = \frac{1}{r}\frac{\partial}{\partial r}\left(r\frac{\partial E_i}{\partial r}\right) + \frac{1}{r^2}\frac{\partial^2 E_i}{\partial \phi^2} + \frac{\partial^2 E_i}{\partial z^2} - \gamma^2 E_i = 0 \quad i = r, \phi, z$$
$$\qquad (2.94)$$
$$\nabla^2 H = \frac{1}{r}\frac{\partial}{\partial r}\left(r\frac{\partial H_i}{\partial r}\right) + \frac{1}{r^2}\frac{\partial^2 H_i}{\partial \phi^2} + \frac{\partial^2 H_i}{\partial z^2} - \gamma^2 H_i = 0$$

The solution to Equation 2.94 may be written as $E = f(r)g(\phi)h(z)$. By applying E to Equation 2.94 and dividing by $f(r)g(\phi)h(z)$ we get Equation 2.95.

$$\nabla^2 E = \frac{1}{rf}\frac{\partial}{\partial r}\left(r\frac{\partial f}{\partial r}\right) + \frac{1}{r^2 g}\frac{\partial^2 g}{\partial \phi^2} + \frac{\partial^2 h}{h\partial z^2} - \gamma^2 = 0 \qquad (2.95)$$

f is a function that varies in the r direction, g is a function that varies in the ϕ direction and h is a function that varies in the z direction. The sum of f, g and h may be equal

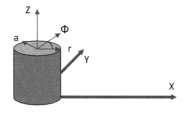

FIGURE 2.13   Circular waveguide structure.

to zero only if they are equal to a constant. The solution for h is written in Equation 2.96. The propagation constant is γ_g.

$$\frac{\partial^2 h}{h \partial z^2} - \gamma_g^2 = 0$$

$$h = A e^{-\gamma_g z} + B e^{\gamma_g z}$$

(2.96)

$$\frac{r}{f}\frac{\partial}{\partial r}\left(r\frac{\partial f}{\partial r}\right) + \frac{1}{g}\frac{\partial^2 g}{\partial \phi^2} - \left(\gamma^2 - \gamma_g^2\right) = 0$$

(2.97)

The solution for g is written in Equation 2.98.

$$\frac{\partial^2 g}{g \partial \phi^2} + n^2 = 0$$

$$g = A_n \sin n\phi + B_n \cos n\phi$$

(2.98)

$$r\frac{\partial}{\partial r}\left(r\frac{\partial f}{\partial r}\right) + \left(\left(k_c r\right)^2 - n^2\right)f = 0$$

$$k_c^2 + \gamma^2 = \gamma_g^2$$

(2.99)

Equation 2.99 is a Bessel equation. The solution of this equation is written in Equation 2.100. $J_n(k_c r)$ is a Bessel equation with order n and represents a standing wave that varies as a cosine function in the circular waveguide. $N_n(k_c r)$ is a Bessel equation with order n and represents a standing wave that varies as a sine function in the circular waveguide.

$$f = C_n J_n\left(k_c r\right) + D_n N_n\left(k_c r\right)$$

(2.100)

The general solution for the electric fields in the circular waveguide is given in Equation 2.101. For $r = 0$, $N_n(k_c r)$ goes to infinity, so $D_n = 0$.

$$E\left(r,\phi,z\right) = \left(C_n J_n\left(k_c r\right) + D_n N_n\left(k_c r\right)\right)\left(A_n \sin n\phi + B_n \cos n\phi\right)e^{\pm\gamma_g z}$$

(2.101)

$$A_n \sin n\phi + B_n \cos n\phi = \sqrt{A_n^2 + B_n^2}\cos\left(n\phi + \tan^{-1}\left(\frac{A_n}{B_n}\right)\right) = F_n \cos n\phi$$

(2.102)

The general solution for the electric fields in the circular waveguide is given in Equation 2.103.

$$E\left(r,\phi,z\right) = E_0\left(J_n\left(k_c r\right)\right)\left(\cos n\phi\right)e^{\pm\gamma_g z} \quad \text{if } \alpha = 0$$

$$E\left(r,\phi,z\right) = E_0\left(J_n\left(k_c r\right)\right)\left(\cos n\phi\right)e^{\pm\beta_g z}$$

$$\beta_g = \pm\sqrt{\omega^2 \mu\varepsilon - k_c^2}$$

(2.103)

2.11.1 TE WAVES IN CIRCULAR WAVEGUIDE

In TE waves $e_z = 0$, H_z is a solution of Equation 2.103. The solution to Equation 2.104 may be written as given in Equation 2.105.

$$\nabla^2 H_z = \gamma^2 H_z \tag{2.104}$$

$$H_z = H_{0z}\left(J_n\left(k_c r\right)\right)\left(\cos n\phi\right)e^{\pm j\beta_g z} \tag{2.105}$$

The electric and magnetic fields are solutions of Maxwell equations as written in Equations 2.106 and 2.107.

$$\nabla X E = -j\omega\mu H \tag{2.106}$$

$$\nabla \times H = j\omega\varepsilon E \tag{2.107}$$

Field variation in the z direction may be written as $e^{-j\beta z}$. The derivative of this expression in the direction may be written as $-j\beta e^{j\beta z}$. The electric and magnetic fields' components are solutions of Equations 2.108 and 2.109.

$$E_r = -\frac{j\omega\mu}{k_c^2}\frac{1}{r}\left(\frac{\partial H_z}{\partial \phi}\right)$$

$$E_\phi = \frac{j\omega\mu}{k_c^2}\left(\frac{\partial H_z}{\partial r}\right) \tag{2.108}$$

$$H_\phi = -\frac{-j\beta_g}{k_c^2}\frac{1}{r}\left(\frac{\partial H_z}{\partial \phi}\right)$$

$$H_r = \frac{-j\beta_g}{k_c^2}\left(\frac{\partial H_z}{\partial r}\right) \tag{2.109}$$

H_z, H_r, and E_ϕ boundary conditions are written in Equation 2.110.

$$\frac{\partial H_z}{\partial r} = 0 \quad r = a$$

$$H_r = 0 \quad r = a \tag{2.110}$$

$$E_\phi = 0 \quad r = a$$

By applying the boundary conditions to Equation 2.109 we get the relations written in Equation 2.111. The solutions of Equation 2.111 are listed in Table 2.17.

$$\frac{\partial H_z}{\partial r}\mid r = a = H_{0z}\left(J_n'\left(k_c a\right)\right)\left(\cos n\phi\right)e^{-\beta_g z} = 0$$

$$J_n'\left(k_c a\right) = 0 \tag{2.111}$$

TABLE 2.17
Circular waveguide transverse electric modes

p	$(n = 0)$ X'_{np}	$(n = 1)$ X'_{np}	$(n = 2)$ X'_{np}	$(n = 3)$ X'_{np}	$(n = 4)$ X'_{np}	$(n = 5)$ X'_{np}
1	3.832	1.841	3.054	4.201	5.317	6.416
2	7.016	5.331	6.706	8.015	9.282	10.52
3	10.173	8.536	9.969	11.346	12.682	13.987
4	13.324	11.706	13.170			

The wave number at cutoff is $k_{c,np}$. $k_{c,np}$ depends on the waveguide dimensions. The propagation constant $\gamma_{g,np}$ is given in Equation 2.112.

$$\gamma_{g,np} = j\beta_{g,np} = j\left(k_0^2 - k_c^2\right)^{1/2} =$$
$$= j\left[\left(\frac{2\pi}{\lambda_0}\right)^2 - \left(\frac{X'_{np}}{a}\right)^2\right]^{1/2} \tag{2.112}$$

For $k_0 > k_{c,nm}$, β is real and the wave will propagate. For $k_0 < k_{c,nm}$, β is imaginary and the wave will decay rapidly. Frequencies that define propagating and decaying waves are called cutoff frequencies. We may calculate cutoff frequencies by using Equation 2.113.

$$f_{c,nm} = \frac{cX'_{np}}{2\pi a} \tag{2.113}$$

We get the field's components by solving Equations 2.108 and 2.109. The field's components are written in Equation 2.114.

$$H_z = H_{0z}\left(J_n\left(\frac{X'_{np}r}{a}\right)\right)(\cos n\phi)e^{-j\beta_g z}$$

$$H_\phi = \frac{E_{0r}}{Z_g}\left(J_n\left(\frac{X'_{np}r}{a}\right)\right)(\sin n\phi)e^{-j\beta_g z}$$

$$H_r = \frac{E_{0\phi}}{Z_g}\left(J'_n\left(\frac{X'_{np}r}{a}\right)\right)(\cos n\phi)e^{-j\beta_g z}$$

$$E_\phi = E_{0\phi}\left(J'_n\left(\frac{X'_{np}r}{a}\right)\right)(\cos \phi)e^{-j\beta_g z} \tag{2.114}$$

$$E_r = E_{0r}\left(J_n\left(\frac{X'_{np}r}{a}\right)\right)(\sin \phi)e^{-j\beta_g z}$$

The impedance of the *np* modes is written in Equation 2.115.

$$Z_{g,np} = \frac{E_r}{H_\phi} = \frac{\omega\mu}{\beta_{g,np}} = \frac{\eta}{\sqrt{1-\left(\frac{f_c}{f}\right)^2}} \tag{2.115}$$

2.11.2 TM Waves in Circular Waveguide

In TM waves $h_z = 0$, e_z is a solution of Equation 2.116. The solution to Equation 2.116 is written in Equation 2.117.

$$\nabla^2 E_z = \gamma^2 E_z \tag{2.116}$$

$$E_z = E_{0z}\left(J_n(k_c r)\right)\left(\cos n\phi\right)e^{\pm j\beta_g z} \tag{2.117}$$

The electric and magnetic fields are solutions of Maxwell equations as written in Equations 2.118 and 2.119.

$$\nabla X E = -j\omega\mu H \tag{2.118}$$

$$\nabla \times H = j\omega\varepsilon E \tag{2.119}$$

Field variation in the z direction may be written as $e^{-j\beta z}$. The derivative of this expression in the direction may be written as $-j\beta e^{j\beta z}$. The electric and magnetic fields' components are solutions of Equations 2.120 and 2.121.

$$H_r = \frac{j\omega\varepsilon}{k_c^2}\frac{1}{r}\left(\frac{\partial H_z}{\partial \phi}\right)$$
$$H_\phi = \frac{-j\omega\varepsilon}{k_c^2}\left(\frac{\partial H_z}{\partial r}\right) \tag{2.120}$$

$$E_\phi = -\frac{-j\beta_g}{k_c^2}\frac{1}{r}\left(\frac{\partial E_z}{\partial \phi}\right)$$
$$E_r = \frac{-j\beta_g}{k_c^2}\left(\frac{\partial E_z}{\partial r}\right) \tag{2.121}$$

E_r boundary condition is written in Equation 2.122.

$$E_z = 0 \quad \text{at } r = a \tag{2.122}$$

By applying the boundary conditions to Equation 2.121 we get the relations written in Equation 2.123. The solutions of Equation 2.123 are listed in Table 2.18.

$$E_z|(r = a)) = H_{0z}\left(J_n(k_c a)\right)\left(\cos n\phi\right)e^{-j\beta_g z} = 0$$
$$J_n(k_c a) = 0 \tag{2.123}$$

The wave number at cutoff is $k_{c,np}$. $k_{c,np}$ depends on the waveguide dimensions. The propagation constant $\gamma_{g,np}$ is given in Equation 2.124.

TABLE 2.18

Circular waveguide transverse magnetic modes

p	$n = 0, X_{np}$	$n = 1, X_{np}$	$n = 2, X_{np}$	$n = 3, X_{np}$	$n = 4, X_{np}$	$n = 5, X_{np}$
1	2.405	3.832	5.136	6.38	7.588	8.771
2	5.52	7.106	8.417	9.761	11.065	12.339
3	8.645	10.173	11.62	13.015	14.372	
4	11.792	13.324	14.796			

$$\gamma_{g,np} = j\beta_{g,np} = j\left(k_0^2 - k_c^2\right)^{1/2} =$$
$$= j\left[\left(\frac{2\pi}{\lambda_0}\right)^2 - \left(\frac{X_{np}}{a}\right)^2\right]^{1/2} \quad (2.124)$$

For $k_0 \rangle k_{c,nm}$, β is real and the wave will propagate. For $k_0 \langle k_{c,nm}$, β is imaginary and the wave will decay rapidly. Frequencies that define propagating and decaying waves are called cutoff frequencies. We may calculate cutoff frequencies by using Equation 2.125.

$$f_{c,nm} = \frac{cX_{np}}{2\pi a} \quad (2.125)$$

We get the field's components by solving Equations 2.120 and 2.121. The field's components are written in Equation 2.126.

$$E_z = E_{0z}\left(J_n\left(\frac{X_{np}r}{a}\right)\right)(\cos n\phi)e^{-j\beta_g z}$$

$$H_\phi = \frac{E_{0r}}{Z_g}\left(J_n'\left(\frac{X_{np}r}{a}\right)\right)(\cos n\phi)e^{-j\beta_g z}$$

$$H_r = \frac{E_{0\phi}}{Z_g}\left(J_n\left(\frac{X_{np}r}{a}\right)\right)(\sin n\phi)e^{-j\beta_g z} \quad (2.126)$$

$$E_\phi = E_{0\phi}\left(J_n\left(\frac{X_{np}r}{a}\right)\right)(\sin \phi)e^{-j\beta_g z}$$

$$E_r = E_{0r}\left(J_n'\left(\frac{X_{np}r}{a}\right)\right)(\cos \phi)e^{-j\beta_g z}$$

The impedance of the np modes is written in Equation 2.127.

$$Z_{g,np} = \frac{E_r}{H_\phi} = \frac{\beta_{g,np}}{\omega\varepsilon} = \eta\sqrt{1 - \left(\frac{f_c}{f}\right)^2} \quad (2.127)$$

The mode with the lowest cutoff frequency in circular waveguide is TE11. TE11 fields in circular waveguide are shown in Figure 2.14.

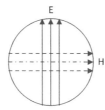

FIGURE 2.14   Transverse electric TE11 fields in circular waveguide.

REFERENCES

[1] W. Ramo, V. Duzer (1994). *Fields and Waves in Communication Electronics*, 3rd edition, John Wiley, New York.

[2] R. E. Collin (1966). *Foundation for Microwave Engineering*, McGraw-Hill, New York.

[3] C.A. Balanis (1996). *Antenna Theory: Analysis and Design*, 2nd edition, Wiley, Hoboken, NJ.

[4] A. Sabban (2015). *Low Visibility Antennas for Communication Systems*, Taylor & Francis Group, Boca Raton, FL.

[5] A. Sabban (2016). *Wideband RF Technologies and Antenna in Microwave Frequencies*, John Wiley & Sons, New York.

[6] L. C. Godara Ed. (2002). *Handbook of Antennas in Wireless Communications*, CRC Press LLC, Boca Raton, FL.

[7] J. D. Kraus, R. J. Marhefka (2002). *Antennas for All Applications*, 3rd edition, McGraw Hill, New Delhi, India.

[8] F. T. Ulaby (2005). *Electromagnetics for Engineers*, Pearson Education, Upper Saddle River, NJ.

[9] J. R. James, P. S Hall, C. Wood (1981). *Microstrip Antenna Theory and Design*, Peregrinus on behalf of the Institution of Electrical Engineers, London.

[10] A. Sabban (2014). *RF Engineering, Microwave and Antennas*, Saar Publication, Israel.

[11] *ITU-R Recommendation V.431: Nomenclature of the Frequency and Wavelength Bands Used in Telecommunications*. International Telecommunication Union, Geneva, 2015.

[12] *IEEE Standard 521-2002: Standard Letter Designations for Radar-Frequency Bands*, 2002, E-ISBN: 0-7381-3356-6.

[13] AFR 55-44/AR 105-86/OPNAVINST 3430.9A/MCO 3430.1, 27 October 1964 superseded by AFR 55-44/AR 105-86/OPNAVINST 3430.1A/MCO 3430.1A, 6 December 1978: *Performing Electronic Countermeasures in the United States and Canada*, Attachment 1, ECM Frequency Authorizations.

[14] L. A. Belov, S. M. Smolskiy, V. N. Kochemasov (2012). *Handbook of RF, Microwave, and Millimeter-Wave Components*, Artech House, Boston, pp. 27–28, ISBN: 978-1-60807-209-5.

[15] L. C. Chirwa, P. A. Hammond, S. Roy, D. R. S. Cumming (2003). "Electromagnetic radiation from ingested sources in the human intestine between 150 MHz and 1.2 GHz," *IEEE Trans. Biomed. Eng.*, 50(4), 484–492.

[16] D. Werber, A. Schwentner, E. M. Biebl (2006). "Investigation of RF transmission properties of human tissues," *Adv. Radio Sci.*, 4, 357–360.

[17] A. Sabban (2017). *Novel Wearable Antennas for Communication and Medical Systems*, Taylor & Francis Group, Boca Raton, FL.

3 Antennas for Wearable 5G Communication and Medical Systems

Albert Sabban

CONTENTS

3.1 INTRODUCTION TO ANTENNAS

Antennas are major components in communication systems [1–17]. Mobile antenna systems are presented in [10]. An antenna is used to efficiently radiate electromagnetic energy in desired directions. Antennas match radio frequency systems, sources of electromagnetic energy, to space. All antennas may be used to receive or radiate energy. Antennas transmit or receive electromagnetic waves. Antennas convert electromagnetic radiation into an electric current, or vice versa. Antennas transmit and receive electromagnetic radiation at radio frequencies. Antennas are commonly employed in air or outer space, but can also be operated under water, on and inside the human body or even through soil and rock at low frequencies for short distances. Physically, an antenna is an arrangement of one or more conductors. In transmitting mode, an alternating current is created in the elements by applying a voltage at the antenna terminals, causing the elements to radiate an electromagnetic field. In receiving mode, an electromagnetic field from another source induces an alternating current in the elements and a corresponding voltage at the antenna's terminals. Some receiving antennas (such as parabolic and horn) incorporate shaped reflective surfaces to receive the radio waves striking them and direct or focus them onto the actual conductive elements.

3.2 ANTENNA: DEFINITIONS

Antenna: An antenna efficiently radiates electromagnetic energy in desired directions. Antennas match radio frequency systems, sources of electromagnetic energy, to space.

Radiator: This is the basic element of an antenna. An antenna can be made up of multiple radiators.

Radiation pattern: A radiation pattern is the antenna radiated field as a function of the direction in space. It is a way of plotting the radiated power from an antenna. This power is measured at various angles at a constant distance from the antenna.

Main beam: Main beam is the region around the direction of maximum radiation, usually the region that is within 3 dB of the peak of the main lobe.

Beam width: Beam width is the angular range of the antenna pattern in which at least half of the maximum power is emitted. This angular range, of the major lobe, is defined as the points at which the field strength falls around 3 dB with regard to the maximum field strength.

Side lobes: Side lobes are smaller beams that are away from the main beam. Side lobes present radiation in undesired directions. The side lobe level is a parameter used to characterize the antenna radiation pattern. It is the maximum value of the side lobes away from the main beam and is usually expressed in decibels.

Bore sight: The direction in space to which the antenna radiates maximum electromagnetic energy.

Range: Antenna range is the radial range from an antenna to an object in space.

Azimuth (AZ): The angle from left to right from a reference point, from $0°$ to $360°$.

Elevation (EL): The EL angle is the angle from horizontal (x, y) plane, from $-90°$ (down) to $+90°$ (up).

Radiated power: Total radiated power when the antenna is excited by a current or voltage of known intensity.

Isotropic radiator: Theoretical lossless radiator that radiates, or receives, equal electromagnetic energy in free space to all directions.

Wearable antenna: An antenna worn on a human body.

Directivity: The ratio between the amounts of energy propagating in a certain direction compared to the average energy radiated to all directions over a sphere.

$$D = \frac{P(\theta,\phi)\,\text{maximal}}{P(\theta,\phi)\,\text{average}} = 4\pi \frac{P(\theta,\phi)\,\text{maximal}}{P\,\text{rad}} \tag{3.1}$$

$$P(\theta,\phi)\,\text{average} = \frac{1}{4\pi}\iint P(\theta,\phi)\sin\theta\, d\theta\, d\phi = \frac{P\,\text{rad}}{4\pi} \tag{3.2}$$

$$D \sim \frac{4\pi}{\theta E \times \theta H} \tag{3.3}$$

- θE – Beam width *in radian in EL plane*
- θH – Beam width *in radian in AZ plane*

Antenna effective area (Aeff): The antenna area which contributes to the antenna directivity.

$$D = \frac{4\pi\,\text{Aeff}}{\lambda^2} \tag{3.4}$$

Antenna gain (G): The ratio between the amounts of energy propagating in a certain direction compared to the energy that would be propagating in the same direction if the antenna were not directional, isotropic radiator, is known as its gain.

Radiation efficiency (α): Radiation efficiency is the ratio of power radiated to the total input power. The efficiency of an antenna takes into account losses, and is equal to the total radiated power divided by the radiated power of an ideal lossless antenna.

$$G = \alpha D \tag{3.5}$$

$$\text{For small antennas}\left(l < \frac{\lambda}{2}\right)\ G \cong \frac{41,000}{\theta E^\circ \times \theta H^\circ} \tag{3.6a}$$

$$\text{For medium-size antennas}\left(\lambda < l < 8\lambda\right)\ G \cong \frac{31,000}{\theta E^\circ \times \theta H^\circ} \tag{3.6b}$$

$$\text{For big antennas}\left(8\lambda < l\right)\ \ G \cong \frac{27,000}{\theta E^\circ \times \theta H^\circ} \tag{3.6c}$$

Antenna impedance: Antenna impedance is the ratio of voltage at any given point along the antenna to the current at that point. Antenna impedance depends upon the height of the antenna above the ground and the influence of surrounding objects. The impedance of a quarter-wave monopole near a perfect ground is approximately 36 Ω. The impedance of a half-wave dipole is approximately 75 Ω.

Antenna array: An array of antenna elements is a set of antennas used for transmitting or receiving electromagnetic waves.

Active antenna: An active antenna consists of a radiating element and active elements such as amplifiers.

Phased arrays: Phased array antennas are electrically steerable. The physical antenna can be stationary. Phased arrays, or smart antennas, incorporate active components for beam steering.

3.2.1 STEERABLE ANTENNAS

- Arrays with switchable elements and partially mechanically and electronically steerable arrays.
- Hybrid antenna systems – to fully electronically steerable arrays. Such systems can be equipped with phase and amplitude shifters for each element, or the design can be based on digital beam forming (DBF).
- This technique, in which the steering is performed directly on a digital level, allows the most flexible and powerful control of the antenna beam.

3.2.2 TYPES OF ANTENNAS

3.2.2.1 Small Antennas for Wearable Communication Systems

Monopole: Quarter-wavelength wire antenna.

Dipole: Dipole antenna consists of two quarter-wavelength wires. A dipole is a half-wavelength wire antenna. Monopole and dipole couple to the electric field of the electromagnetic wave in the region near the antenna.

Slot antenna: Half-wavelength slot. A slot antenna consists of a metal surface with a *slot* cut out.

Bi-conical antennas: Bi-conical half-wavelength wire antenna. A bi-conical antenna is a wideband antenna made of two conical conductive objects.

Loop antennas: Loop wire antennas. The small loop antenna is known as a magnetic loop. The loop antenna behaves as an inductor. It couples to the magnetic field of the electromagnetic wave in the region near the antenna, in contrast to monopole and dipole antennas which couple to the electric field of the wave.

Helical antennas: Helical wire antennas.

Printed antennas: Antennas printed on a dielectric substrate, usually have low efficiency due to losses [17, 18].

3.2.2.2 Aperture Antennas for Base Station Communication Systems

- Horn and open waveguide.
- Reflector antennas.
- Arrays and phased arrays.

TABLE 3.1
Antenna directivity versus antenna gain

Antenna type	Directivity (dBi)	Gain (dBi)
Isotropic radiator	0	0
Dipole λ/2	2	2
Dipole above ground plane	6–4	6–4
Microstrip antenna	7–8	6–7
Yagi antenna	6–18	5–16
Helix antenna	7–20	6–18
Horn antenna	10–30	9–29
Reflector antenna	15–60	14–58

- Microstrip and printed antenna arrays.
- Slot antenna arrays.

A comparison of directivity and gain values for several antennas is given in Table 3.1.

3.3 DIPOLE ANTENNA

A dipole antenna is a small wire antenna. A dipole antenna consists of two straight conductors excited by a voltage fed via a transmission line as shown in Figure 3.1. Each side of the transmission line is connected to one of the conductors. The most common dipole is the half-wave dipole, in which each of the two conductors is approximately a quarter-wavelength long, so the length of the antenna is half-wavelength.

We can calculate the fields radiated from the dipole by using the potential function. The electric potential function is φ_l. The electric potential function is A. The potential function is given in Equation 3.7.

$$\phi_l = \frac{1}{4\pi\varepsilon_0} \int_c \frac{\rho_l e^{j(\omega t - \beta R)}}{R} \, dl$$

$$A_l = \frac{\mu_0}{4\pi} \int_c \frac{i e^{j(\omega t - \beta R)}}{R} \, dl \tag{3.7}$$

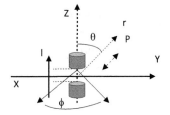

FIGURE 3.1   Dipole antenna.

3.3.1 Radiation from Small Dipole

The length of a small dipole is small compared to wavelength and is called elementary dipole. We may assume that the current along the elementary dipole is uniform. We can solve the wave equation in spherical coordinates by using the potential function given in Equation 3.7. The electromagnetic fields in a point $P(r, \theta, \varphi)$ is given in Equation 3.8. The electromagnetic fields in Equation 3.8 vary as $1/(r)^3$ for $r \ll 1$. The dominant component of the field varies as $1/(r)^3$ and is given in Equation 3.9. These fields are the dipole near fields. In this case the waves are standing waves and the energy oscillates in the antenna near zone, and is not radiated to the open space. The real part of the Poynting vector is equal to zero. For $r \gg 1$, the dominant component of the field varies as $1/r$ as given in Equation 3.10. These fields are the dipole far fields.

$$E_r = \eta_0 \frac{lI_0 \cos\theta}{2\pi r^2}\left(1 - \frac{j}{\beta r}\right)e^{j(\omega t - \beta r)}$$

$$E_\theta = j\eta_0 \frac{\beta lI_0 \sin\theta}{4\pi r}\left(1 - \frac{j}{\beta r} - \frac{1}{(\beta r)^2}\right)e^{j(\omega t - \beta r)}$$

$$H_\phi = j\frac{\beta lI_0 \sin\theta}{4\pi r}\left(1 - \frac{j}{\beta r}\right)e^{j(\omega t - \beta r)}$$

$$H_r = 0 \quad H_\theta = 0 \quad E_\varphi = 0$$

$$I = I_0 \cos \omega t$$

(3.8)

$$E_r = -j\eta_0 \frac{lI_0 \cos\theta}{2\pi\beta r^3}e^{j(\omega t - \beta r)}$$

$$E_\theta = -j\eta_0 \frac{lI_0 \sin\theta}{4\pi\beta r^3}e^{j(\omega t - \beta r)}$$

$$H_\phi = \frac{lI_0 \sin\theta}{4\pi r^2}e^{j(\omega t - \beta r)}$$

(3.9)

$$E_r = 0$$

$$E_\theta = j\eta_0 \frac{l\beta I_0 \sin\theta}{4\pi r}e^{j(\omega t - \beta r)}$$

$$H_\phi = j\frac{l\beta I_0 \sin\theta}{4\pi r}e^{j(\omega t - \beta r)}$$

(3.10)

$$\frac{E_\theta}{H_\phi} = \eta_0 = \sqrt{\frac{\mu_0}{\varepsilon_0}}$$

(3.11)

   In the far fields the electromagnetic fields vary as sin θ. Wave impedance in free space is given in Equation 3.11.

3.3.1.1 Dipole Radiation Pattern

The antenna radiation pattern represents the radiated fields in space at a point $P(r, \theta, \varphi)$ as a function of θ, φ. The antenna radiation pattern is three dimensional. When φ is constant and θ varies we get the E plane radiation pattern. When φ varies and θ is constant, usually $\theta = \pi/2$, we get the H plane radiation pattern.

3.3.1.2 Dipole E Plane Radiation Pattern

The dipole E plane radiation pattern is given in Equation 3.12 and presented in Figure 3.2.

$$|E_\theta| = \eta_0 \frac{l\beta I_0 |\sin\theta|}{4\pi r} \tag{3.12}$$

At a given point $P(r, \theta, \varphi)$ the dipole E plane radiation pattern is given in Equation 3.13.

$$
\begin{aligned}
|E_\theta| &= \eta_0 \frac{l\beta I_0 |\sin\theta|}{4\pi r} = A|\sin\theta| \\
Choose \quad & A = 1 \\
|E_\theta| &= |\sin\theta|
\end{aligned}
\tag{3.13}
$$

   The dipole E plane radiation pattern in spherical coordinate system is shown in Figure 3.3.

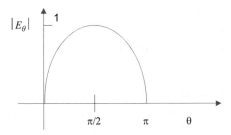

FIGURE 3.2   Dipole E plane radiation pattern.

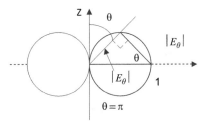

FIGURE 3.3   Dipole E plane radiation pattern in spherical coordinate system.

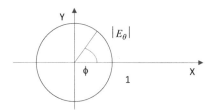

FIGURE 3.4   Dipole H plane radiation pattern for $\theta = \pi/2$.

3.3.1.3 Dipole H Plane Radiation Pattern

For $\theta = \pi/2$ the dipole H plane radiation pattern is given in Equation 3.14 and presented in Figure 3.4.

$$|E_\theta| = \eta_0 \frac{l\beta I_0}{4\pi r} \tag{3.14}$$

At a given point $P(r, \theta, \varphi)$ the dipole H plane radiation pattern is given in Equation 3.15.

$$
\begin{aligned}
|E_\theta| &= \eta_0 \frac{l\beta I_0 |\sin\theta|}{4\pi r} = A \\
Choose \quad & A = 1 \\
|E_\theta| &= 1
\end{aligned}
\tag{3.15}
$$

The dipole H plane radiation pattern in x-y plane is a circle with $r = 1$.

The radiation pattern of a vertical dipole is omnidirectional. It radiates equal power in all azimuthal directions perpendicular to the axis of the antenna. A dipole H plane radiation pattern in spherical coordinate system is shown in Figure 3.4.

3.3.1.4 Antenna Radiation Pattern

A typical antenna radiation pattern is shown in Figure 3.5. The antenna −3dB beam width is measured between the points with intensity of 0.707E from the maximum

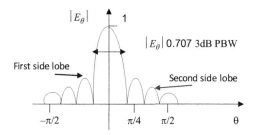

FIGURE 3.5   Antenna typical radiation pattern.

relative field intensity E. Half of the radiated power is concentrated in the antenna main beam. The antenna main beam is called 3 dB beam width. Radiation to undesired direction is concentrated in the antenna side lobes.

For a dipole, the power intensity varies as $(\sin^2 \theta)$. At $\theta = 45°$ and $\theta = 135°$ the radiated power is equal to half the power radiated toward $\theta = 90°$. The dipole beam width is $\theta = (135 - 45) = 90°$.

3.3.1.5 Dipole Directivity

Directivity is defined as the ratio between the amounts of energy propagating in a certain direction compared to the average energy radiated to all directions over a sphere as written in Equations 3.16 and 3.17.

$$D = \frac{P(\theta,\phi)\text{maximal}}{P(\theta,\phi)\text{average}} = 4\pi \frac{P(\theta,\phi)\text{maximal}}{P\,\text{rad}} \tag{3.16}$$

$$P(\theta,\phi)\text{average} = \frac{1}{4\pi}\iint P(\theta,\phi)\sin\theta\, d\theta\, d\phi = \frac{P\,\text{rad}}{4\pi} \tag{3.17}$$

The radiated power from a dipole is calculated by computing the Poynting vector P as given in Equation 3.18.

$$P = 0.5\left(E \times H^*\right) = \frac{15\pi I_0^2 l^2 \sin^2\theta}{r^2 \lambda^2}$$

$$W_T = \int_s P \cdot ds = \frac{15\pi I_0^2 l^2}{\lambda^2} \int_0^\pi \sin^3\theta\, d\theta \int_0^{2\pi} d\phi = \frac{40\pi^2 I_0^2 l^2}{\lambda^2} \tag{3.18}$$

The overall radiated energy is *WT*. *WT* is computed by integration of the power flow over an imaginary sphere surrounding the dipole. The power flow of an isotropic radiator equal to W_T divided by the surrounding area of the sphere, $4\pi r^2$, is given in Equation 3.19. The dipole directivity at $\theta = 90°$ is 1.5 or 1.76 dB as shown in Equation 3.20.

$$\oint_s ds = r^2 \int_0^\pi \sin\theta\, d\theta \int_0^{2\pi} d\varphi = 4\pi r^2$$

$$P_{iso} = \frac{W_T}{4\pi r^2} = \frac{10\pi I_0^2 l^2}{r^2 \lambda^2} \tag{3.19}$$

$$D = \frac{P}{P_{iso}} = 1.5\sin^2\theta$$

$$G_{dB} = 10\log_{10} G = 10\log_{10} 1.5 = 1.76 dB \tag{3.20}$$

For small antennas or for antennas without losses, $D = G$, losses are negligible. For a given θ and φ for small antennas the approximate directivity is given by Equation 3.21.

$$D = \frac{41253}{\theta_{3dB}\varphi_{3dB}}$$

$$G = \xi D \qquad \xi = Efficency$$

(3.21)

Antenna losses degrade the antenna efficiency. Antenna losses consist of conductor loss, dielectric loss, radiation loss and mismatch losses. For resonant small antennas $\xi = 1$. For reflector and horn antennas, the efficiency varies between $\xi = 0.5$ and $\xi = 0.7$. The beam width of a small dipole, 0.1λ long, is around $90°$. The 0.1λ dipole impedance is around 2Ω. The beam width of a dipole, 0.5λ long, is around $80°$. The impedance of a 0.5λ dipole is around 73Ω.

3.3.1.6 Antenna Impedance

Antenna impedance determines the efficiency of transmitting and receiving energy in antennas. The dipole impedance is given in Equation 3.22.

$$R_{rad} = \frac{2W_T}{I_0^2}$$

$$For\ a\ Dipole : R_{rad} = \frac{80\pi^2 l^2}{\lambda^2}$$

(3.22)

3.3.1.7 Impedance of a Folded Dipole

A folded dipole is a half-wave dipole with an additional wire connecting its two ends. If the additional wire has the same diameter and cross section as the dipole, two nearly identical radiating currents are generated. The resulting far-field emission pattern is nearly identical to the one for the single-wire dipole described above, but at resonance its feed point impedance R_{rad-f} is four times the radiation resistance of a dipole. This is because for a fixed amount of power, the total radiating current $I0$ is equal to twice the current in each wire and thus equal to twice the current at the feed point. Equating the average radiated power to the average power delivered at the feed point, we obtain $R_{rad-f} = 4R_{rad} = 300\ \Omega$. The folded dipole has a wider bandwidth than a single dipole.

3.4 MONOPOLE ANTENNA FOR WEARABLE COMMUNICATION SYSTEMS

A monopole antenna is usually a quarter-wavelength conductor mounted above a ground plane or the earth. Based on image theory, the monopole image is located behind the ground plane. The monopole antenna and the monopole image form a dipole antenna.

A monopole antenna is half a dipole that radiates electromagnetic fields above the ground plane. The impedance of a 0.5λ monopole antenna is around 37Ω. The beam

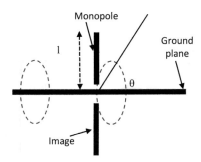

FIGURE 3.6   Monopole antenna.

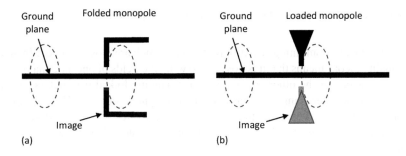

FIGURE 3.7   (a) Inverted monopole antenna, (b) loaded monopole antenna.

width of a monopole, 0.25 λ long, is around 40°. The directivity of a monopole, 0.25 λ long, is around 3–5 dBi. Usually, in wireless communication systems, very short monopole antennas are employed. The impedance of a 0.05 λ monopole antenna is around 1 Ω with capacitive reactance. The beam width of a monopole, 0.05 λ long, is around 45°. The directivity of a monopole, 0.05 λ long, is around 3 dBi.

An inverted monopole antenna is shown in Figure 3.7a. A loaded monopole antenna is shown in Figure 3.7b. Monopole antennas may be printed on a dielectric substrate as part of wearable communication devices.

3.5 LOOP ANTENNAS FOR WIRELESS COMMUNICATION SYSTEMS

Loop antennas are compact, low-profile and low-cost antennas. Loop antennas are employed in wearable wireless communication systems.

3.5.1 DUALITY RELATIONSHIP BETWEEN DIPOLE AND LOOP ANTENNAS

The loop antenna is referred to as the dual of the dipole antenna (see Figure 3.8). A small dipole has magnetic current flowing (as opposed to electric current as in a regular dipole); the fields would resemble that of a small loop. The short dipole has a capacitive impedance (imaginary part of the impedance is negative). The impedance of a small loop is inductive (positive imaginary part).

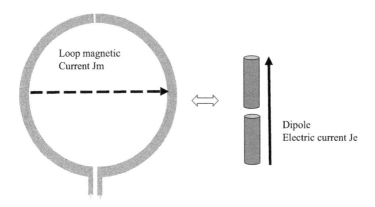

FIGURE 3.8 Duality relationship between dipole and loop antennas.

Duality means combining two different things which are closely linked. In antennas, duality theory means that it is possible to write the fields of one antenna from the field expressions of the other antenna by interchanging linked parameters. Electric current may be interchanged by an equivalent magnetic current. The variation of electromagnetic waves as a function of time may be written as $e^{j\omega t}$. The derivative as function of time is $j\omega e^{j\omega t}$. Maxwell's equations for system 1 may be written as in Equation 3.23:

$$\nabla X E_1 = -j\omega\mu_1 H_1$$
$$\nabla X H_1 = (\sigma + j\omega\varepsilon_1)E = J_e + j\omega\varepsilon_1 E_1 \quad (3.23)$$

System 2 is a dual system to system 1. Maxwell's equations for system 2 may be written as in Equation 3.24:

$$\nabla X H_2 = j\omega\varepsilon_2 E_2$$
$$\nabla X E_2 = -J_m - j\omega\mu_2 H_2 \quad (3.24)$$

System 1 electric current source	System 2 magnetic current source
J_e	J_m
E_1	H_2
μ_1	ε_2
H_1	$-E_2$
ε_1	μ_2

By using the duality principle, the far electromagnetic fields of the loop antenna are given in Equation 3.25. The directivity of a loop antenna with circumference of 0.5 λ long is around 1 dBi. The directivity of a loop antenna with circumference of 1 λ long is around 4 dBi. The H plane 3D radiation pattern of a loop antenna with circumference of 0.45 λ in free space is shown in Figure 3.9.

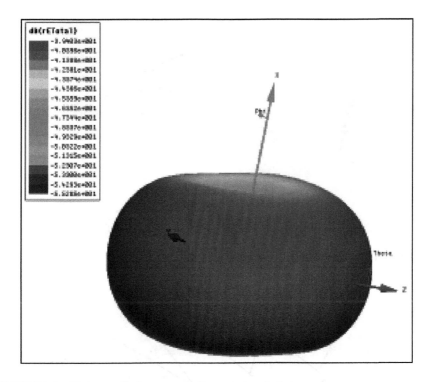

FIGURE 3.9   H plane radiation pattern of loop antenna in free space.

$$E_r = H_\varphi = E_\theta = 0$$

$$H_\theta = j\eta_0 \frac{l\beta I_m \sin\theta}{4\pi r} e^{j(\omega t - \beta r)}$$

$$E_\varphi = -j \frac{l\beta I_m \sin\theta}{4\pi r} e^{j(\omega t - \beta r)}$$

(3.25)

3.5.2 MEDICAL APPLICATIONS OF PRINTED LOOP ANTENNAS

Loop antennas have low radiation resistance and high reactance. It is difficult to match the antenna to a transmitter. Loop antennas are most often used as receive antennas, where impedance mismatch loss can be accepted. Small loop antennas are used in medical devices as field strength probes, in pagers and in wireless measurements.

A BALUN transformer is connected to the loop feed lines. A BALUN transformer is a transformer from a balance transmission line to an unbalance transmission line. The transformer may be used to match the loop antenna to the communication system. A photo of loop antennas is shown in Figure 3.10. The loop antenna may be inserted in a sleeve as shown in Figure 3.10. The electrical properties of the sleeve were chosen to match the loop antenna to the human body. The sleeve also protects the loop antenna from the environment.

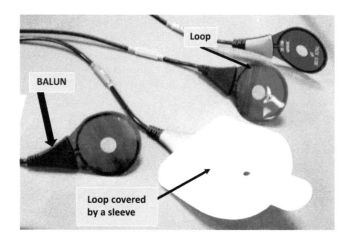

FIGURE 3.10   Photo of loop antennas.

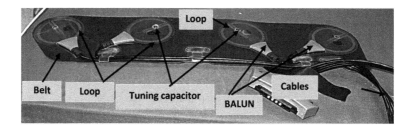

FIGURE 3.11   Photo of loop antenna array inside a belt.

A photo of a four-loops antenna array inside a belt is shown in Figure 3.11. The loop antenna may be tuned by adding a capacitor or a varactor as shown in Figure 3.11.

3.6 WEARABLE LOOP ANTENNAS

Loop antennas are used as receive antennas in wearable communication and medical systems. Loop antennas may be printed on a dielectric substrate or manufactured as a wired antenna. In this section several loop antennas are presented.

3.6.1 SMALL WEARABLE LOOP ANTENNA

A small loop antenna is shown in Figure 3.12. The shape of the loop antenna may be circular or rectangular. These antennas have low radiation resistance and high reactance. It is difficult to match the antenna to a transmitter. Loop antennas are most often used as receive antennas, where impedance mismatch loss can be accepted. Small loop antennas are used as field strength probes, in pagers and in wireless measurements. The loop lies in the x-y plane. The radius a, of the small loop antenna is

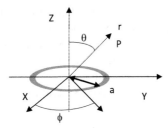

FIGURE 3.12 Small loop antenna.

smaller than a wavelength ($a \ll \lambda$). A loop antenna electric field is given Equation 3.26. A loop antenna magnetic field is given in Equation 3.27.

$$E_\phi = \eta \frac{(ak)^2 I_0 \sin\theta}{4r} e^{j(\omega t - \beta r)} \tag{3.26}$$

$$H_\theta = -\frac{(ak)^2 I_0 \sin\theta}{4r} e^{j(\omega t - \beta r)} \tag{3.27}$$

The variation of the radiation pattern with direction is $\sin\theta$, the same as dipole antennas. The fields of a small loop have the E and H fields switched relative to that of a short dipole. The E field is horizontally polarized in the x-y plane.

The small loop is often referred to as the dual of the dipole antenna, because if a small dipole had a magnetic current flowing (as opposed to an electric current as in a regular dipole), its fields would be similar to the fields of a small loop. The short dipole has a capacitive impedance (imaginary part of the impedance is negative). The impedance of a small loop is inductive (positive imaginary part). The radiation resistance (and ohmic loss resistance) can be increased by adding more turns to the loop. If there are N turns of a small loop antenna, each with a surface area S, the radiation resistance for small loops can be approximated as given in Equation 3.28.

$$R_{rad} = \frac{31,329 N^2 S^2}{\lambda^4} \tag{3.28}$$

For a small loop, the reactive component of the impedance can be determined by finding the inductance of the loop. For a circular loop with radius a, and wire radius r, the reactive component of the impedance is given by Equation 3.29.

$$X = 2\pi a f \mu \left(\ln\left(\frac{8a}{r}\right) - 1.75 \right) \tag{3.29}$$

Loop antennas behave better in the vicinity of the human body than dipole antennas. The reason is that the electric near fields in a dipole antenna are very strong. For $r \ll 1$, the dominant component of the field varies as $1/r^3$. These fields are the dipole near fields. In this case, the waves are standing waves and the energy oscillates in the antenna near zone, and the waves are not radiated to the open space. The real part of the Poynting vector is equal to zero. Near the body, the electric fields decay rapidly. However, the magnetic fields near the body are not affected. The magnetic fields are strong in the near field of the loop antenna. These magnetic fields give rise to the loop antenna radiation. The loop antenna radiation near the human body is stronger than the dipole radiation near the human body. Several loop antennas are used as "wearable antennas".

3.6.2 WEARABLE PRINTED LOOP ANTENNA

The diameter of a printed loop antenna is around half-wavelength. A loop antenna is dual to a half-wavelength dipole. Several loop antennas were designed for medical systems at a frequency range between 400 MHz and 500 MHz. In Figure 3.13a a printed loop antenna is presented. A photo of a printed loop antenna with a BALUN transformer is presented in Figure 3.13b.

A BALUN transformer is a transformer from a balance transmission line to an unbalance transmission line. The loop may be attached to a human body or inserted inside a wearable belt. The antenna was printed on flame retardant-4 (FR-4) with 0.5 mm thickness. The loop diameter is 45 mm. The loop antenna voltage standing wave ratio (VSWR) is around 4:1. The printed loop antenna radiation pattern at 435 MHz is shown in Figure 3.14.

The loop antenna gain is around 1.8 dBi. An antenna with a tuning capacitor is shown in Figure 3.15.

The loop antenna VSWR without the tuning capacitor was 4:1. This loop antenna may be tuned by adding a capacitor or varactor as shown in Figure 3.16.

Matching stubs are employed to tune the antenna to the resonant frequency. Tuning the antenna allows us to work in a wider bandwidth as shown in Figure 3.16. Loop antennas are used as receive antennas in medical systems. The loop antenna radiation pattern on a human body is shown in Figure 3.17.

The computed 3D radiation pattern and the coordinate used in this chapter are shown in Figure 3.18.

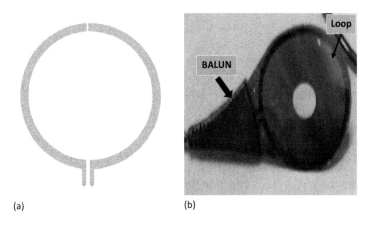

(a)   (b)

FIGURE 3.13   (a) Printed loop antenna, (b) photo of a printed loop antenna.

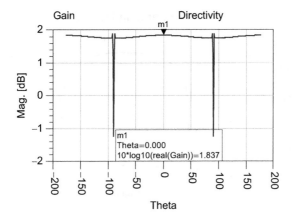

FIGURE 3.14 Printed loop antenna radiation pattern at 435 MHz.

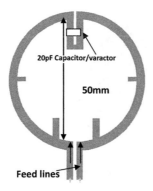

FIGURE 3.15 Tunable loop antenna without ground plane.

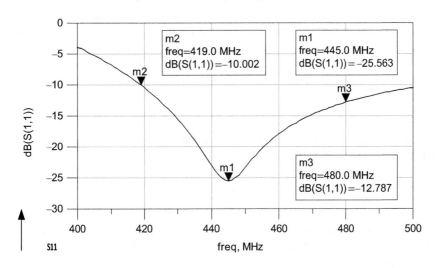

FIGURE 3.16 Computed S11 of loop antenna, without ground plane, with a tuning capacitor.

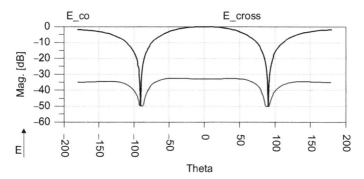

FIGURE 3.17   Radiation pattern of loop antenna on human body.

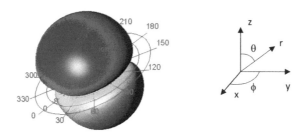

FIGURE 3.18   Loop antenna three-dimensional radiation pattern.

3.6.3 WIRED LOOP ANTENNA

A wire seven-turn loop antenna is shown in Figure 3.19. The antenna length $l =$ 4.5 mm. The loop diameter is 3.5 mm. Equation 3.30 is an approximation to calculate the inductance value for an air coil loop antenna with N turns, diameter r and length l.

$$L(nH) = \frac{3.94 r_{mm} N^2}{0.9 \dfrac{l}{r} + 1} = \frac{0.1 r_{mil} N^2}{0.9 \dfrac{l}{r} + 1} \tag{3.30}$$

The inductance value for an air coil loop antenna with N turns, diameter r and length l can also be written as Equation 3.31.

$$L(nH) \approx \frac{r_{mm}^2 N^2}{2l + r} \tag{3.31}$$

For length $l = 4.5$ mm, wire diameter 0.6 mm, loop diameter 3.5 mm and $N = 7$ the inductance is around $L = 52$ nH. The quality factor of this seven-turn wire loop is around 100.

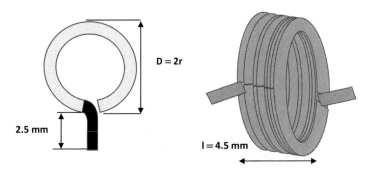

FIGURE 3.19 Wire seven-turn loop antenna.

For length $l = 1.7$ mm, wire diameter 0.6 mm, loop diameter 6.5 mm and $N = 2$ the inductance is around $L = 45$nH.

For length $l = 2$ mm, wire diameter 0.6 mm, loop diameter 3.62 mm and $N = 2$ the inductance is around $L = 42.1$nH.

For length $l = 0.5$ mm, wire diameter 0.6 mm, loop diameter 7 mm and $N = 2$ the inductance is around $L = 87$nH.

For length $l = 2.5$ mm, wire diameter 0.6, loop diameter 5 mm and $N = 2$ the inductance is around $L = 20.7$nH.

The wire loop antenna has very low radiation efficiency. The amount of power radiated is a small fraction of the input power. The antenna efficiency is around 0.01%, −41 dB. The ratio between the antenna's dimension and wavelength is around 1:100. Small antennas are characterized by radiation resistance Rr. The radiated power is given by I^2Rr, where I is the current through the coil. Remember that the current through the coil is Q times the current through the antenna. For example, for current of 2 mA and Q of 20, if $Rr = 10^{-3}$ Ω, the radiated power is around −32 dBm.

Figure 3.20 presents a wire loop antenna with 2.5 turns on a printed circuit board (PCB). Figure 3.21 presents a wire loop antenna with seven turns on a PCB.

FIGURE 3.20 Wire loop antenna on printed circuit board.

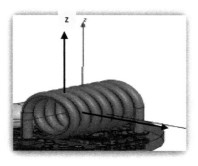

FIGURE 3.21   Seven-turn loop antenna on printed circuit board.

3.7 WEARABLE LOOP ANTENNAS WITH GROUND PLANE

A new loop antenna with ground plane has been designed on Kapton substrates with relative dielectric constant of 3.5 and thickness of 0.25 mm. The antenna is presented in Figure 3.22.

Matching stubs are employed to tune the antenna to the resonant frequency. The loop antenna with ground plane diameter is 45 mm. The antenna was designed by

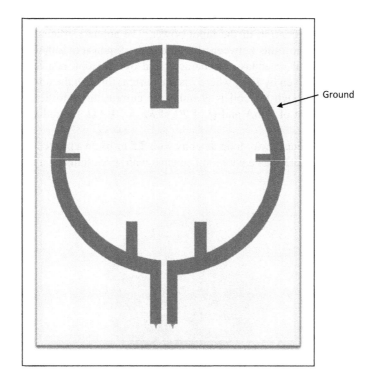

FIGURE 3.22   Printed loop antenna with ground plane.

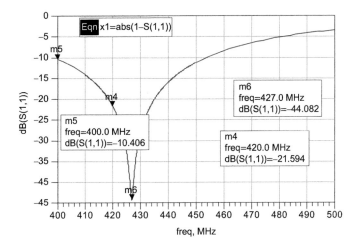

FIGURE 3.23 Computed S11 of loop antenna with ground plane.

employing 3D full-wave software [19]. The antenna center frequency is 427 MHz. The antenna bandwidth for VSWR better than 2:1 is around 12% as shown in Figure 3.23.

The printed loop antenna radiation pattern at 435 MHz is shown in Figure 3.24. The loop antenna gain is around 1.8 dBi. The loop antenna with ground plane beam width is around 100°.

A printed loop antenna with ground plane with shorter tuning stubs is shown in Figure 3.25. The diameter the loop antenna with ground, with shorter tuning stubs plane, is 45 mm. The antenna center frequency is 438 MHz.

The antenna bandwidth for VSWR better than 2:1 is around 12% as shown in Figure 3.26.

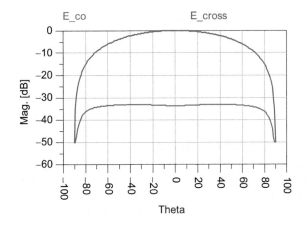

FIGURE 3.24 Radiation pattern of loop antenna with ground plane.

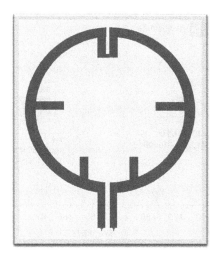

FIGURE 3.25   Printed loop antenna with ground plane and short tuning stubs.

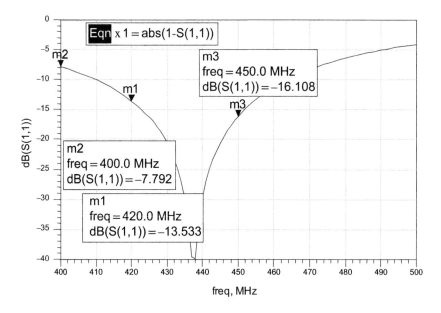

FIGURE 3.26   Computed S11 of loop antenna with ground plane and short tuning stubs.

The printed loop antenna with shorter tuning stubs radiation pattern at 435 MHz is shown in Figure 3.27.

Table 3.2 Comparison of the electrical performance of a loop antenna with ground plane, GND, with a loop antenna without ground plane

The loop antenna gain is around 1.8 dBi. The loop antenna with ground plane beam width is around 100°. The printed loop antenna with shorter tuning stubs 3D radiation pattern at 430 MHz is shown in Figure 3.28.

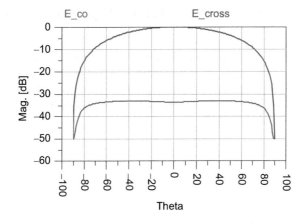

FIGURE 3.27   Radiation pattern of loop antenna with ground plane and short tuning stubs.

TABLE 3.2
Electrical performance of several loop antenna configurations

Antenna with no tuning capacitor	Beam width 3 dB	Gain dBi	VSWR
Loop with no ground, GND	100°	0	4:1
Loop with tuning capacitor (no GND)	100°	0	2:1
Wearable loop with GND	100°	0–2	2:1
Loop with GND in free space	100°–110°	−3	5:1

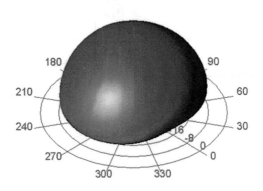

FIGURE 3.28   A three-dimensional radiation pattern of loop antenna with ground plane.

3.8 RADIATION PATTERN OF A LOOP ANTENNA NEAR A METAL SHEET

E and H plane 3D radiation pattern of a wire loop antenna in free space is shown in Figure 3.29.

E and H plane 3D radiation pattern of a wire loop antenna 30 cm from a metal sheet as shown in Figure 3.30 was computed by employing High Frequency Structure Simulator (HFSS) software [20].

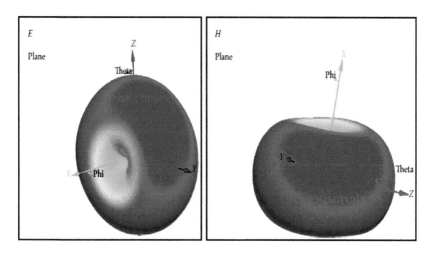

FIGURE 3.29   E and H plane radiation pattern of loop antenna in free space.

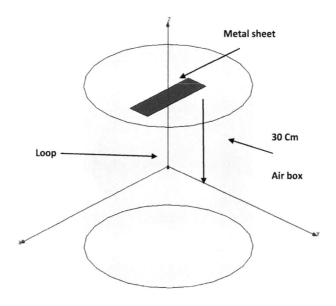

FIGURE 3.30   Loop antenna located near a metal sheet.

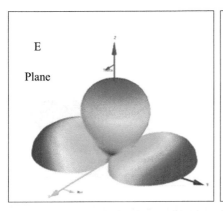
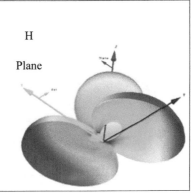

FIGURE 3.31   E and H plane radiation pattern of a loop antenna for distance of 30 cm from a metal sheet.

Figure 3.31 presents the E and H plane radiation pattern of loop antenna for distance of 30 cm from a metal sheet. We can see that a metal sheet in the vicinity of an antenna splits the main beam and creates holes of around 20 dB in the radiation pattern.

E and H plane 3D radiation pattern of a wire loop antenna 10 cm from a metal sheet as shown in Figure 3.32 was computed by employing HFSS software [20].

Figure 3.33 presents the E and H plane radiation pattern of a loop antenna located 10 cm from a metal sheet. We can see that a metal sheet in the vicinity of the antenna splits the main beam and creates holes up to 15 dB in the radiation pattern.

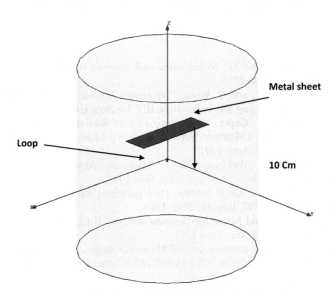

FIGURE 3.32   Loop antenna located 10 cm from a metal sheet.

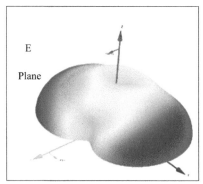
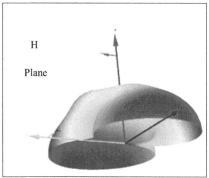

FIGURE 3.33 E and H plane radiation pattern of a loop antenna for distance of 10 cm from a metal sheet.

3.9 CONCLUSIONS

This chapter presents several wideband wearable antennas with high efficiency for medical and sport applications. Design considerations and computed and measured results of several wearable printed antennas are presented in this chapter. Several compact loop antennas for 5G, IOT and medical applications are presented in this chapter. Several loop antennas were designed for medical systems at a frequency range between 400 MHz and 500 MHz. The loop antenna VSWR without the tuning capacitor or a varactor is around 4:1. The loop antenna bandwidth with a tuning capacitor is 15% for VSWR better than 2:1.

A capacitor or a varactor may be employed to tune and compensate variations in the antenna resonant frequency at different locations on the human body.

REFERENCES

[1] A. Sabban, *"Wideband RF Technologies and Antenna in Microwave Frequencies,"* Wiley Sons, New York, 2016.

[2] J.R. James, P.S. Hall, and C. Wood, *"Microstrip Antenna Theory and Design,"* The Institution of Engineering and Technology, IET, London, UK, 1981.

[3] A. Sabban and K.C. Gupta, "Characterization of Radiation Loss from Microstrip Discontinuities Using a Multiport Network Modeling Approach," IEEE Transactions on MTT, 39(4), 705–712, April 1991.

[4] A. Sabban, "A New Wideband Stacked Microstrip Antenna," IEEE Antenna and Propagation Symp., Houston, Texas, USA, June 1983.

[5] U.S Patent, Inventors: Albert Sabban, Dual polarized dipole wearable antenna, U.S Patent number: 8203497, June 19, 2012, USA

[6] A. Sabban, "Wideband Microstrip Antenna Arrays," IEEE Antenna and Propagation Symp MELCOM., Tel-Aviv, June 1981.

[7] A. Sabban, "Microstrip Antenna Arrays," *Microstrip Antennas*, Nasimuddin Nasimuddin, InTech, 361–384, 2011, ISBN: 978-953-307-247-0 , http://www.intechopen.com/articles/show/title/microstrip-antenna-arrays.

[8] L.C. Chirwa, P.A. Hammond, S. Roy, and D.R.S. Cumming, "Electromagnetic Radiation from Ingested Sources in the Human Intestine between 150 MHz and 1.2 GHz," *IEEE Transaction on Biomedical Engineering*, 50(4), 484–492, April 2003.

[9] D. Werber, A. Schwentner, and E. M. Biebl, "Investigation of RF Transmission Properties of Human Tissues," *Advanced Radio Science*, 4, 357–360, 2006.

[10] K. Fujimoto and J.R. James, Eds., *Mobile Antenna Systems Handbook*, Artech House, Boston, MA, USA, 1994.

[11] B. Gupta, S. Sankaralingam, and S. Dhar, "Development of Wearable and Implantable Antennas in the Last Decade," Microwave Mediterranean Symposium (MMS), Guzelyurt, Turkey, 251–267, August 2010.

[12] T. Thalmann, Z. Popovic, B.M. Notaros, and J.R. Mosig, "Investigation and Design of a Multi-band Wearable Antenna," 3rd European Conference on Antennas and Propagation, EuCAP 2009, Berlin, Germany, 462–465, 2009.

[13] P. Salonen, Y. Rahmat-Samii, and M. Kivikoski, "Wearable Antennas in the Vicinity of Human Body," IEEE Antennas and Propagation Society Symposium, Monterey, USA, Vol. 1, 467–470, 2004.

[14] T. Kellomaki, J. Heikkinen, and M. Kivikoski, "Wearable Antennas for FM Reception," First European Conference on Antennas and Propagation, EuCAP 2006, Hague, Netherlands, 1–6, 2006.

[15] A. Sabban, "Wideband printed antennas for medical applications," APMC 2009 Conference, Singapore, 12/2009.

[16] Y. Lee, "*Antenna Circuit Design for RFID Applications*," Microchip Technology Inc., Microchip, AN, 710c.

[17] A. Sabban, "*Low Visibility Antennas for Communication Systems*," Taylor & Francis Group, New York, 2015.

[18] A. Sabban, "PhD Thesis, Multiport Network Model for Evaluating Radiation Loss and Coupling Among Discontinuities in Microstrip Circuits," University of Colorado at Boulder, January 1991.

[19] ADS Software, Keysightt, http://www.keysight.com/en/pc-1297113/advanced-design-system-ads?cc=IL&lc=eng.

[20] HFSS Software, ANSYS HFSS, https://www.ansys.com/products/electronics/ansys-hfss.

4 Wideband Wearable Antennas for 5G Communication Systems, IOT and Medical Systems

Albert Sabban

CONTENTS

4.1 INTRODUCTION

Microstrip antennas are widely employed in communication systems and seekers. Microstrip antennas possess attractive features that are crucial to communication and medical systems. The features of microstrip antennas are low profile, flexibility, light weight and low production cost. In addition, the benefit of a compact low-cost feed network is attained by integrating the radio frequency (RF) front end with the radiating elements on the same substrate. Printed antennas have been widely presented in books and papers in the last decade [1–9]. The RF transmission properties of human tissues have been investigated in several articles [10, 11]. However, the effect of the human body on the electrical performance of wearable antennas has not been presented [14, 15]. Several wearable antennas have been presented in the last decade [1–20]. A review of wearable and body-mounted antennas designed and developed for various applications at different frequency bands over the last decade can be found in [13]. In [14], meander line wearable antennas close to the human body are presented in the frequency range between 800 and 2700 MHz. In [15] the performance of a textile antenna in the vicinity of the human body is presented at 2.4 GHz. In [16] the effect of the human body on wearable 100 MHz portable radio antennas is studied. In [16] the authors concluded that wearable antennas need to be shorter by 15%–25% from the antenna length in free space. Measurement of the antenna gain in [16] shows that a wide dipole (116 × 10 cm) has −13 dBi gain. The antennas presented in [12–16] were developed mostly for cellular applications. Requirements and the frequency range for medical applications differ from those for cellular applications. Medical wearable sensors are presented in [22–49]. Wearable technology will support the development of individualized treatment systems with real-time feedback to help promote patient health. Wearable medical systems and sensors can perform gait analysis and can measure body temperature, heart rate, blood pressure, sweat rate and other physiological parameters of the person wearing the medical device as presented in [22–49]. In this chapter, a new class of wideband compact wearable antennas for communication and medical RF systems is presented. Numerical results in free space and in the presence of the human body are discussed.

4.2 PRINTED WEARABLE ANTENNAS

Wearable antennas should be compact, flexible, light weight and low cost. Microstrip antennas, printed dipoles, slot antennas and loop antennas are compact, flexible, light weight and have low volume. These antennas are a good choice to be employed as wearable antennas.

Applications of wearable antennas:

- medical
- wireless communication
- wireless local area network (WLAN)

- High Performance Radio local area network (HiperLAN)
- Global Positioning System (GPS)
- military applications.

4.2.1 Double-Layer Printed Wearable Dipole Antennas

Single-layer printed dipole antennas have a narrow bandwidth between 0.5% and 1% for voltage standing wave ratio (VSWR) better than 2:1. The length of dipole may be between a quarter wavelength and half a wavelength. The antenna directivity is around 0 dBi and the beam width is around 100°. The antenna bandwidth may be improved by printing the antenna feed network on a dielectric substrate and printing the radiating dipole on a second layer. The electromagnetic fields are coupled from the feed lines to the radiating dipole. The bandwidth of the double-layer printed dipole is between 1% and 5% for VSWR better than 2:1 as a function of the dipole configuration and the layers' thickness. The printed dipole antenna consists of two layers. The first layer consists of RO3035 0.8 mm dielectric substrate. The second layer consists of RT-Duroid 5880 0.8 mm dielectric substrate. The substrate thickness determines the antenna bandwidth. However, thinner antennas are flexible. The antenna dimensions may be optimized to operate on the human body by full wave electromagnetic software [21]. A double-layer dipole antenna is shown in Figure 4.1.

The directivity of the antenna at 420 MHz is around 4 dBi as presented in Figure 4.2.

A double-layer 460 MHz dipole antenna is shown in Figure 4.3. The antenna dimensions are 20 × 4 cm. The directivity of the antenna at 460 MHz is around 5 dBi as presented in Figure 4.4. The antenna beam width is around 120°.

4.2.2 Printed Wearable Dual Polarized Dipole Antennas

In several communication and medical systems, the polarization of the received signal is not known. The polarization of the received signal may be vertical, horizontal or circular. In these systems it is crucial to use dual polarized receiving antennas. Two wearable antennas are presented in this section: the first is a dual polarized printed dipole, the second antenna is a dual polarized folded printed microstrip dipole. The compact microstrip loaded dipole antenna has been designed to provide horizontal polarization. The antenna's dimensions have been optimized to operate on the human body by employing Keysight Advanced Design System (ADS) software [21]. The antenna consists of two layers. The first layer consists of RO3035 0.8 mm dielectric substrate.

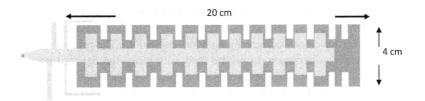

FIGURE 4.1   Wearable double-layer 420 MHz printed dipole antenna.

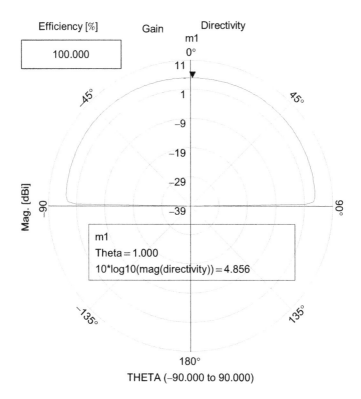

FIGURE 4.2   Radiation pattern of a wearable double-layer printed dipole antenna.

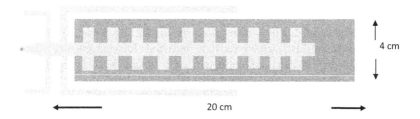

FIGURE 4.3   Wearable double-layer 460 MHz printed dipole antenna.

The second layer consists of RT-Duroid 5880 0.8 mm dielectric substrate. The substrate thickness determines the antenna bandwidth. However, thinner antennas are flexible. Thicker antennas have been designed with wider bandwidth. The printed slot antenna provides a vertical polarization. In several medical systems the required polarization may be vertical or horizontal. The proposed antenna is dually polarized. The printed dipole and the slot antenna provide dual orthogonal polarizations. The dimensions and current distribution of the dual polarized wearable antenna are presented in Figure 4.5.

The antenna dimensions are $26 \times 6 \times 0.16$ cm. The antenna may be used as a wearable antenna on a human body. The antenna may be attached to a patient's shirt,

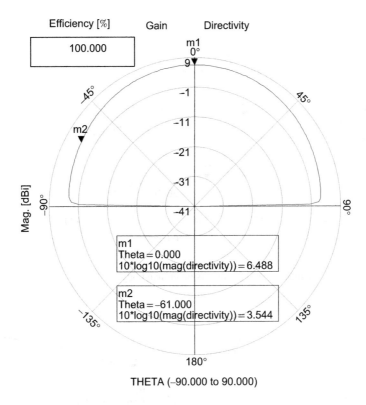

FIGURE 4.4   Radiation pattern of a wearable double-layer printed dipole at 460 MHz.

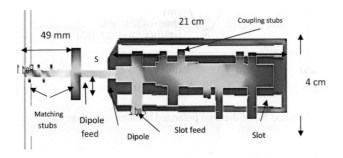

FIGURE 4.5   Current distribution of the wearable antenna.

patient's stomach or in the back zone. The antenna has been analyzed by using Keysight Momentum software [21]. There is a good agreement between measured and computed results. The antenna bandwidth is around 10% for VSWR better than 2:1. The antenna beam width is around 100°. The antenna gain is around 2 dBi. The computed S_{11} and S_{22} parameters are presented in Figure 4.6. Figure 4.7 presents the antenna's measured S_{11} parameters. The computed radiation patterns are shown in Figure 4.8.

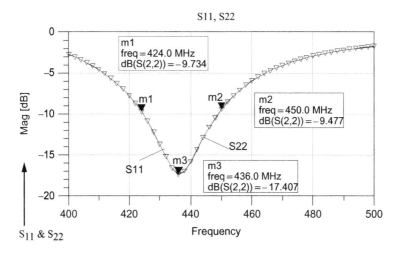

FIGURE 4.6 Computed S_{11} and S_{22} results on human body.

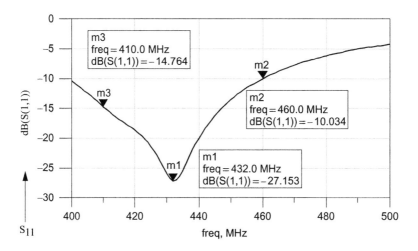

FIGURE 4.7 Measured S_{11} of the dual polarized antenna on human body.

The co-polar radiation pattern belongs to the *yz* plane. The cross-polar radiation pattern belongs to the *xz* plane. The antenna's cross-polarized field strength may be adjusted by varying the slot feed location. The dimensions and current distribution of the folded dually polarized antenna are presented in Figure 4.9. The antenna's dimensions are $6 \times 5 \times 0.16$ cm. Figure 4.10 presents the antenna's computed S_{11} and S_{22} parameters. The computed radiation patterns of the folded dipole are shown in Figure 4.11. The antenna's radiation characteristics on a human body have been measured by using a phantom. The phantom's electrical characteristics represent the human body's electrical characteristics.

The phantom has a cylindrical shape with a 40 cm diameter and a length of 1.5 m. The phantom contains a mix of 55% water, 44% sugar and 1% salt. The antenna under

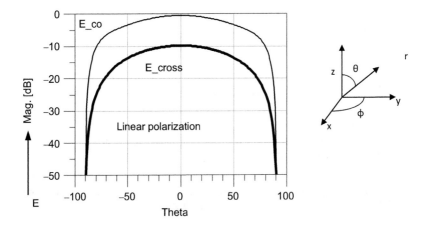

FIGURE 4.8 Wearable antenna, antenna shown in Figure 4.1, radiation patterns.

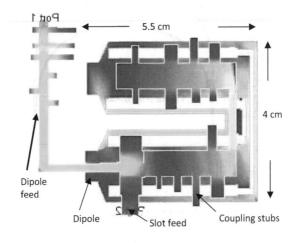

FIGURE 4.9 Current distribution of the folded dipole antenna, $6 \times 5 \times 0.16$ cm.

test was placed on the phantom during the measurements of the antenna's radiation characteristics. S_{11} and S_{12} parameters were measured directly on the human body by using a network analyzer. The measured results were compared to a known reference antenna.

4.3 PRINTED WEARABLE LOOP ANTENNA

Wearable loop antennas with ground plane are presented in this section. A wearable loop antenna with ground plane has been designed on Kapton substrates with thickness of 0.25 and 0.4 mm. An antenna without ground plane is shown in Figure 4.12a. A photo of wearable antennas is presented in Figure 4.12b. The loop antenna VSWR

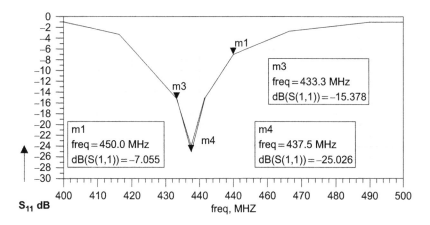

FIGURE 4.10   Folded antenna computed S_{11} and S_{22} results on human body.

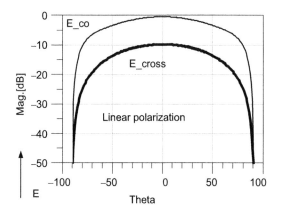

FIGURE 4.11   Folded antenna radiation patterns.

without the tuning capacitor was 4:1. This loop antenna may be tuned by adding a capacitor or varactor as shown in Figure 4.13. Tuning the antenna allows us to work in a wider bandwidth. Figure 4.14 presents the computed S_{11} of loop antenna with ground plane on human body. There is a good agreement between measured and computed results for several loop antennas' electrical parameters on human body.

The results presented in Table 4.1 indicate that the match of the loop antenna with ground plane to the human body's environment, without the tuning capacitor, is better than the match of the loop antenna without ground plane. A comparison of the electrical performance of several wearable printed antennas is given in Table 4.1.

Computed S_{11} parameters of the loop antenna with ground, in free space, are shown in Figure 4.15. Loop antenna with ground radiation pattern on human body is presented in Figure 4.16. The computed 3D radiation pattern and the coordinate used in this chapter are shown in Figure 4.17. Computed S_{11} of the loop antenna with a tuning capacitor is given in Figure 4.18. Figure 4.19 presents the radiation pattern of

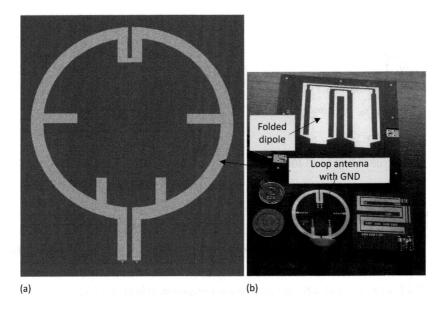

(a)   (b)

FIGURE 4.12   (a) Printed loop antenna with ground, (b) photo of wearable antennas.

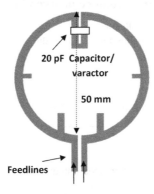

FIGURE 4.13   Tunable loop antenna without ground plane.

loop antenna without ground on human body. Figure 4.20 presents a loop antenna with ground plane printed on 0.4 mm thick substrate. Figure 4.21 presents the loop antenna's computed S_{11} on human body. Loop antennas printed on thicker substrate have a wider bandwidth as presented in Figure 4.21. Figure 4.22 presents the loop antenna's, printed on 0.4 mm thick substrate, radiation pattern.

4.4 WEARABLE MICROSTRIP ANTENNAS

Printed antennas possess attractive features such as low profile, flexibility, light weight, small volume and low production cost. Printed antennas may be used as

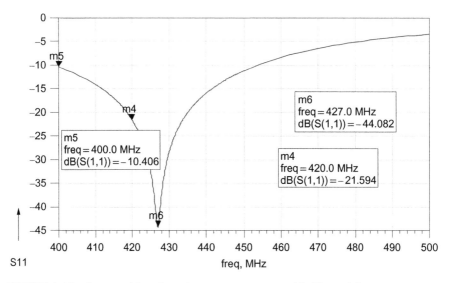

FIGURE 4.14   Computed S_{11} of new loop antenna, presented in Figure 4.8a.

TABLE 4.1
Comparison of electrical performance of several printed antennas

Parameter	Beam width 3 dB	Gain (dBi)	Voltage standing wave ratio	Dimensions (cm)
Printed loop antenna	100°	0	4:1	6 × 5 × 0.05
Printed microstrip dipole	100°	2	2:1	26 × 6 × 0.16
Wearable loop with ground	100°	0–2	2:1	6 × 5 × 0.05
Folded printed microstrip dipole	100°	2 to 3	2:1	6 × 5 × 0.16

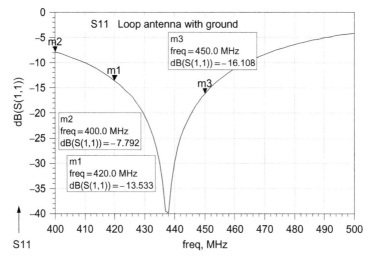

FIGURE 4.15   Computed S_{11} of the loop antenna with ground.

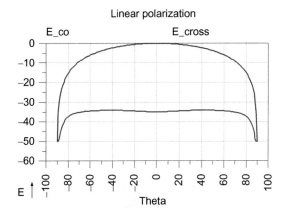

FIGURE 4.16   Loop antenna radiation pattern on human body, presented in Figure 4.8a.

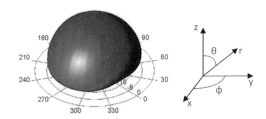

FIGURE 4.17   Loop antenna with ground, three-dimensional radiation pattern.

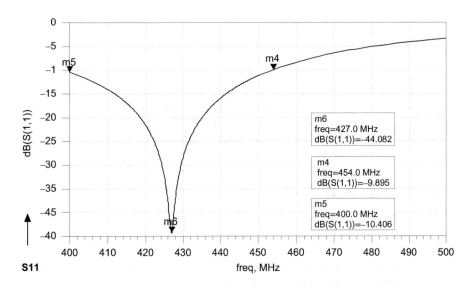

FIGURE 4.18   Computed S_{11} of a tuned loop antenna, without ground plane.

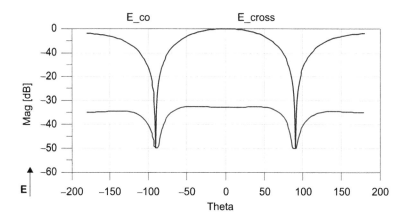

FIGURE 4.19   Radiation pattern of printed loop antenna on human body.

FIGURE 4.20   Loop antenna with ground plane printed on 0.4 mm thick substrate.

wearable antennas. Printed antennas have been widely presented in books and papers in the last decade [1–19]. The most popular type of printed antenna is the microstrip antenna. However, planar inverted-F antenna (PIFA), slot and dipole printed are widely used in communication systems. Printed antennas may be employed in communication links, seekers and in medical systems.

4.4.1 WEARABLE MICROSTRIP ANTENNAS

Microstrip antennas are printed on a dielectric substrate with low dielectric losses. A cross section of a microstrip antenna is shown in Figure 4.23. Microstrip antennas are thin patches etched on a dielectric substrate er. The substrate thickness, H, is less than 0.1λ. Microstrip antennas are widely presented in [1–7]. The printed antenna may be attached to a human body or inserted inside a wearable belt.

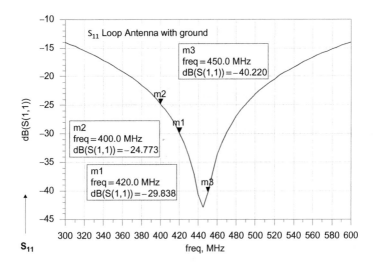

FIGURE 4.21   Computed S_{11} of a loop antenna printed on 0.4 mm thick substrate.

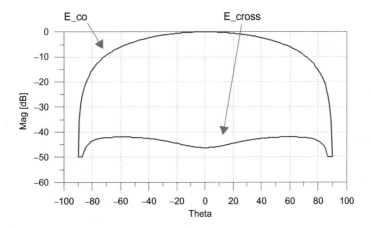

FIGURE 4.22   Loop antenna radiation pattern, printed on 0.4 mm thick Kapton substrate.

Advantages of microstrip antennas:

- Light weight and low volume.
- Flexible, conformal structures are possible.
- Low cost to fabricate.
- Easy to form large uniform arrays and phased arrays.

These features are very important for wearable communication systems.

Disadvantages of microstrip antennas:

- Limited bandwidth (usually 1%–5%; however, a wider bandwidth is possible with increased antenna structure complexity).

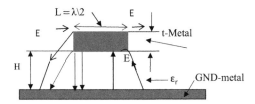

FIGURE 4.23   Microstrip antenna cross section.

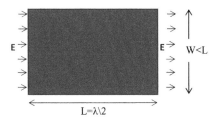

FIGURE 4.24   Rectangular microstrip antenna.

- Low power handling up to 50 W.
- Limited gain up to 30 dBi.
- High feed network losses at high frequencies, above 18 GHz.

The electric field along the radiating edges is shown in Figure 4.24.

The magnetic field is perpendicular to the E field according to Maxwell's equations. At the edge of the strip ($X/L = 0$ and $X/L = 1$) the H field drops to zero, because there is no conductor to carry the RF current, it is maximum in the center. The E field intensity is at maximum magnitude (and opposite polarity) at the edges ($X/L = 0$ and $X/L = 1$) and zero at the center. The ratio of E to H field is proportional to the impedance that we see when we feed the patch. Microstrip antennas may be fed by a microstrip line or by a coaxial line or probe feed. By adjusting the location of the feed point between the center and the edge, we can get any impedance, including 50 Ω. The shape of microstrip antennas may be square, rectangular, triangle, circle or any arbitrary shape as shown in Figure 4.25.

The dielectric constant that controls the resonance of the antenna is the effective dielectric constant of the microstrip line. The antenna dimension W is given by Equation 4.1.

$$W = \frac{c}{2f\sqrt{\epsilon_{\text{eff}}}} \tag{4.1}$$

The antenna bandwidth is given in Equation 4.2.

$$BW = \frac{H}{\sqrt{\epsilon_{\text{eff}}}} \tag{4.2}$$

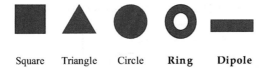

Square   Triangle   Circle   **Ring**   **Dipole**

FIGURE 4.25   Microstrip antenna shapes.

The gain of a microstrip antenna is between 0 and 7 dBi. The microstrip antenna gain is a function of the antenna's dimensions and configuration. We may increase printed antenna gain by using microstrip antenna's array configuration. In a microstrip antenna array, the benefit of a compact low-cost feed network is attained by integrating the RF feed network with the radiating elements on the same substrate. Microstrip antenna feed networks are presented in Figure 4.26. Figure 4.26a presents a parallel feed network. Figure 4.26b presents a parallel series feed network.

4.4.2 TRANSMISSION LINE MODEL OF MICROSTRIP ANTENNAS

In the transmission line model (TLM) of patch microstrip antennas, the antenna is represented as two slots connected by a transmission line. A TLM is presented in Figure 4.27. The TLM is not an accurate model; however, it gives a good physical understanding of patch microstrip antennas. The electric field along and underneath the patch depends on the z coordinate, see Equation 4.3. In the design of a wearable

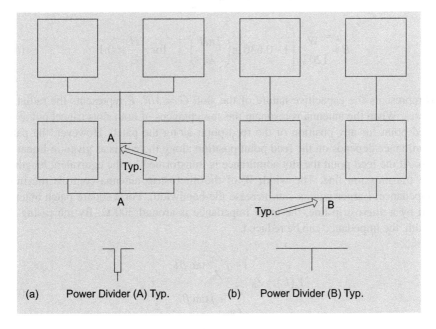

FIGURE 4.26   Configuration of microstrip antenna array. (a) Parallel feed network, (b) parallel series feed network.

a. Parallel feed network. b. Parallel series feed network.

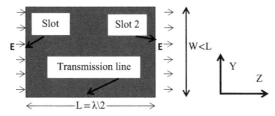

FIGURE 4.27   Transmission line model of patch microstrip antennas.

patch antenna, the human body's electrical parameter should be considered in the design.

$$E_x \sim \cos\left(\frac{\pi z}{L_{eff}}\right) \tag{4.3}$$

At $z = 0$ and $z = L_{eff}$ the electric field is maximum. At $z = \dfrac{L_{eff}}{2}$ the electric field equals zero. For $\dfrac{H}{\lambda_0} < 0.1$ the electric field distribution along the x-axis is assumed to be uniform. The slot admittance may be written as given in Equations 4.4 and 4.5.

$$G = \frac{W}{120\lambda_0}\left[1 - \frac{1}{24}\left(\frac{2\pi H}{\lambda_0}\right)^2\right] \quad \text{for} \quad \frac{H}{\lambda_0} < 0.1 \tag{4.4}$$

$$B = \frac{W}{120\lambda_0}\left[1 - 0.636\ln\left(\frac{2\pi H}{\lambda_0}\right)^2\right] \quad \text{for} \quad \frac{H}{\lambda_0} < 0.1 \tag{4.5}$$

B represents the capacitive nature of the slot. $G = 1/R$. R represents the radiation losses. When the antenna is resonant the susceptances of both slots cancel out at the feed point for any position of the feed point along the patch. However, the patch admittance depends on the feed point position along the z-axis as given in Equation 4.6. At the feed point the slot admittance is transformed by the equivalent length of the transmission line. The width W of the microstrip antenna controls the input impedance. Larger widths can increase the bandwidth. For a square patch antenna fed by a microstrip line, the input impedance is around 300 Ω. By increasing the width, the impedance can be reduced.

$$Y(l_1) = Z_0 \frac{1 + j\dfrac{Z_L}{Z_0}\tan\beta l_1}{\dfrac{Z_L}{Z_0} + j\tan\beta l_1} = Y_1 \tag{4.6}$$

$$Y_{in} = Y_1 + Y_2$$

4.4.3 Higher-Order Transmission Modes in Microstrip Antennas

In order to prevent higher-order transmission modes we should limit the thickness of the microstrip substrate to 10% of a wavelength. The cut-off frequency of the higher-order mode is given in Equation 4.7.

$$f_c = \frac{c}{4H\sqrt{\varepsilon - 1}}$$
(4.7)

4.4.4 Effective Dielectric Constant

Part of the fields in the microstrip antenna structure exists in air and the other part of the fields exists in the dielectric substrate. The effective dielectric constant is somewhat less than the substrate's dielectric constant. The effective dielectric constant of the microstrip line may be calculated by Equations 4.8a and 4.8b as a function of W/H:

$$\text{For } \left(\frac{W}{H}\right) < 1 \quad \varepsilon_e = \frac{\varepsilon_r + 1}{2} + \frac{\varepsilon_r - 1}{2}\left[\left(1 + 12\left(\frac{H}{W}\right)\right)^{-0.5} + 0.04\left(1 - \left(\frac{W}{H}\right)\right)^2\right]$$
(4.8a)

$$\text{For } \left(\frac{W}{H}\right) \geq 1 \quad \varepsilon_e = \frac{\varepsilon_r + 1}{2} + \frac{\varepsilon_r - 1}{2}\left[\left(1 + 12\left(\frac{H}{W}\right)\right)^{-0.5}\right]$$
(4.8b)

This calculation ignores strip thickness and frequency dispersion, but their effects are negligible.

4.4.5 Losses in Microstrip Antennas

Losses in microstrip line are due to conductor loss, radiation loss and dielectric loss.

4.4.5.1 Conductor Loss

Conductor loss may be calculated by using Equation 4.9.

$$\alpha_c = 8.686\log\left(R_S/(2WZ_0)\right) \quad \text{dB/length}$$
$$R_S = \sqrt{\pi f \mu \rho} \quad \text{Skin resistance}$$
(4.9)

Conductor losses may also be calculated by defining an equivalent loss tangent δ_c, given by $\delta_c = \delta_s/h$, where $\delta_s = \sqrt{\dfrac{2}{\omega\mu\sigma}}$, where σ is the strip conductivity, h is the substrate height and μ is the free space permeability.

FIGURE 4.28   Coordinate system.

4.4.5.2 Dielectric Loss

Dielectric loss may be calculated by using Equation 4.10.

$$\alpha_d = 27.3 \frac{\varepsilon_r}{\sqrt{\varepsilon_{eff}}} \frac{\varepsilon_{eff} - 1}{\varepsilon_r - 1} \frac{tg\delta}{\lambda_0} \quad \text{dB/cm}$$

(4.10)

$$tg\delta = \text{dielectric loss coefficient}$$

4.4.6 Patch Radiation Pattern

The patch width, W, controls the antenna's radiation pattern. The coordinate system is shown in Figure 4.28. The normalized radiation pattern is approximately given by Equations 4.11 and 4.12:

$$E_\theta = \frac{\sin\left(\dfrac{k_0 W}{2} \sin\theta \sin\varphi\right)}{\dfrac{k_0 W}{2} \sin\theta \sin\varphi} \cos\left(\dfrac{k_0 L}{2} \sin\theta \cos\varphi\right) \cos\varphi$$

(4.11)

$$k_0 = \frac{2\pi}{\lambda}$$

$$E_\varphi = \frac{\sin\left(\dfrac{k_0 W}{2} \sin\theta \sin\varphi\right)}{\dfrac{k_0 W}{2} \sin\theta \sin\varphi} \cos\left(\dfrac{k_0 L}{2} \sin\theta \cos\varphi\right) \cos\theta \sin\varphi$$

(4.12)

$$k_0 = \frac{2\pi}{\lambda}$$

The magnitude of the fields is given by Equation 4.13:

$$f(\theta,\varphi) = \sqrt{E_\theta^2 + E_\varphi^2}$$

(4.13)

4.5 TWO-LAYER WEARABLE STACKED MICROSTRIP ANTENNAS

Two-layer microstrip antennas were first presented in [1–7]. The major disadvantage of a single-layer microstrip antenna is narrow bandwidth. By designing a

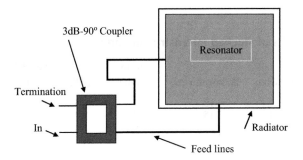

FIGURE 4.29 Circular polarized microstrip stacked patch antenna.

double-layer microstrip antenna we can get a wider bandwidth. Two-layer microstrip antennas may be the best antenna choice for wideband wearable systems.

In the first layer, the antenna feed network and a resonator are printed. In the second layer, the radiating element is printed. The electromagnetic field is coupled from the resonator to the radiating element. The shape of the resonator and the radiating element may be square, rectangular, triangle, circle or any arbitrary shape. The distance between the layers is optimized to get maximum bandwidth with the best antenna efficiency. The spacing between the layers may be air or a foam with low dielectric losses.

A circular polarization double-layer antenna was designed at 2.2 GHz. The resonator and the feed network were printed on a substrate with relative dielectric constant of 2.5 with thickness of 1.6 mm. The resonator is a square microstrip resonator with dimensions $W = L = 45$ mm. The radiating element was printed on a substrate with relative dielectric constant of 2.2 with thickness of 1.6 mm. The radiating element is a square patch with dimensions $W = L = 48$ mm. The patch was designed as a circular polarized antenna by connecting a 3 dB 90° branch coupler to antenna feed lines, as shown in Figure 4.29.

The antenna bandwidth is 10% for VSWR better than 2:1. The measured antenna beam width is around 72°. The measured antenna gain is 7.5 dBi. This antenna may be applied in wideband wireless wearable systems. Measured results of stacked microstrip wearable antennas are listed in Table 4.2. The antennas listed in Table 4.2 may be used in wearable communication systems. Results in Table 4.2 indicated that the bandwidth of two-layer microstrip antennas may be around 9%–15% for VSWR better than 2:1. In Figure 4.30 a stacked microstrip antenna is shown.

The antenna feed network is printed on flame retardant-4 (FR-4) dielectric substrate with dielectric constant of 4 and 1.6 mm thick. The radiator is printed on RT-Duroid 5880 dielectric substrate with dielectric constant of 2.2 and thickness of 1.6 mm. The antenna's electrical parameters were calculated and optimized by using ADS software. The dimensions of the microstrip stacked patch antenna shown in Figure 4.30 are $33 \times 20 \times 3.2$ mm. The computed S_{11} parameters are presented in Figure 4.31. The radiation pattern of the microstrip stacked patch is shown in Figure 4.32. The antenna bandwidth is around 5% for VSWR better than 2.5:1.

TABLE 4.2
Measured results of stacked microstrip antennas

Antenna	F (GHz)	Bandwidth (%)	Beam width (°)	Gain (dBi)	Side lobe (dB)	Polarization
Square	2.2	10	72	7.5	−22	Circular
Circular	2.2	15	72	7.9	−22	Linear
Annular disc	2.2	11.5	78	6.6	−14	Linear
Rectangular	2.0	9	72	7.4	−25	Linear
Circular	2.4	9	72	7	−22	Linear
Circular	2.4	10	72	7.5	−22	Circular

FIGURE 4.30   A microstrip stacked patch antenna.

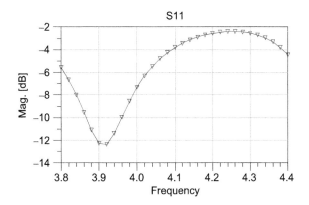

FIGURE 4.31   Computed S_{11} of the microstrip stacked patch.

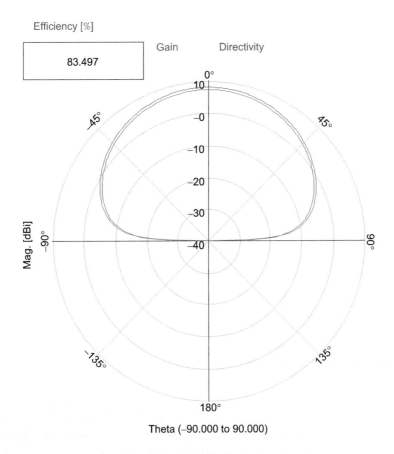

Efficiency [%]

83.497

Gain   Directivity

FIGURE 4.32   Radiation pattern of the microstrip stacked patch.

The antenna bandwidth is improved to 10% for VSWR better than 2.0:1 by adding 8 mm air spacing between the layers. The antenna beam width is around 72°. The antenna gain is around 7 dBi.

4.6 STACKED MONO-PULSE KU BAND PATCH ANTENNA

A mono-pulse double-layer antenna was designed at 15 GHz. The mono-pulse dou-ble-layer antenna consists of four circular patch antennas as shown in Figure 4.33. The resonator and the feed network were printed on a substrate with relative dielec-tric constant of 2.5 with thickness of 0.8 mm. The resonator is a circular microstrip resonator with diameter, $a = 4.2$ mm. The radiating element was printed on a sub-strate with relative dielectric constant of 2.2 with thickness of 0.8 mm. The radiating element is a circular microstrip patch with diameter, $a = 4.5$ mm. The four circular patch antennas are connected to three 3 dB 180° rat-race couplers via the antenna feed lines, as shown in Figure 4.33. The comparator consists of three stripline 3 dB 180° rat-race couplers printed on a substrate with relative dielectric constant of 2.2

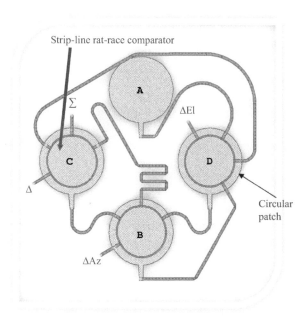

FIGURE 4.33 A microstrip stacked mono-pulse antenna.

with thickness of 0.8 mm. The comparator has four output ports: a sum port Σ, difference port Δ, elevation difference port ΔEl and azimuth difference port ΔAz as shown in Figure 4.33. The antenna bandwidth is 10% for VSWR better than 2:1. The antenna beam width is around 36°. The measured antenna gain is around 10 dBi. The comparator losses around 0.7 dB.

4.6.1 RAT-RACE COUPLER

A rat-race coupler is shown in Figure 4.34. The rat-race circumference is 1.5 wavelengths. The distance from A to Δ port is $3\lambda/4$. The distance from A to Σ port is $\lambda/4$.

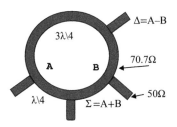

FIGURE 4.34 Rat-race coupler.

For an equal-split rat-race coupler, the impedance of the entire ring is fixed at $1.41 \times Z0$, or $70.7\ \Omega$ for $Z0 = 50\ \Omega$. For an input signal V, the outputs at ports 2 and 4 are equal in magnitude, but $180°$ out of phase.

4.7 WEARABLE PIFA

PIFAs possess attractive features such as low profile, small size and low fabrication costs [16–18]. A PIFA's bandwidth is higher than the bandwidth of a conventional patch antenna, because the antenna thickness of a PIFA is higher than the thickness of patch antennas. The conventional antenna of a PIFA is a grounded quarter-wavelength patch antenna. The antenna consists of ground plane, a top plate radiating element, feed wire and a shorting plate or via holes that connect the radiating element to the ground plane as shown in Figure 4.35.

The patch is shorted to the ground at the patch edge. The fringing fields, which are responsible for radiated fields, are shorted on the far end; so, only the fields nearest the transmission line radiate. Consequently, the gain is reduced, but the patch antenna maintains the same basic properties as a half-wavelength patch. However, the antenna length is reduced by 50%. The feed location may be placed between the open and shorted end. The feed location controls the antenna's input impedance.

4.7.1 Grounded Quarter-Wavelength Patch Antenna

A grounded quarter-wavelength patch antenna was designed on FR-4 substrate with relative dielectric constant of 4.5 and 1.6 mm thickness at 3.85 GHz. The antenna is shown in Figure 4.36. The antenna was designed by using Momentum software. The antenna's dimensions are $34 \times 17 \times 1.6$ mm. S_{11} results of the antenna are shown in Figure 4.37. The antenna bandwidth is around 6% for VSWR better than 3:1 without a matching network. The radiation pattern is shown in Figure 4.38. The antenna beam width is around $76°$. The grounded quarter-wavelength patch antenna gain is around 6.7 dBi. The antenna efficiency is around 92%. The compact antenna of a PIFA may be attached to a human body or inserted inside a wearable belt.

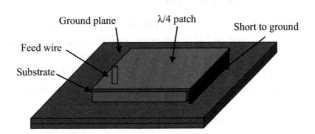

FIGURE 4.35   Conventional planar inverted-F antenna.

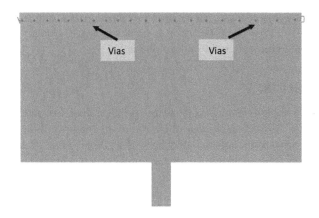

FIGURE 4.36 Grounded quarter-wavelength patch antenna.

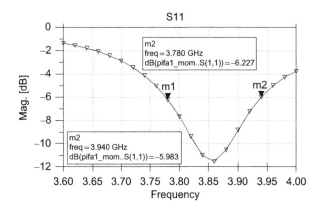

FIGURE 4.37 Grounded quarter-wavelength patch antenna S_{11} results.

4.7.2 A WEARABLE DOUBLE-LAYER PIFA

A new double-layer PIFA was designed. The first layer is a grounded quarter-wavelength patch antenna printed on FR-4 substrate with relative dielectric constant of 4.5 and thickness of 1.6 mm. The second layer is a rectangular patch antenna printed on Duroid substrate with relative dielectric constant of 2.2 and thickness of 1.6 mm. The antenna is shown in Figure 4.39. The antenna was designed by employing ADS software. The antenna's dimensions are $34 \times 17 \times 3.2$ mm. S_{11} results of the antenna are shown in Figure 4.40. The antenna is a dual band antenna. The first resonant frequency is 3.48 GHz. The second resonant frequency is 4.02 GHz. The radiation pattern is shown in Figure 4.41. The antenna beam width is around 74°. The antenna gain is around 7.4 dBi. The antenna efficiency is around 83.4%.

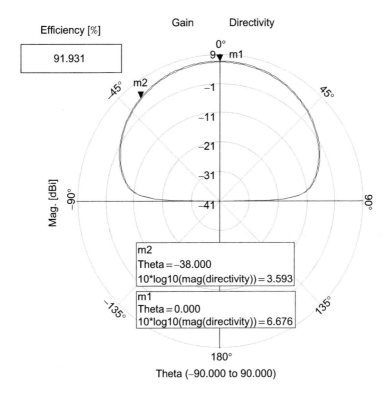

FIGURE 4.38   Grounded quarter-wavelength patch antenna radiation pattern.

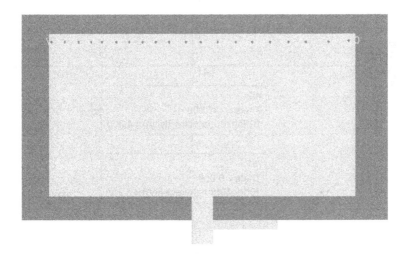

FIGURE 4.39   Double-layer planar inverted-F antenna.

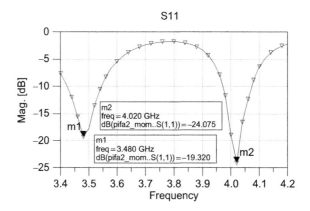

FIGURE 4.40   Double-layer planar inverted-F antenna S_{11} results.

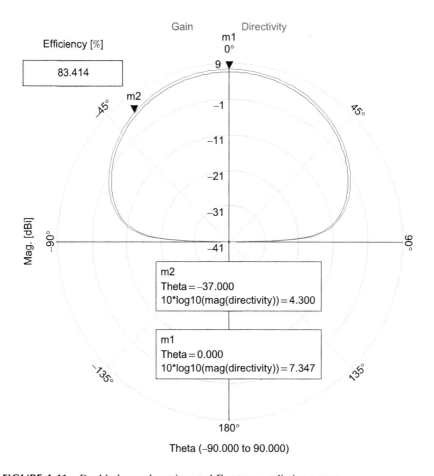

FIGURE 4.41   Double-layer planar inverted-F antenna radiation pattern.

4.8 CONCLUSIONS

This chapter presents several wideband wearable antennas with high efficiency for medical, Internet of Things and sport applications. Wearable technology provides a powerful new tool for medical and surgical rehabilitation services. A wearable body area network is emerging as an important option for medical centers and patients. Wearable technology provides a convenient platform that may quantify the long-term context and physiological response of individuals. Wearable technology will support the development of individualized treatment systems with real-time feedback to help promote patient health. Wearable medical systems and sensors can perform gait analysis and can measure body temperature, heart rate, blood pressure, sweat rate and other physiological parameters of the person wearing the medical device.

Design considerations and computed and measured results of several wearable printed antennas are presented in this chapter. The antenna dimensions may vary from $26 \times 6 \times 0.16$ cm to $5 \times 5 \times 0.05$ cm according to the medical system specification. The antenna's bandwidth is around 10% for VSWR better than 2:1. The antenna beam width is around $100°$. The antenna's gain varies from 0 to 5 dBi. A wideband tunable microstrip antenna with high efficiency for medical applications has been presented in this chapter.

A varactor is employed to compensate variations in the antenna's resonant frequency at different locations on the human body.

REFERENCES

[1] Sabban, A. *Wideband RF Technologies and Antenna in Microwave Frequencies*. Wiley Sons, New York, July 2016.

[2] Sabban, A. *Novel Wearable Antennas for Communication and Medical Systems*, Taylor & Francis Group, Florida, USA, October 2017.

[3] Sabban, A. *Low Visibility Antennas for Communication Systems*. Taylor & Francis Group, Florida USA, 2015.

[4] Sabban, A. Small wearable meta materials antennas for medical systems. *Appl. Comput. Electromagn. Soc. J.*, 31(4), 434–443, April 2016.

[5] Sabban, A. *Microstrip Antenna Arrays*; Nasimuddin, N. Ed. *Microstrip Antennas*, InTech, Croatia, 2011, pp. 361–384, ISBN: 978-953-307-247-0 . http://www.intechopen. com/articles/show/title/microstrip-antenna-arrays.

[6] Sabban, A. New wideband printed antennas for medical applications. *IEEE Trans. Antennas Propag.*, 61(1), 84–91, January 2013.

[7] Sabban, A. Dual polarized dipole wearable antenna. U.S. Patent number: 8203497, June 19, 2012.

[8] Sabban, A. A new wideband stacked microstrip antenna. In *IEEE Antenna and Propagation Symposium*, Houston, TX, June 1983.

[9] Sabban, A. Wideband microstrip antenna arrays. In *IEEE Antenna and Propagation Symposium MELCOM*, Tel-Aviv, June 1981.

[10] Chirwa, L.C.; Hammond, P.A.; Roy, S.; Cumming, D.R.S. Electromagnetic radiation from ingested sources in the human Intestine between 150 MHz and 1.2 GHz. *IEEE Trans. Biomed. Eng.*, 50(4), 484–492, April 2003.

[11] Werber, D.; Schwentner, A.; Biebl, E.M. Investigation of RF transmission properties of human tissues. *Adv. Radio Sci.*, 4, 357–360, 2006.

[12] Fujimoto, K.; James, J.R. Eds. *Mobile Antenna Systems Handbook*, Artech House, Boston, MA, 1994.

[13] Gupta, B., Sankaralingam, S., Dhar, S. Development of wearable and implantable antennas in the last decade. In *Microwave Mediterranean Symposium (MMS)*, Guzelyurt, Turkey, August 2010; pp. 251–267.

[14] Thalmann, T.; Popovic, Z.; Notaros, B.M.; Mosig, J.R. Investigation and design of a multi-band wearable antenna. In *3rd European Conference on Antennas and Propagation, EuCAP 2009*, Berlin, Germany, 2009, pp. 462–465.

[15] Salonen, P.; Rahmat-Samii, Y.; Kivikoski, M. Wearable antennas in the vicinity of human body. In *IEEE Antennas and Propagation Society Symposium*, Monterey, CA, 2004, Vol.1, pp. 467–470.

[16] Kellomaki, T.; Heikkinen, J.; Kivikoski, M. Wearable antennas for FM reception. In *First European Conference on Antennas and Propagation, EuCAP 2006*, Hague Netherlands, 2006, pp. 1–6.

[17] Sabban, A. Wideband printed antennas for medical applications. In *APMC 2009 Conference*, Singapore, December 2009.

[18] Lee, Y. Antenna Circuit Design for RFID Applications. Microchip Technology Inc., Microchip AN 710c., Arizona, USA.

[19] Sabban, A.; Gupta, K.C. Characterization of radiation loss from microstrip discontinuities using a multiport network modeling approach. *IEEE Trans. Microwave Theory Tech.*, 39(4), 705–712, April 1991.

[20] Sabban, A. *Multiport Network Model for Evaluating Radiation Loss and Coupling among Discontinuities in Microstrip Circuits*, PhD thesis, University of Colorado at Boulder, January 1991.

[21] Software, Keysight, http://www.keysight.com/en/pc-1297113/advanced-design-system-ads?cc=IL&lc=eng.

[22] Mukhopadhyay, S.C. Ed. *Wearable Electronics Sensors*, Springer, Cham, 2015.

[23] Bonfiglio, A.; De Rossi, D. Eds. *Wearable Monitoring Systems*, Springer, New York, 2011.

[24] Gao, T.; Greenspan, D.; Welsh, M.; Juang, R.R.; Alm, A. Vital signs monitoring and patient tracking over a wireless network. In *Proceedings of IEEE-EMBS 27th Annual International Conference of the Engineering in Medicine and Biology*, Shanghai, China, September 1–5, 2005; pp. 102–105.

[25] Otto, C.A.; Jovanov, E.; Milenkovic, E.A. WBAN-based system for health monitoring at home. In *Proceedings of IEEE/EMBS International Summer School, Medical Devices and Biosensors*, Boston, MA, September 4–6, 2006; pp. 20–23.

[26] Zhang, G.H.; Poon, C.C.Y.; Li, Y.; Zhang, Y.T. A biometric method to secure telemedicine systems. In *Proceedings of the 31st Annual International Conference of the IEEE Engineering in Medicine and Biology Society*, Minneapolis, MN, September 2009; pp. 701–704.

[27] Srinivasan, V.; Stankovic, J.; Whitehouse, K. Protecting your daily in home activity information from a wireless snooping attack. In *Proceedings of the 10th International Conference on Ubiquitous Computing*, Seoul, Korea, September 21–24, 2008; pp. 202–211.

[28] Casas, R.; Blasco, M.R.; Robinet, A.; Delgado, A.R.; Yarza, A.R.; Mcginn, J.; Picking, R.; Grout, V. User modelling in ambient intelligence for elderly and disabled people. In *Proceedings of the 11th International Conference on Computers Helping People with Special Needs*, Linz, Austria, July 2008; pp. 114–122.

[29] Jasemian, Y. Elderly comfort and compliance to modern telemedicine system at home. In *Proceedings of the Second International Conference on Pervasive Computing Technologies for Healthcare*, Tampere, Finland, January 30–February 1, 2008; pp. 60–63.

[30] Atallah, L.; Lo, B.; Yang, G.Z.; Siegemund, F. Wirelessly accessible sensor populations (WASP) for elderly care monitoring. In *Proceedings of the Second International Conference on Pervasive Computing Technologies for Healthcare*, Tampere, Finland, January 30–February 1, 2008; pp. 2–7.

[31] Hori, T.; Nishida, Y.; Suehiro, T.; Hirai, S. SELF-Network: Design and implementation of network for distributed embedded sensors. In *Proceedings of IEEE/RSJ International Conference on Intelligent Robots and Systems*, Takamatsu, Japan, October 30–November 5, 2000; pp. 1373–1378.

[32] Mori, Y.; Yamauchi, M.; Kaneko, K. Design and implementation of the Vital Sign Box for home healthcare. In *Proceedings of IEEE EMBS International Conference on Information Technology Applications in Biomedicine*, Arlington, VA, November 2000; pp. 104–109.

[33] Lauterbach, C.; Strasser, M.; Jung, S.; Weber, W. Smart clothes self-powered by body heat. In *Proceedings of Avantex Symposium*, Frankfurt, Germany, May 2002; pp. 5259–5263.

[34] Marinkovic, S.; Popovici, E. Network coding for efficient error recovery in wireless sensor networks for medical applications. In *Proceedings of International Conference on Emerging Network Intelligence*, Sliema, Malta, October 11–16, 2009; pp. 15–20.

[35] Schoellhammer, T.; Osterweil, E.; Greenstein, B.; Wimbrow, M.; Estrin, D. Lightweight temporal compression of microclimate datasets. In *Proceedings of the 29th Annual IEEE International Conference on Local Computer Networks*, Tampa, FL, November 16–18, 2004; pp. 516–524.

[36] Barth, A.T.; Hanson, M.A.; Powell, H.C., Jr; Lach, J. Tempo 3.1: A body area sensor network platform for continuous movement assessment. In *Proceedings of the 6th International Workshop on Wearable and Implantable Body Sensor Networks*, Berkeley, CA, June 2009; pp. 71–76.

[37] Gietzelt, M.; Wolf, K.H.; Marschollek, M.; Haux, R. Automatic self-calibration of body worn triaxial-accelerometers for application in healthcare. In *Proceedings of the Second International Conference on Pervasive Computing Technologies for Healthcare*, Tampere, Finland, January 2008; pp. 177–180.

[38] Gao, T.; Greenspan, D.; Welsh, M.; Juang, R.R.; Alm, A. Vital signs monitoring and patient tracking over a wireless network. In *Proceedings of the 27th Annual International Conference of the IEEE EMBS*, Shanghai, China, September 1–4, 2005; pp. 102–105. Purwar, A.; Jeong, D.U.; Chung, W.Y. Activity monitoring from realtime triaxial accelerometer data using sensor network. In *Proceedings of International Conference on Control, Automation and Systems*, Hong Kong, March 21–23, 2007; pp. 2402–2406.

[39] Baker, C.; Armijo, K.; Belka, S.; Benhabib, M.; Bhargava, V.; Burkhart, N.; Der Minassians, A.; Dervisoglu, G.; Gutnik, L.; Haick, M.; Ho, C.; Koplow, M.; Mangold, J.; Robinson, S.; Rosa, M.; Schwartz, M.; Sims, C.; Stoffregen, H.; Waterbury, A.; Leland, E.; Pering, T.; Wright, P. Wireless sensor networks for home health care. In *Proceedings of the 21st International Conference on Advanced Information Networking and Applications Workshops*, Niagara Falls, Canada, May 21–23, 2007; pp. 832–837.

[40] Aziz, O.; Lo, B.; King, R.; Darzi, A.; Yang, G.Z. Pervasive body sensor network: An approach to monitoring the postoperative surgical patient. In *Proceedings of International Workshop on Wearable and implantable Body Sensor Networks (BSN 2006)*, Cambridge, MA; 2006; pp. 13–18.

[41] Kwon, D.Y.; Gross, M. Combining body sensors and visual sensors for motion training. In *Proceedings of the 2005 ACM SIGCHI International Conference on Advances in Computer Entertainment Technology*, Valencia, Spain, June 15–17, 2005; pp. 94–101.

[42] Boulgouris, N.K.; Hatzinakos, D.; Plataniotis, K.N. Gait recognition: A challenging signal processing technology for biometric identification. *IEEE Signal Process. Mag.*, 22, 78–90, 2005.

[43] Kimmeskamp, S.; Hennig, E.M. Heel to toe motion characteristics in parkinson patients during free walking. *Clin. Biomech.*, 16, 806–812, 2001.

[44] Turcot, K.; Aissaoui, R.; Boivin, K.; Pelletier, M.; Hagemeister, N.; de Guise, J.A. New accelerometric method to discriminate between asymptomatic subjects and patients with medial knee osteoarthritis during 3-D gait. *IEEE Trans. Biomed. Eng.*, 55, 1415–1422, 2008.

[45] Bamberg, S.J.M.; Benbasat, A.Y.; Scarborough, D.M.; Krebs, D.E.; Paradiso, J.A. Gait analysis using a shoe-integrated wireless sensor system. *IEEE Trans. Inf. Technol. Biomed.*, 12, 413–423, 2008.

[46] Choi, J.H.; Cho, J.; Park, J.H.; Eun, J.M.; Kim, M.S. An efficient gait phase detection device based on magnetic sensor array. In *Proceedings of the 4th Kuala Lumpur International Conference on Biomedical Engineering*, Kuala Lumpur, Malaysia, June 25–28, 2008; Vol. 21, pp. 778–781.

[47] Hidler, J. Robotic assessment of walking in individuals with gait disorders. In *Proceedings of the 26th Annual International Conference of the IEEE Engineering in Medicine and Biology Society*, San Francisco, CA, September 1–5, 2004; Vol. 7, pp. 4829–4831.

[48] Wahab, Y.; Bakar, N.A. Gait analysis measurement for sport application based on ultrasonic system. In *Proceedings of the 2011 IEEE 15th International Symposium on Consumer Electronics*, Singapore, June 14–17, 2011; pp. 20–24.

[49] ElSayed, M.; Alsebai, A.; Salaheldin, A.; El Gayar, N.; ElHelw, M. Ambient and wearable sensing for gait classification in pervasive healthcare environments. In *Proceedings of the 12th IEEE International Conference on e-Health Networking Applications and Services (Healthcom)*, Lyon, France, July 1–3, 2010; pp. 240–245.

5   Small Wearable Antennas
Experimental Case Studies

Ely Levine

CONTENTS

5.1 INTRODUCTION

In most cases, an antenna designer assumes that the antenna is surrounded by free space (air). The current or voltage sources and the electromagnetic fields are solved numerically and the radiation intensity can be defined in the free space. The space in the vicinity of the antenna can be divided into three regions [1–2]: (a) reactive near field, (b) radiating near field (Fresnel) and (c) far field (Fraunhofer). The reactive near zone is defined as [1] "that portion of the near-field region, immediately surrounding the antenna, wherein the reactive field predominates." For small antennas (in terms of wavelengths), the outer boundary of this region is commonly taken to exist at a distance R from the antenna surface, as shown in Figure 5.1.

$$R < \lambda/2\pi \qquad (5.1)$$

which is referred to as the "radian distance." In this distance, the energy is basically stored as electrical power or as magnetic power. The radiative near field zone is defined as [1] "that region of the field of an antenna between the reactive near field region and the far field region wherein radiation fields predominate and wherein the angular field distribution is dependent upon the distance from the antenna." If the antenna has a maximum dimension that is not large compared to the wavelength, this region may not exist. For an antenna focused at infinity, the radiating near field has an inner distance R known as *Fresnel* distance:

$$R > 0.62 \sqrt{D^3/\lambda} \qquad (5.2)$$

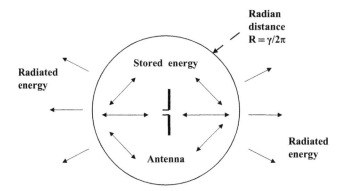

FIGURE 5.1   The reactive near field (less than the radian distance).

where D is the largest dimension of the antenna and also large compared to the wavelength (D >λ). The outer boundary of the near field is the *Fraunhofer* distance R:

$$R > 2D^2/\lambda. \tag{5.3}$$

The far field zone is defined as [1] "that region of the field of the antenna where the angular field distribution is essentially independent of the distance from the antenna." The boundaries between the zones depicted in Figure 5.2 are not sharp and definitive.

In cases in which the antenna is surrounded with objects or materials, these boundaries are changed. For example, the maximal size D may be a function of the

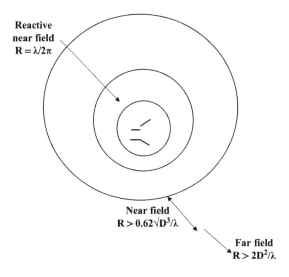

FIGURE 5.2   Three regions of the fields surrounding the antenna. Reactive near field R<λ/2π. Radiative near field $0.62\sqrt{D^3/\lambda} < R < 2D^2/\lambda$. Far field R > 2D²/λ.

materials/tissues and not the pure size of the antenna itself. We conclude that a close distance between the antenna and the body may change the impedance of the antenna due to its reactive nature and the radiation pattern, with and without actual shadowing of the tissues. The design of wearable antennas can be divided into cases of weak interaction with the body, where some changes in the antenna can adjust to the specific environment, and into cases of strong interactions, where new modeling or testing should be done. In most practical cases, and especially for small antennas, one can use phantoms like fruits or meat to emulate the effects of the body.

The design of wearable antennas should take into account the effects of the human body, either by simulations or by measurements. The antennas must be small, low-weight and flexible to allow convenient movement. They must meet the safety standards of non-ionized radiation and be user-friendly. Many cases and examples are described in the literature. We mention here a few of them, mainly reviews and books [3–12].

This chapter brings readers to the actual world of small wearable antenna design and applications. The small antennas under consideration were operated in the vicinity of a human body (hand, head, chest etc.) at different frequencies, licensed and unlicensed, from VHF to S band. This chapter describes effects of the human body in several aspects: detuning of the working frequency, additional loss by the live tissue, effects on the radiation patterns and on the gain, and possible polarization rotation. We emphasize actual realization through detailed examples with measured performances and essential "tips" from an experimentalist point of view. The five design cases presented here range from wearable antennas with minimal interaction with the body (mounted on the head or on the chest with a metallic ground plane) to wrist-type antennas that are indeed influenced by the hand.

- Antenna on helmet with high front to back (F/B) ratio.
- Wearable antenna for high power cellular jammer.
- Radio-frequency identification (RFID) UHF reader antenna in the pocket.
- Small helical antenna for a personal location beacon.
- VHF antenna for personal communications.

5.2 ANTENNA ON HELMET WITH HIGH F/B RATIO

Soldiers, police officers and security agents carry on their body a variety of communication devices and antennas. An optimal location for RFID active tags is the helmet, which enjoys minimal shadowing and absorption by the body. We present here a specific application of a small printed antenna, at 915 MHz, which is mounted on the back side of a standard helmet. The main concern in this case was to reduce the F/B (in this case, the back to front) ratio in the radiation pattern. The value of the F/B ratio is largely dependent on the size of the ground plane in terms of wavelength. For the industrial, scientific and medical (ISM) frequencies 915 MHz or 868 MHz the ground plane, allowed on the human body, is necessarily smaller than the wavelength, hence the F/B ratio is not high (10–15 dB).

FIGURE 5.3   A helmet covered with thin copper layer.

The design starts with covering the helmet with thin metal strips (40 μm) as shown in Figure 5.3. The cross section of the helmet is about 20 × 20 cm. In order to reduce the antenna size and keep the antenna small in comparison to the ground, we have chosen a narrow quarter-wave radiator. The microstrip element is built on a flexible foam substrate with size 3 × 7 cm and thickness 6 mm as shown in Figure 5.4.

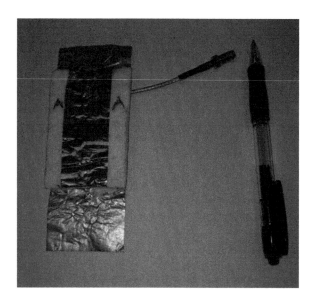

FIGURE 5.4   A quarter-wave microstrip element (3 × 7 cm).

FIGURE 5.5   A quarter-wave microstrip element (3 × 7 cm) on a ground plane 16 × 12 cm.

Four samples are tested: (a) ground 20 × 20 cm, (b) ground plane 16 × 16 cm, (c) ground plane 12 × 16 cm and (d) ground plane is the helmet. In all cases, the bandwidth is 4% which covers the 902–928 MHz band and the gain is 4–5 dBi. The beamwidths vary between 85° and 110°. Radiation patterns in E and H planes for different ground planes (20 × 20 cm, 16 × 16 cm and 12 × 16 cm) are shown in Figures 5.5–5.11.

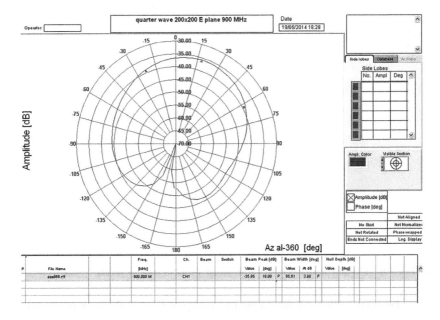

FIGURE 5.6   Ground plane 20 × 20 cm, E plane. Front to back = 17 dB.

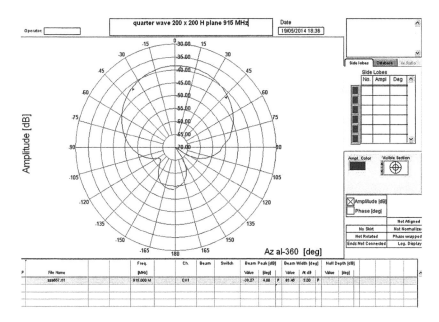

FIGURE 5.7   Ground plane 20 × 20 cm, E plane. Front to back = 15 dB.

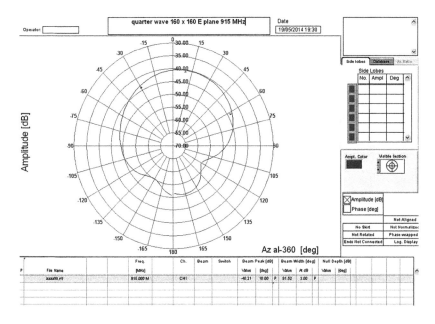

FIGURE 5.8   Ground plane 16 × 16 cm, E plane. Front to back = 12 dB.

As a last step, we have mounted the microstrip element to the back of the helmet

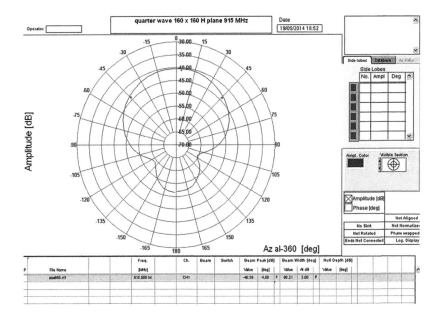

FIGURE 5.9 Ground plane 16 × 16 cm, H plane. Front to back = 12 dB.

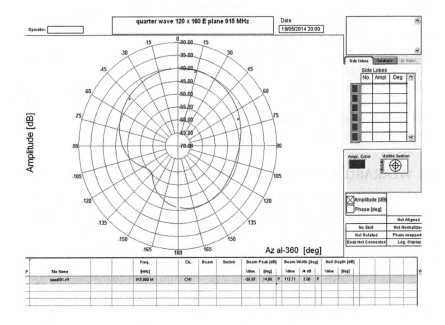

FIGURE 5.10 Ground plane 12 × 16 cm, E plane. Front to back = 6 dB.

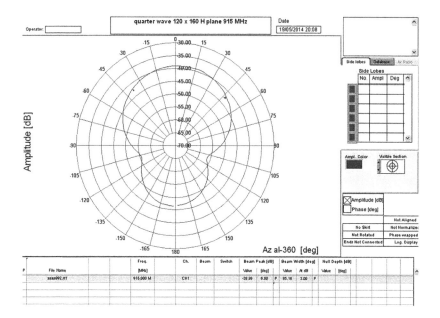

FIGURE 5.11   Ground plane 12×16 cm, H plane. Front to back = 8 dB.

as shown in Figure 5.12. Two cases were compared: (a) vertical mounting (as shown in Figure 5.12) and (b) horizontal mounting, with horizontal polarization.

The patterns of the two cases are shown in Figures 5.13 and 5.14. It is shown that the F/B with helmet is 13–14 dB, somewhat lower than F/B with planar ground of the same size (15–17 dB). The beamwidths in the two polarizations are 106° and 121°.

A case study of a small microstrip element (3×7 cm) operating at 902–928 MHz while mounted on a battle helmet is presented. This case can be considered a wearable antenna with minimal interaction with the body. The relevant parameter of F/B ratio depends on the size of the ground plane and its specific shape. The work was supported by Virtual Extension Ltd. [13].

5.3 WEARABLE ANTENNA FOR HIGH POWER CELLULAR JAMMER

Wearable antennas made of thin layers of metals and foams have attracted a lot of attention in recent years. A subset of such antennas is focused on cellular applications in the frequency range of 700–2700 MHz [14–19]. Here we are interested in high power applications, typically 10 W continuous wave (CW), for Long Term Evolution (LTE) jamming, deception and manipulations [20]. The required gain of the antenna is 5–7 dBi and the permitted area while mounted on the chest of the carrier is 30×25 cm. The requirements for broadband, directive and flat antenna led to the solution of a printed monopole, mounted above a ground plane. The monopole has a trapezoidal shape to allow the proper bandwidth. However, the LTE band of frequency has been covered by two separated elements, as shown in Figure 5.15. Another antenna embedded in a box (case 2) is shown in Figure 5.16.

FIGURE 5.12 A quarter-wave microstrip element (3 × 7 cm) mounted on a metallic helmet.

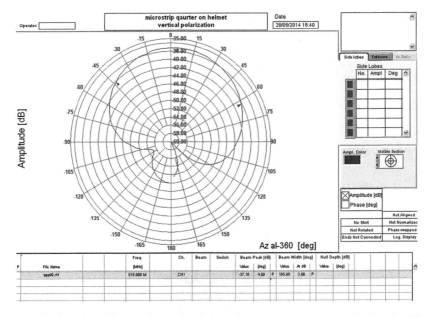

FIGURE 5.13 Ground plane is metallic helmet. The antenna is mounted vertically. F/B = 14 dB.

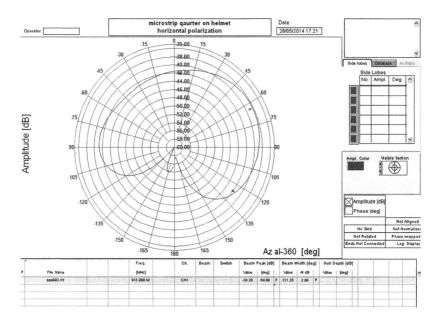

FIGURE 5.14 Ground plane is metallic helmet. The antenna is mounted horizontally. F/B = 13 dB.

FIGURE 5.15 A photograph of a two-element flat antenna.

FIGURE 5.16   Photograph of the two elements within a box.

The S11 graphs of the two elements (case 1) are shown in Figure 5.17 and Figure 5.18.

The return loss at 700–960 MHz, 1700–2200 MHz and 2500–2700 MHz is better than −5 dB. The S11 graphs of the two elements in the box (case 2) are shown in Figure 5.19 and Figure 5.20. The return loss at 700–960 MHz, at 1700–2200 MHz and at 2500–2700 MHz is better than −5 dB. The gain of each element is 6–8 dBi at 700–960 MHz, 7–9 dBi at 1700–2170 MHz and 7–9 dBi at 2500–2700 MHz, all in vertical polarization.

The larger monopole dimensions are: 23 cm length, 14 cm width and 4 cm thickness. The smaller monopole dimensions are: 9 cm length, 6 cm width and 2 cm thickness. The total area of the antenna is 30×25 cm and the overall thickness is 4 cm. The antenna is fed by two coaxial ports: the first port for 700–960 MHz and the second port for 1700–2700 MHz. The fractional bandwidth of the larger monopole is $260/830 = 31\%$ and of the smaller monopole $1000/2200 = 45\%$. The antenna layers are made from flexible materials (thin metals and foams). The ground of the monopole in this case is the vertical wall of the box. The elements are somewhat smaller and the F/B ratio is better.

The radiation exposure for the carrier of such an antenna is naturally of major concern. The international standard IEEE ANSI C.95 [21] for uncontrolled environment defines permitted average power density of f/300 mW/cm^2 where f is the frequency (between 300 and 3000 MHz). The power densities are therefore 2 mW/cm^2 (@700 MHz) to 7 mW/cm^2 (@2200 MHz). The exposure is defined over a period of 6 minutes. Other standards of ICNIRP and ETSI allow close values [22].

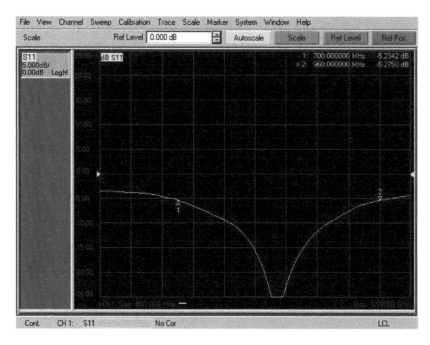

FIGURE 5.17   S11 of the larger monopole (case 1).

FIGURE 5.18   S11 of the smaller monopole (case 1).

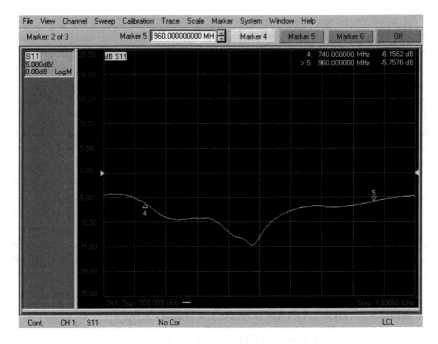

FIGURE 5.19 S11 of the larger monopole (case 2).

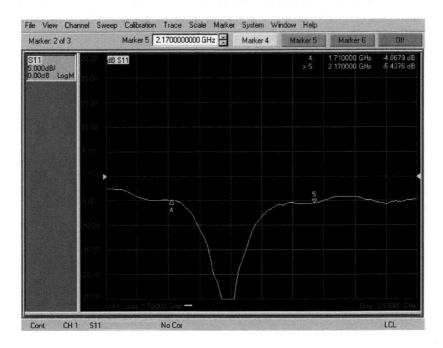

FIGURE 5.20 S11 of the smaller monopole (case 2).

We tested the power density of the antenna at a distance of 10 cm *in front* of the aperture. For input power of 10 mW it was found to be between 0.01 and 0.02 mw/cm². The F/B ratio, as seen from typical azimuthal patterns in Figures 5.21–5.26, is 10 dB.

Hence, for input power of 10 W, the maximal power density at the *back* of the antenna would be between 1 and 2 mW/cm² as allowed by the safety standards. It can also be argued that the worst case scenario will be if 10 W is radiated with gain of 5 towards a sphere of radius 10 cm. The maximal radiation density in front will be $PtGt/4\pi R^2 = 50$ mW/cm² and in the back 5 mW/cm², also within the safety standard. As far as the Specific Absorption Rate (SAR) definition applies, we can expect that 2 mW/cm² will be spread on an area of 25 × 30 cm, hence no more than 1.5 W can reach the body.

In conclusion, we have presented a wearable antenna that covers the cellular bands 700–960 MHz and 1700–2700 MHz with directive gain of 6–9 dBi and typical beamwidths 50°–70°. The overall area is 30 × 25 cm, fitted to be carried on the chest of the operator, as shown in Figure 5.27 (flexible laboratory prototype). The nature of the ground causes differences between the patterns of the H-plane and the E-plane. An estimate of the exposure by the carrier to input power of 10 W, shows that the level of radiation is below international standards.

5.4 RFID READER UHF ANTENNA IN THE POCKET

RFID, known now as Internet of Things (IOT), plays a major role in monitoring people, animals and assets. The following application is concentrated on the design of a high quality antenna reader (not a tag by a person). The objective was to insert a circular polarized antenna into the pocket without any degradation in antenna

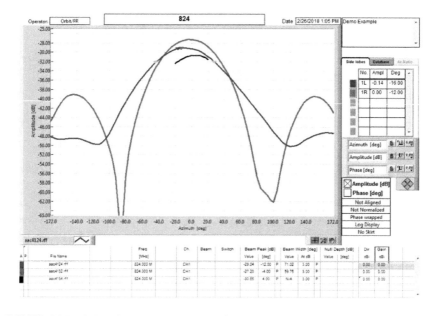

FIGURE 5.21 Azimuth (H-plane) and elevation (E-plane) radiation patterns at 824 MHz (case 1). Reference antenna is 5–6 dBi.

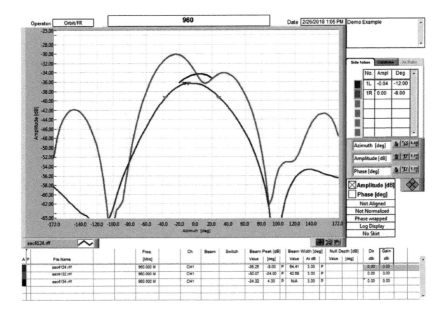

FIGURE 5.22 Azimuth (H-plane) and elevation (E-plane) radiation patterns at 960 MHz (case 1). Reference antenna is 5–6 dBi.

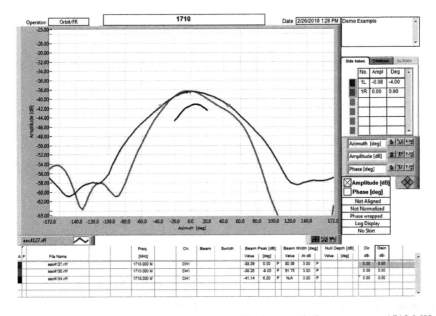

FIGURE 5.23 Azimuth (H-plane) and elevation (E-plane) radiation patterns at 1710 MHz (case 1). Reference antenna is 5–6 dBi.

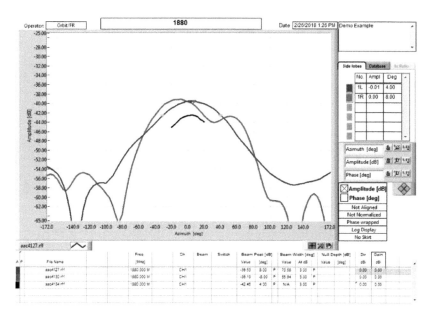

FIGURE 5.24 Azimuth (H-plane) and elevation (E-plane) radiation patterns at 1880 MHz (case 1). Reference antenna is 5–6 dBi.

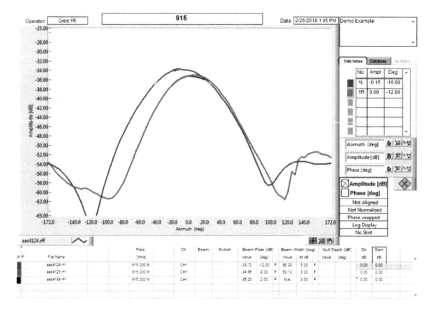

FIGURE 5.25 Azimuth (H-plane) Radiation Patterns at 915 MHz (case 1 and case 2). Reference Antenna is 5–6 dBi. Beamwidths are 60° and 66° respectively.

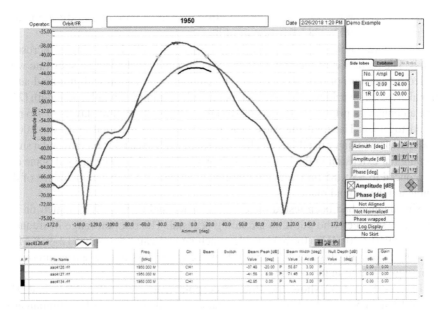

FIGURE 5.26 Azimuth (H-plane) radiation patterns at 1950 MHz (case 1 and case 2). Reference antenna is 5–6 dBi. Beamwidths are 71° and 59°, respectively.

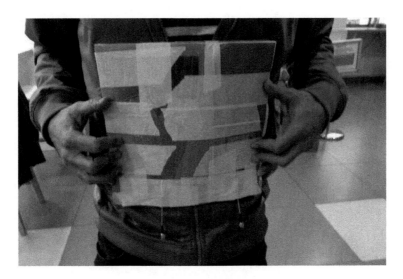

FIGURE 5.27 A wearable antenna for 700–2700 MHz bands.

performance. The required gain was 5–6 dBic with axial ratio of 1–2 dB at the frequency range of 902–928 MHz or part of it.

The natural choice for a thin antenna with gain 5–7 dBi is the single microstrip patch. The reduction of the area to $0.25\lambda \times$ is possible by choosing a substrate with dielectric constant of 6 or 10 [23, 24]. The circular polarization with minimal additional area can be achieved by one feed point with asymmetric width and length [25]. The antenna was printed on Rogers 6010 (Dk = 10), made by three layers of 1.6 mm glued together. The total thickness was h = 5 mm and the total area was 100 × 100 mm. The radiator itself has an elliptical shape with axes 65 × 68 mm and a small rectangular extension in the left side, as shown in Figure 5.28. The ground plane can be reduced to 85 × 85 mm, leaving margins of at least 3 h between the edges of the patch and the ground plane.

The antenna had been tested once in free space and then while mounted on the arm of a volunteer who put it on a rotating pedestal. The relevant range of frequencies was 910–920 MHz (for reasonable matching and gain) but only 912–918 MHz for pure axial ratio. Figures 5.29–5.30 show the return loss in free space and on the arm (equivalent to in the pocket).

Figures 5.31–5.32 show the free space polar patterns at 915 MHz and 917 MHz. Figures 5.33–5.34 show the polar patterns in the presence of the human arm at 915 MHz and 917 MHz.

In conclusion, a compact microstrip element was adapted to be carried on the human body without degradation of the performance, in accordance to Table 5.1. The work was supported by Virtual Extension Ltd [26].

FIGURE 5.28 An elliptical microstrip patch for pocket UHF reader. The patch sizes are 65 × 68 mm and the ground is 100 × 100 mm.

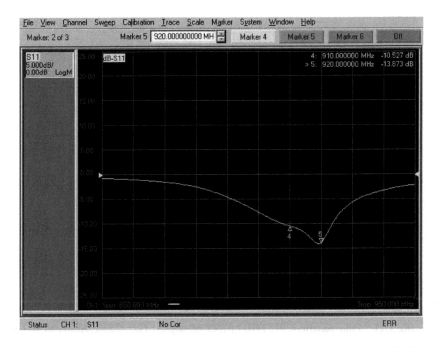

FIGURE 5.29   Return loss in free space. Voltage standing wave ratio = 2 at 910–920 MHz.

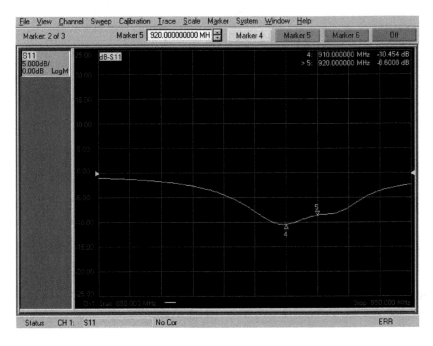

FIGURE 5.30   Return loss on the arm. Voltage standing wave ratio = 2.2 at 910–920 MHz.

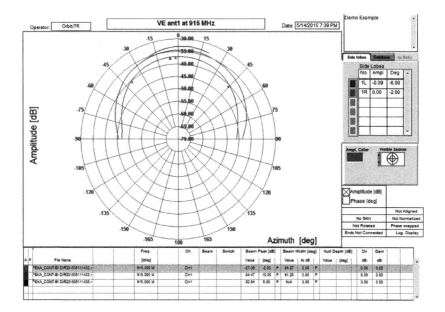

FIGURE 5.31   Polar patterns at 915 MHz (at four roll angles). Gain 6 dBic. Axial ratio = 2 dB.

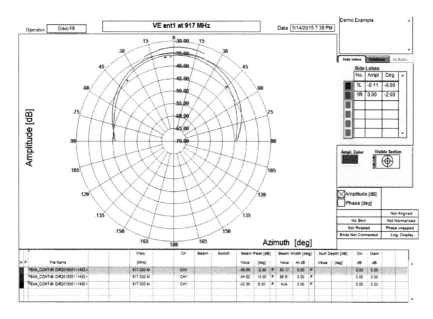

FIGURE 5.32   Polar patterns at 917 MHz (at four roll angles). Gain 6 dBic. Axial ratio = 1 dB.

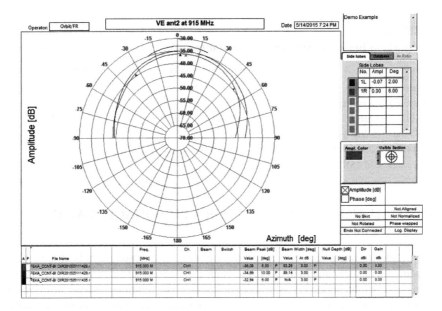

FIGURE 5.33 Polar patterns at 915 MHz in the pocket. Gain 5.5 dBic. Axial ratio = 1 dB.

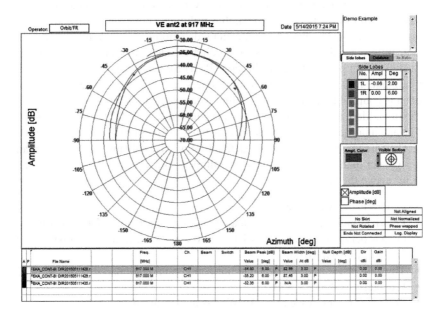

FIGURE 5.34 Polar patterns at 917 MHz in the pocket. Gain 5.5 dBic. Axial ratio = 0.5 dB.

TABLE 5.1

Comparison between antenna performance in free space and in the pocket

Parameter	Antenna in free space	Antenna in pocket
Frequency	910–920 MHz	910–920 MHz
Voltage standing wave ratio	2.0	2.2
Gain (915–917 MHz)	6.0 dBic	5.5 dBic
Axial ratio (915–917 MHz)	2 dB	1 dB
Beamwidths at 915 MHz	$84° \times 91°$	$83° \times 89°$
Beamwidths at 917 MHz	$83° \times 90°$	$83° \times 87°$

5.5 SMALL HELICAL ANTENNA FOR A PERSONAL LOCATOR BEACON

A portable personal communication device with an extendable helical antenna is presented. This helical antenna is made of an elastic conductive spring, configured to change its height, mounted over a ground plane. The antenna has two positions as shown in Figure 5.35: (a) a stowed position where the antenna is pressed down between the ground plane and a rigid cover, achieving a low profile; and (b) an operational position where the rigid cover is removed and the antenna is extended by its own spring force. In this position, the device is seen as a sport wrist watch.

Normally, a planar antenna may be placed over the same ground plane, not exceeding the footprint of the helical antenna, thus utilizing a compactly small volume and yet achieving a considerable electromagnetic decoupling between the antennas, due to their different radiation patterns. According to one embodiment, these antennas are installed in a personal locator beacon (PLB) for search and rescue of people in distress, like mariners, aviators, hikers, skiers and off-road drivers. The planar antenna is coupled to a navigation satellite receiver, like Global Positioning System (GPS) or Galileo, and the helical antenna is coupled to VHF/UHF transmit radio.

The antenna specifications are defined by the CosPas-SarSat satellite constellation [27]. The antenna characteristics are defined for all azimuth angles and for elevation angles greater than 5° and lower than 60° (although the tests are done at elevations of 10°–50° above the horizon).

- Pattern is hemispherical.
- Polarization is circular (right-handed circular polarization; RHCP) or linear.
- Gain is between −3 dBi and 4 dBi over 90% of the elevation sector between 10° and 50° above horizon.
- The gain is a sum of vertical and horizontal polarizations.
- Frequency is 406 ± 2 MHz.
- Voltage standing wave ratio (VSWR) is not greater than 1.5:1.

In early experiments we compared several alternatives for such a hand-held antenna: monopole wire perpendicular to the hand, printed horizontal strip attached to the hand and a coil above a small ground plane. The alternatives were tested in the presence of a 1.5 liter plastic bottle, filled with water or with slices of sausage. The

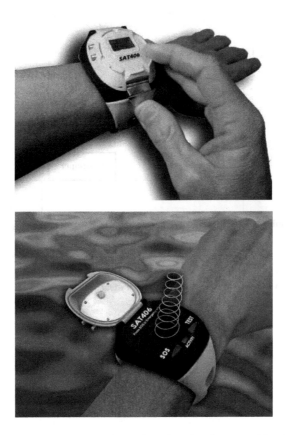

FIGURE 5.35   Helical antenna in a wrist device (closed and open positions).

bottle emulates the presence of the hand, or the body, or water of the sea. The radiation patterns were measured in an un-echoic chamber as shown in Figure 5.36.

The monopole height was 20 cm and its ground plane was a disc with diameter of 5 cm. The maximal gain of the monopole in free space was 2–3 dBi and the gain within the relevant coverage zone was −1 to 0 dBi. The monopole antenna in vicinity of a bottle with meat at distance of 1 cm had excessive loss of 4–7 dB. The monopole antenna in vicinity of a bottle with water at distance of 1 cm had excessive loss of 1–18 dB. The elevation patterns of the monopole at 400 MHz and at 410 MHz are shown in Figures 5.37 and 5.38.

The printed antenna shown in Figure 5.39 was 110 × 40 mm and it was attached to the bottle (with meat or with water). While isolated, the maximal gain was 0–1 dBi and within the required coverage zone the gain was −4 to −3 dBi. The printed antenna in vicinity of a bottle with meat at distance of 1 cm had excessive loss of 4–7 dB. The monopole antenna in vicinity of a bottle with water at distance of 1 cm had excessive loss of 1–18 dB. The elevation patterns of the printed antenna at 400 MHz and at 410 MHz are shown in Figures 5.40 and 5.41.

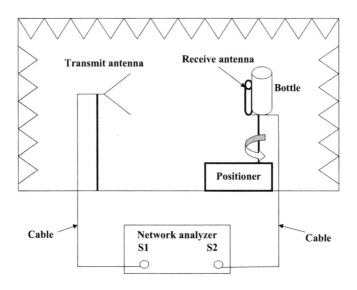

FIGURE 5.36 A schematic diagram of a pattern measurement.

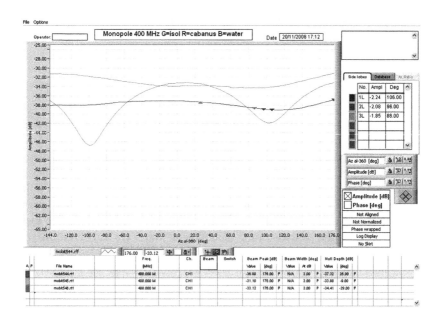

FIGURE 5.37 Elevation patterns of a monopole at 400 MHz. Green = isolated (maximal gain 3 dBi). Red = near a bottle with meat (losses of 3–6 dB). Blue = near a bottle with water (losses of 0–14 dB).

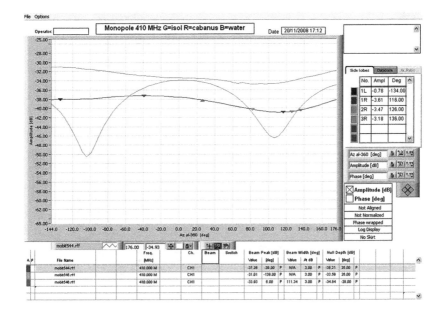

FIGURE 5.38 Elevation patterns of a monopole at 410 MHz. Green = isolated (maximal gain 3 dBi). Red = near a bottle with meat (losses of 4–7 dB). Blue = near a bottle with water (losses of 1–18 dB).

Both the vertical monopole and the ground strips are large in their dimensions and are much influenced by the presence of the human body. The size limitations of the product and the required isolation from the body dictate a small wire antenna mounted above a small ground. We have faced the challenge of building such a short coil (up to 7 cm) mounted on a wrist whose diameter is 5 cm. Also, in the center of the wrist there is a ceramic GPS antenna whose size is 18 × 18 mm or close. As mentioned, the novel solution was a helical spring that functions as an antenna while it is opened.

The normal-mode helix, mounted above a small ground plane, is a well-known antenna [28–31] that founded many applications throughout the years [32, 33]. The current may be assumed to be of uniform magnitude and with a constant phase along

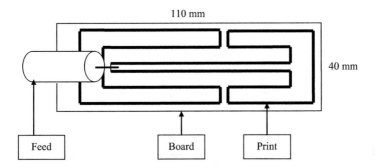

FIGURE 5.39 A scheme of a printed antenna at 406 MHz.

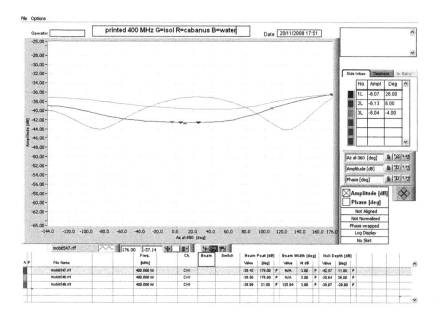

FIGURE 5.40   Elevation patterns of a printed antenna at 400 MHz. Green = isolated (maximal gain 1 dBi). Red = near a bottle with meat (losses of 0–3 dB). Blue = near a bottle with water (losses of +2 dB (gain) to 6 dB).

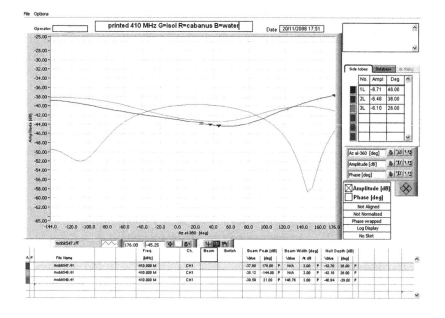

FIGURE 5.41   Elevation patterns of a printed antenna at 410 MHz. Green = isolated (maximal gain 1 dBi). Red = near a bottle with meat (losses of 0–2 dB). Blue = near a bottle with water (losses of +3 dB (gain) to 18 dB).

the helix. It can also be seen as a short dipole and a small loop combined together with elliptical polarization. The azimuthal cut is nearly omnidirectional and the elevation cut is sensitive to the overall height and the presence of the ground plane. This antenna enjoys the presence of two polarizations: at low elevation angles above the horizon the vertical polarization is dominant and at high elevation angles above the horizon the horizontal polarization is dominant.

Early experiments with a normal-mode helix antenna, with an example shown in Figure 5.42, show absolute gain in the desired coverage between −6 dBi to 2 dBi. The lower values of the gain occur at high elevation angles (40°–50° above the horizon). Table 5.2 shows five models, matched at 406 MHz. Figures 5.43 and 5.44 show the elevation cut in two polarizations for models 3 and 5. The defined absolute gain allows to add the power of the two orthogonal polarizations.

After an optimization process of the antenna dimensions we have chosen a helix with eight turns whose height is 7 cm. The spring has a slight conical nature for better storage conditions inside the beacon. The main characteristic parameter to be validated is the absolute gain. The pre-compliance tests were done at 3 m open antenna range in two test conditions. In the first test condition the PLB is mounted on a 125 cm radius ground plane as shown in Figures 5.45 and 5.46. In the second test condition the PLB is mounted on an absorber 45 cm above the floor as shown in

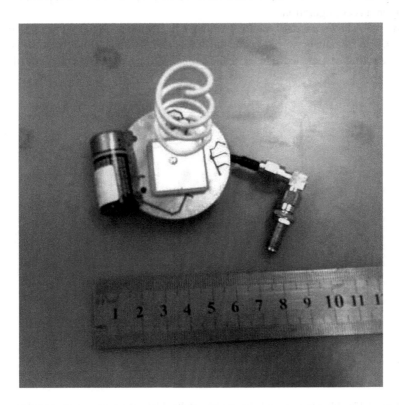

FIGURE 5.42 An example of a short helix with four turns, mounted on a 50 mm ground plane. A Global Positioning System antenna is shown as well.

TABLE 5.2
Five models of helix antenna over a ground plane with diameter 50 mm

Model	Height (mm)	Base (mm)	Turns
1	3	25	2
2	40	36	3
3	42	15	4
4	76	16	6
5	94	12	11

Figures 5.47 and 5.48. The PLB is allowed to transmit between 3 W and 7 W (35–38 dBm) and the antenna gain is given by the effective isotropic radiated power (EIRP) – Pt + Cable Loss.

The compliance results of the gain are shown in Tables 5.3 and 5.4 [34]. The results in Table 5.3 seem to comply with the requirement of minimal gain of −3 dBi and the results in Table 5.4 seem not to comply. However, in the compliance tests held in Europe, the antenna passed successfully, since a waiver of 3 dB had been given to the second test configuration and because only 80% of the test points should apply in this configuration.

We have shown a wrist-type PLB with a helical antenna in the shape of a conical spring. The antenna is stowed inside the device and operates while the cover is open and the spring is stretched. The helical antenna, whose height is 7 cm, made of eight

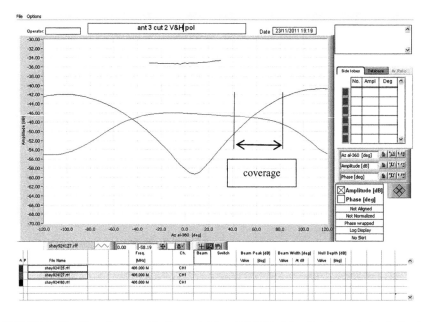

FIGURE 5.43 Elevation cut of helix model 3 at 406 MHz. Blue line is vertical polarization; red line is horizontal polarization. Black line is a 4 dBi reference antenna. Absolute gain values in the required coverage zone are between −6 dBi and −2 dBi.

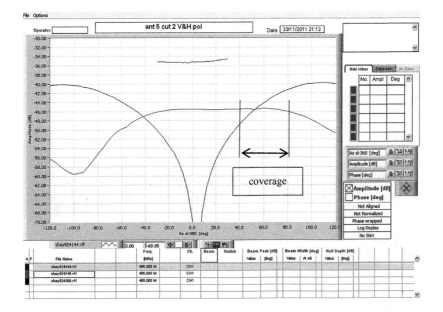

FIGURE 5.44   Elevation cut of helix model 5 at 406 MHz. Blue line is vertical polarization; red line is horizontal polarization. Black line is a 4 dBi reference antenna. Absolute gain values are between −5 dBi and −1 dBi.

turns, had been tested and approved by CosPas-SarSat global constellation. The minimal gain of the antenna (summed at both polarizations) was higher than −3 dBi while mounted on a large ground plane and was higher than −6 dBi (80% of the points) while mounted on an absorber. The antenna had been developed for Mobit Telecom Ltd [35] and is patented.

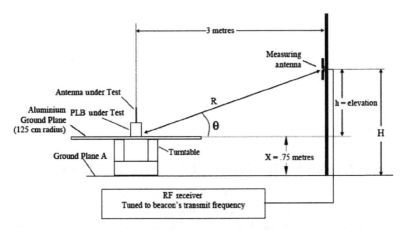

FIGURE 5.45   Test structure personal locator beacon on a 125 cm radius ground plane.

FIGURE 5.46 Outdoor range with the personal locator beacon on a ground plane.

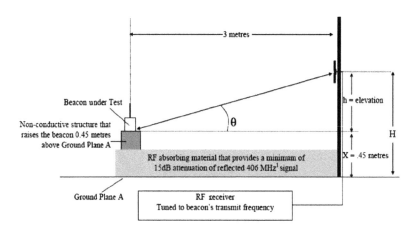

FIGURE 5.47 Test structure personal locator beacon on an absorber, 45 cm above the floor.

5.6 VHF ANTENNA FOR PERSONAL COMMUNICATIONS

A small antenna, made of a compact coil, mounted on the hand is presented. The antenna serves a personal communication link at 162 ± 1 MHz. The antenna should have omnidirectional coverage in mixed polarizations, with a typical size of $d = 60$ mm and thickness of 20 mm. At this frequency, where the size of the antenna

FIGURE 5.48 Outdoor range with the personal locator beacon on an absorber.

TABLE 5.3
Measured gain of a small helix antenna mounted on a wrist device

	EL = 10° (dBi)	EL = 20° (dBi)	EL = 30° (dBi)	EL = 40° (dBi)	EL = 50° (dBi)
AZ = 0°	−1.8	3.2	2.9	−0.3	−2.1
AZ = 90°	−1.4	3.7	2.7	0.6	−2.0
AZ = 180°	−2.8	3.9	2.5	−1.6	−1.8
AZ = 270°	−1.7	3.2	2.1	−1.0	−0.7

Antenna height is 7 cm; ground plane diameter is 5 cm. Test condition includes a 125 cm radius ground plane.

TABLE 5.4
Measured gain of a small helix antenna mounted on a wrist device

	EL = 10° (dBi)	EL = 20° (dBi)	EL = 30° (dBi)	EL = 40° (dBi)	EL = 50° (dBi)
AZ = 0°	−1.7	−1.3	−1.8	−5.5	−9.0
AZ = 90°	−1.4	−1.0	−1.4	−4.5	−7.1
AZ = 180°	−1.7	−1.1	−1.1	−3.7	−5.1
AZ = 270°	−2.0	−1.2	−1.5	−4.3	−6.8

Antenna height is 7 cm; ground plane diameter is 5 cm. Test condition includes an absorber 45 cm above the floor.

is 0.03λ, it is very difficult to isolate the antenna from the body but, nevertheless, it had been decided to use a small ground plane. In order to allow for a reasonable bandwidth and to reduce the weight, it was also decided not to use dielectric materials with high dielectric constant.

The early model of the antenna was built from several coils, arranged in different directions at a total area of 60 × 60 mm; it is shown in Figure 5.49. The total number of turns N and the distance to the ground plane H were optimized to achieve the center frequency of 162 MHz.

Four out of many cases are presented in Figures 5.50–5.53. The best working point was N = 37 and H = 15 mm.

A second series of samples was built on a circular ground plane with diameter of 55 mm (fitted to look like a wrist watch) and spacers of 20 mm made of Delrin, as shown in Figure 5.54. The final antenna is made of five coils, each of which has six turns, arranged in a circle as shown in Figure 5.55.

The antenna is well matched between 155 and 165 MHz and has almost the same patterns at the range 154–158 MHz as shown in Figures 5.56–5.58. The maximal gain is −4 dBi and the gain in all azimuthal coverage is above −16 dB. While keeping in mind that the size of the antenna is 0.03λ, this value is totally reasonable and enables sufficient link margin within a few kilometers. A small trimming in one of the coils, reducing N to 28, will shift the center frequency to 162 ± 2 MHz. The antenna was developed for Thea Marine Safety Ltd.

In conclusion, small VHF antennas were tested while mounted on the arm. The measured gain at 154–158 MHz was −4 dBi to −16 dBi at all polarizations. In this chapter it has been demonstrated that an antenna's characteristics depend on the interaction of the antenna with the human body. The key factors in the design are:

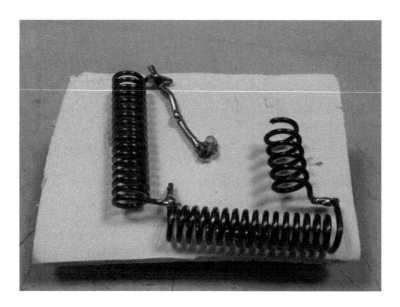

FIGURE 5.49   Early model of a coil antenna with N = 38 turns.

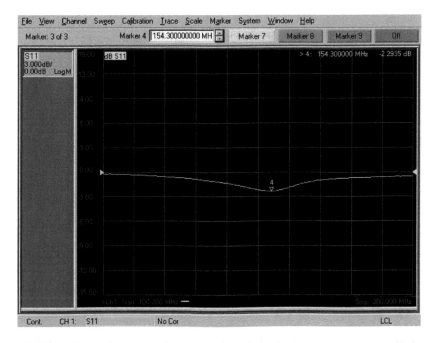

FIGURE 5.50   Return loss of **case 1**. N = 38 and H = 5 mm. Center frequency **154 MHz**.

FIGURE 5.51   Return loss of **case 2**. N = 38 and H = 10 mm. Center frequency **171 MHz**.

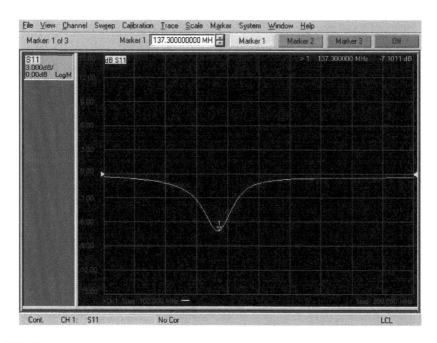

FIGURE 5.52   Return loss of **case 3**. N = 48 and H = 15 mm. Center frequency **137 MHz**.

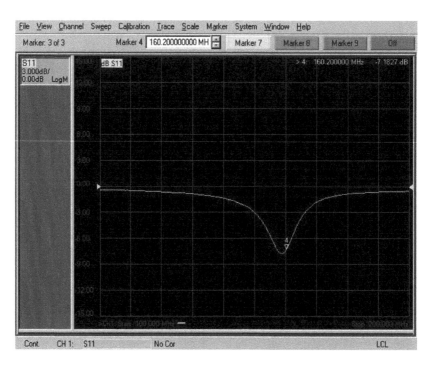

FIGURE 5.53   Return loss of **case 4**. N = 37 and H = 15 mm. Center frequency **158 MHz**.

FIGURE 5.54   Ground planes and Delrin spacers. H = 15 mm.

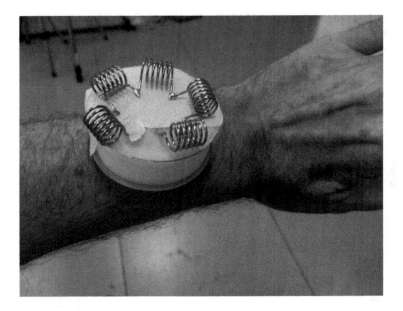

FIGURE 5.55   The final antenna made of five coils arranged in a circle (N=30) mounted on hand.

reducing the size of the antenna, introducing metallic separation from the body, use of absorbing materials in the edges of the antenna and on the cables, intensive use of BALUNs on the feeds/cables and use of phantoms or live tissues for final optimization. Needless to say, all safety standards concerning exposure to radiation should be met during experiments and in regular operation.

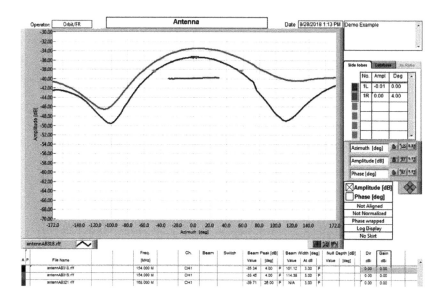

FIGURE 5.56   Azimuthal cut at **154 MHz**. The blue line is a calibrated gain antenna with −10 dBi. Two polarizations are shown. The gain over azimuth is −4 dBi to −16 dBi in vertical polarization and −6 dBi to −20 dBi in horizontal polarization.

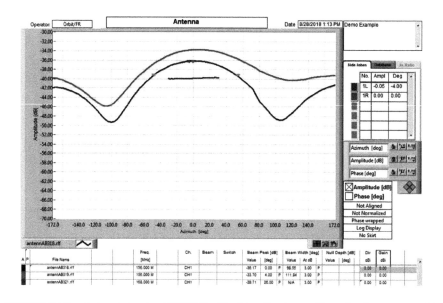

FIGURE 5.57   Azimuthal cut at **156 MHz**. The blue line is a calibrated gain antenna with −10 dBi. Two polarizations are shown. The gain over azimuth is −4 dBi to −16 dBi in vertical polarization and −6 dBi to −19 dBi in horizontal polarization.

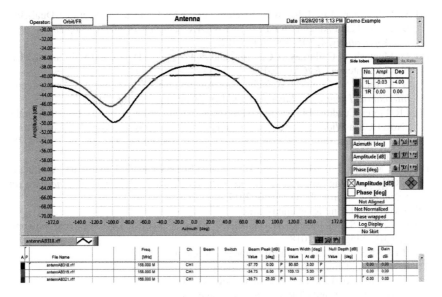

FIGURE 5.58   Azimuthal cut at **158 MHz**. The blue line is a calibrated gain antenna with −10 dBi. Two polarizations are shown. The gain over azimuth is −5 dBi to −16 dBi in vertical polarization and −8 dBi to −20 dBi in horizontal polarization.

REFERENCES

[1] C.A. Balanis, *Antenna Theory Analysis and Design*, third edition, Wiley Interscience, 2005.

[2] J.D. Kraus and R.J. Marhefka, *Antennas for All Applications*, third edition, McGraw Hill, 2002.

[3] N.H.M. Raiset al., "A Review of Wearable Antenna," *LAPC, Loughborough*, 2009.

[4] B. Gupta, S. Sankaralingam, and S. Dhar, "Development of Wearable and Implanted Antennas in the Last Decade," *10th Mediterranean Microwave Systems Conference (MMS)*, 2010.

[5] H. Khaleel, *Innoation in Wearable and Flexible Antennas*, WIT Press, 2015.

[6] A. Sabban, *Low Visibility Antennas for Communications*, Taylor & Francis Group, 2015.

[7] K. Paracha et al., "Wearable Antennas : A Review of Materials, Structures and Inoative Features for Autonomous Communications and Sensing," *ResearchGate* 224089551, 2016.

[8] S. F. Sabri et al., "Review of the Current design on Wearable Antenna in Medical Field and its Challenges," *Jurnal Teknologi (Sciences & Engineering)*, 111–117, 2016.

[9] T. Patel and M. Sahoo, "A Technical Review: Design Issues of Wearable Antennas," *IJARIIE(O)-2395-4396*, Vol. 2, Issue-3, 2016.

[10] A. Sabban, "Wearable Antenna Measurements in Vicinity of Human Body," *Wireless Engineering and Technolgy*, 7, 97–104, 2016.

[11] J.C. Wang, E.G. Lun, M. Leach, Z. Wang, and L. Man, "Review of Wearable Antennas for WBAN Applications," *IAENG International Journal of Computer Science*, 43(4), 2016.

[12] O.A. Saraereh, I. Khan, B. M. Lee, and A.K.S. Al-Bayati, "Modeling and Analysis of Wearable Antennas," *Electronics*, 8(7), 2019.

[13] www.virtual-extension.com

[14] M.A. Soliman, T.E. Taha, W.E. Swelam, and A.M. Gomaa, "A Wearable Dual-Band Dielectric Patch Antenna for LTE and WLAN," *Journal of Electromagnetic Analysis and Applications*, 4, 305–309, July 2012.

[15] M. Mantash, S. Collardey, A.C. Tarot, and A. Presse, "A Dual Band WiFi and 4G LTE Flexible Antenna," *EuCAP*, 2013.

[16] H. K. Raad, "A Compact Wearable Slot Antenna for LTE and WLAN Applications," *International Journal of Electrical and Computer Engineering*, 11(2), 2017.

[17] J.R. Flores-Cuadras, J.L. Medina-Monroy, R.A. Chavez-Perez, and H. Lobato-Morales, "Novel Ultra Wideband Flexible Antenna for Wearable Wrist Worn Devices with 4G LTE Communications," *Microwave and Optical Technology Letters*, 59(4), 777–783, April 2017.

[18] M.V. Kavitha, I. F. Ahamed, and H. Ayman, "Body-Wearable Flexible RF Antenna for 3 GHz Jamming Applications," *International Journal of Current Engineering and Scientific Research*, 4(11),16–19, 2017.

[19] www.Pharad.com/lte-antennas.html

[20] E. Levine, S. Kahlon and H. Matzner, "Safety Aspects of LTE Wearable Antenna," *APEMC*, Singapore, 2018.

[21] IEEE ANSI C.95.1, 2019, http://standards.ieee.org

[22] ICNIRP Guidelines, 2019, www.icnirp.org

[23] Rogers RO-3010 and RO-3006, www.rogers.com

[24] Taconic RF-60A and CER-10, www.4taconic.com

[25] S. Gao, Q. Luo, and F. Zhu, *Circularly Polarized Antennas*, Wiley and IEEE Press, 2014.

[26] www.virtual-extension.com

[27] CosPas SarSat Specification Document, T series T7NOV1_07, www.cospas-sarsat.org

[28] G.F. Levy, "Loop Antenna for Aircraft," *Proceedings of the Institutional Radio Engineers*, 31, 56, 1943.

[29] J.B. Sherman, "Circular Loop Antenna at Ultra High Frequencies," *Proceedings of the Institutional Radio Engineers*, 32, 534, 1944.

[30] H.A. Wheeler, "A Helical Antenna for Circular Polarization," *Electronics*, 20, 109–111, 1947.

[31] J.D. Kraus, "Helical Beam Antenna," *Proceedings of the IRE* 35, 1484–1488, 1947.

[32] W.G. Hong, Y. Yamada, and N. Michishita, "Low Profile Small Normal Mode Helical Antenna," *Proceedings of the Asia Pacific Microwave Conference*, 2007.

[33] C. Su, H. Ke, and T. Hubing, "A Simplified Model for Normal Mode Helical Antenna," *25th Annual Review of Progress in Applied Communication Electromagnetics*, 2009.

[34] D. Katz and E. Levine, "Small Helical Antenna for a Personal Communication Device," *Proceedings of the 27th IEEE Convention in Israel*, Eilat, November 2012.

[35] www.mobitcom.com

6 Small Antennas Mounted near the Human Body
Experimental Case Studies

Ely Levine

CONTENTS

6.1 INTRODUCTION

Antennas as part of communication devices, sensors or radars can be mounted in close vicinity to the human body or even inside the body. The body is composed of many materials which affect in different ways the behavior and the performance of the antennas. The general theory of electromagnetic phenomena is based on Maxwell's equations [1, 2], which constitute a set of four coupled first-order partial differential equations, relating the space and time changes of electric and magnetic fields to their scalar source densities (divergence) and vector source densities (curl). For stationary media, Maxwell's equations in the differential form are:

$$\nabla \cdot D(r,t) = \rho(r,t) \quad \left(\text{Gauss's electric law}\right) \tag{6.1}$$

$$\nabla \cdot B(r,t) = 0 \quad \left(\text{Gauss's magnetic law}\right) \tag{6.2}$$

$$\nabla \times E(r,t) = -\partial B(r,t)/\partial t \quad \left(\text{Faraday's law}\right) \tag{6.3}$$

$$\nabla \times H(r,t) = \partial D(r,t)/\partial t + J(r,t) \quad \left(\text{Ampére's law}\right). \tag{6.4}$$

Maxwell's equations involve only macroscopic electromagnetic fields and, explicitly, only macroscopic densities of free charge $\rho(r,t)$ giving rise to the free current density $J(r,t)$. The effect of the macroscopic charges and current densities bounded to the medium's molecules is indicated by auxiliary magnitudes D and H, which are related to the electric and magnetic fields E and H by the so-called constitutive equations that describe the behavior of the medium. In general, the quantities in these equations are functions of the position (r) and the time (t), with the following definitions:

E = electric field intensity (volt/meter)
H = magnetic field intensity (ampere/meter)
D = electric flux density (coulomb/square meter)
B = magnetic flux density (tesla or weber/square meter)
ρ = free electric charge density (coulomb/cubic meter)
J = free electric current density (ampere/square meter).

The power density that the electromagnetic field carries in the free space is defined by Poynting's vector:

$$S = E \times H \left(watt/square\ meter \right) \tag{6.5}$$

which represents the power passing through a unit area perpendicular to the propagation vector r. Inside matter, the power density Sv (watt/cubic meter) is related to the work (or heat) supplied to the charge distribution:

$$Sv = dS/dv = E \cdot J \tag{6.6}$$

known as the point form of Joule's law.

Inside matter, we introduce additional physical quantities to describe the interaction of the waves with the molecules. The electric permittivity ε is connected to the electric field and to the electric flux, but also to a new macroscopic vector P (coulomb/square meter) called the electric polarization vector, such as:

$$D = \varepsilon\,E + P. \tag{6.7}$$

For most materials, called linear isotropic media, especially human body tissues, the vector P can be considered colinear and perpendicular to the electric field applied. Thus we get:

$$P = \varepsilon\,\chi e\,E. \tag{6.8}$$

Where ε is the dielectric constant defined as a complex number:

$$\varepsilon = \varepsilon o\,\left(\varepsilon' - j\varepsilon'' \right) \tag{6.9}$$

and χe is the electric susceptibility of the matter (capability of the matter to be polarized). For most ordinary materials, D can be written as:

$$D = \varepsilon o \, \varepsilon r \, E. \tag{6.10}$$

The magnetic behavior in matter involves a similar treatment with the following relations:

$$H = B/\mu - M \tag{6.11}$$

$$M = \chi m \, H \tag{6.12}$$

$$B = \mu o \, \mu r \, H \tag{6.13}$$

where μ is the magnetic permeability and χm is the magnetic susceptibility. Since the human body has very minor magnetic effects, we can neglect the magnetic nature of the tissues and consider only their electric nature. Very often, the relation between the electric field E and the current density J is given at any specific point by Ohm's law:

$$J(r,t) = \sigma E(r,t) \tag{6.14}$$

where σ (siemens/meter) is called the conductivity of the matter.

Now we consider the configuration of electromagnetic waves in boundaries between air and matter. In free space, the existing fields are pure E and H and the power density obeys Poynting's law with $\varepsilon r = 1$ and $\mu r = 1$. In the human body, the fields are non-magnetic, thus $B = H$ and the local parameters are ε and σ. When a plane wave that propagates in free space hits the human body, part of the energy is reflected back to air and part of the energy penetrates into the body. Some of the penetrating energy continues to propagate inwards into the body and some is absorbed by the tissues and the bones. The absorbed power can be summed as the integral of many local interactions:

$$S\,(\text{absorbed}) = 0.5 \int Pv\,dv = 0.5 \int (E \cdot J)\,dv = 0.5 \int \sigma |E|^2 \, dv = 0.5\omega \int \varepsilon'' |E|^2 \, dv. \tag{6.15}$$

The factor of 0.5 demonstrates the fact that the dissipation is related to average power and not to the peak power. The absorbed power in the human body is often expressed by the term Specific Absorption Rate (SAR) which is:

$$SAR = (\sigma/2\rho)\,|E|^2 \tag{6.16}$$

where σ is the conductivity (mho/meter or siemens/meter) in any specified area and ρ is the mass density (kg/cubic meter) in any specified area. The units of the SAR quantity are therefore watt/kg. SAR is the most commonly used indicator and measure for safety standards in radio frequency (RF) exposure. International values for SAR are 1.6 watt/kg in any 1 g of tissue for a period of 30 minutes (North America) or 2.0 watt/kg in any 10 g of tissue for a period of 6 minutes (Europe).

TABLE 6.1
Electric properties of human body tissues and comparison to water

Tissue	Dielectric constant	Conductivity	Tangent Delta
	εr	σ (S/m)	tanδ
Fat	5–10	0.1	0.15
Muscle	50–60	1.7	0.25
Skin	40–50	1.5	0.5
Bone	5–7	4	2
Chest	50–60	3–4	
Arm	5–6	0.1	
Water	80	0.05	0.2

The design of antennas in the vicinity of the human body requires theoretical or experimental knowledge of the electrical properties of various tissues. Experimental data concerning the dielectric constant and the dielectric loss in human tissues [3–6] show some typical values in the frequency range of 1–6 GHz as depicted in Table 6.1. The variations of data from specimen to specimen and dependence on the water content suggest that each antenna should be designed and optimized for its specific location near or inside the body. The design can include simulations, experiments on phantoms and experiments on live tissues (as long as low power of less than 10 mW is involved).

This chapter brings readers to the actual world of the design and measurement of antennas in close vicinity to the human body. Detailed reviews of such developments, either external or internal to the body (implantable), can be found in references [7–14]. The small antennas under consideration were mounted on a human body (neck, belly, chest, etc.) or inside a phantom of the body, at different frequencies, licensed and unlicensed, from VHF to S band. The chapter describes the effects of the human body in several aspects: detuning of the working frequency, additional loss by the live tissue, effects on the radiation patterns and the gain, and possible polarization rotation. Some commonly used measures to minimize the effects of the body are described and compared: choice of antenna shape, control of the distance to the body, isolating the antennas by a metallic sheet, use of absorbing materials and others. The chapter emphasizes the actual realization of the antennas and naturally it provides detailed examples with essential "tips" from an experimentalist point of view. The five case studies presented here range from antennas attached to the body to antennas inside the body.

- Wearable UHF radio-frequency identification (RFID) tag on the neck.
- Short-range link through the body at 2.4 GHz.
- Measurement of body parameters with electrocardiogram (EKG) pads.
- Cellular antenna on a phantom.
- Small antennas inserted into a phantom.

6.2 WEARABLE UHF RFID TAG ON THE NECK

RFID technology, known now as Internet of Things (IOT), plays a major role in monitoring people, animals and assets. The following application is concentrated on the design of an antenna for a passive tag, attached to the neck of human beings, without the risk that they might take off or change their clothes. The required gain was −3 to 0 dBi at the frequency range of 902–928 MHz and with mixed polarizations. The target cost of the antenna is 1 US dollar. The antenna should be small, flexible and not very sensitive to the influence of any specific body.

As a first step, we built a printed monopole antenna with a folded conductor and a ground plane, shown in Figure 6.1.

The monopole was printed on a thin FR-4 board whose size was 50 × 12 mm. The length of the monopole was 80 mm and reasonable matching (voltage standing wave ratio; VSWR < 3) was demonstrated in the range of 910–1000 MHz (Figure 6.2).

When the antenna was put in the vicinity of a human neck the return loss dramatically changed. Typical measured return loss patterns, as a function of the distance, as also described in [15], are shown in Figures 6.3–6.6.

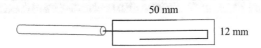

FIGURE 6.1 Schematic structure of a folded monopole.

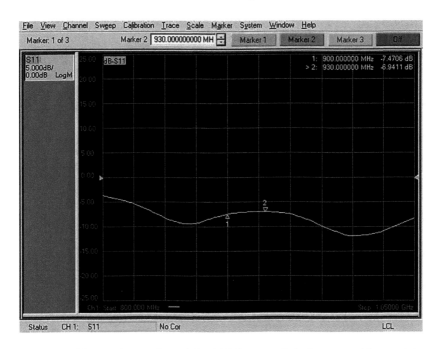

FIGURE 6.2 Measured return loss of the folded monopole in free space. The matching at 910–1000 MHz was better than voltage standing wave ratio = 3.

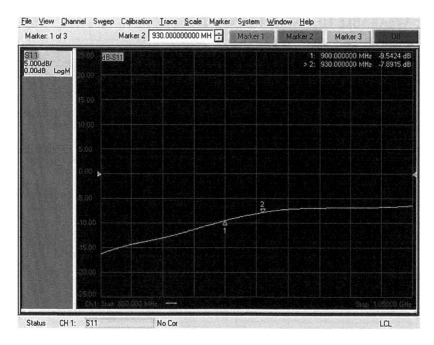

FIGURE 6.3 Measured return loss of the folded monopole at a distance of **2 mm** from the neck. The resonance seems to be at 760 MHz.

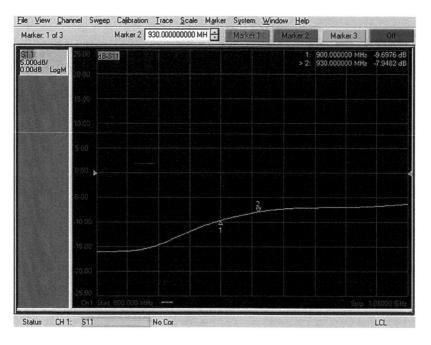

FIGURE 6.4 Measured return loss of the folded monopole at a distance of **4 mm** from the neck. The resonance seems to be at 810 MHz.

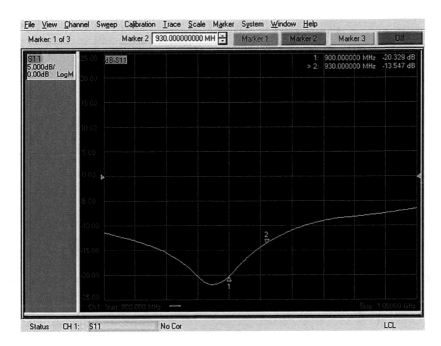

FIGURE 6.5 Measured return loss of the folded monopole at a distance of **6 mm** from the neck. The resonance seems to be at 890 MHz.

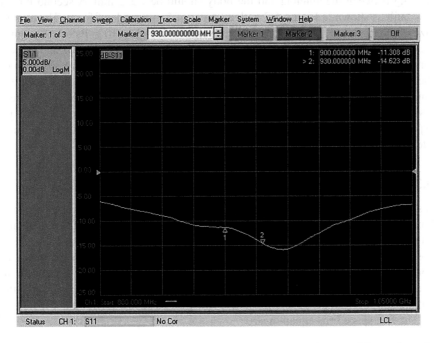

FIGURE 6.6 Measured return loss of the folded monopole at a distance of **8 mm** from the neck. The resonance seems to be at 940 MHz.

FIGURE 6.7   Monopole with foam spacer. Frequency range of operation while mounted on the neck (many cases) was 900–930 MHz.

As a conclusion from this set of measurements, we have decided that the optimal spacing between the antenna and the body should be 7 ± 2 mm. A second model was built as shown in Figure 6.7 with a spacer of 5 mm made of foam. The overall size of the tag is still 50 × 12 mm and its weight is less than 10 g. Pattern measurements of the antenna (without a body) show maximal gain of −8 dBi and typical range between −8 and −16 dBi at a full sphere and in all polarizations. The gain of the antenna in the presence of a body is of course much more complicated to know, but it can be estimated to be lower by 10 to 20 dB than the gain in free space. Field tests show that passive tags like this can maintain a detection range of 2 to 3 m, especially after several readings. The antenna was developed for Virtual Extension Ltd [16].

6.3 SHORT-RANGE LINK THROUGH THE BODY AT 2.4 GHZ

A short-range link was required from a hand-held remote control unit to a therapeutic box attached to a patient's spine. The link passes through the belly of the patient and thus suffers from an excessive loss. The antennas must be small and as isotropic as possible, also from the polarization point of view. The most relevant frequency of operation was in the range 2400–2500 MHz which is allowed for transmission all over the world at output power of 10 mW (10 dBm). As a first step, we measured the radiated power from the transmitter at free space range of 1 m as shown in Figure 6.8.

The free space Friis equation is:

$$Pr/Pt = Gr\ Gt\ \lambda^2/(4\pi)^2\ R^2 L \qquad (6.17)$$

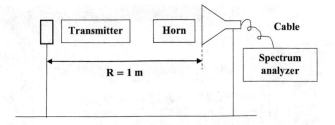

FIGURE 6.8   A test set up for detection of radiated power from a transmitter.

where
  Pr is the received power (measured by a spectrum analyzer)
  Pt is the transmit power
  Gr is the receive antenna gain = 8 dBi
  Gt is the transmit antenna gain
  λ^2 is the wavelength = 0.12^2 m^2 = -18 dB
  $(4\pi)^2 = 157 = 22$ dB
  $R^2 = 1$ m$^2 = 0$ dB
  L is the cable loss = 2 dB.

Since the antenna is embedded in the transmitter and cannot be separated from the source we use the definition of effective isotropic radiated power (EIRP) and find that:

$$\text{EIRP} = \text{Pt Gt} = \text{Pr} \left(4\pi\right)^2 R^2 L / Gr \; \lambda^2 = \text{Pr} + 22 + 0 + 2 - 8 + 18 = \text{Pr} + 34 \, \text{dB}. \quad (6.18)$$

Typical readings of the received power, at different rotations of the transmitter in three angles (azimuth, elevation and roll), were between -50 and -60 dBm. The signal is transmitted in two frequencies and modulated. An example of a reading is shown in Figure 6.9 where Pr = -50 dBm and thus EIRP = -16 dBm. We estimate that the output power in each frequency was 1 mW (0 dBm) and the antenna gain was -16 dBi.

In the second step we tried to find what the loss caused by the human body is at these frequencies. A volunteer patient (the author) sat between a transmit antenna and a receive antenna connected to the S12 parameters of the network analyzer (Figure 6.10).

In order to emulate a small isotropic antenna we have chosen a printed inverted-F antenna (PIFA) with small dimensions and wide spatial coverage [17]. In order to reduce any sensitivity to rotation we chose a circular-shaped conductor, for example the Laird D-Puck antenna [18] shown in Figure 6.11. The antenna sizes are 16×16 mm and the typical gain is 3 dBi while mounted on a 51×62 mm board. With a board of 20×20 mm the gain drops to -8 dBi.

More examples of small antennas with circular polarization, PIFA and others, can be found in [19, 20]. We show a rectangular loop with size of 15×15 mm in Figure 6.12 with gain of -13 to -6 dBi in all polarizations.

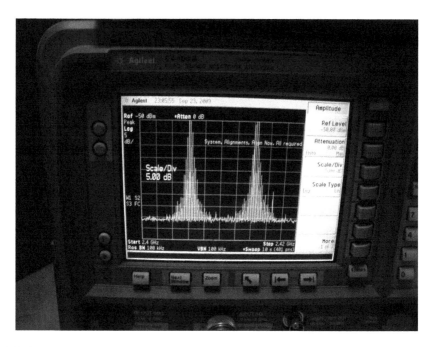

FIGURE 6.9 A graphical presentation of the emitted spectrum. Maximal reading at two frequencies 2406 and 2414 MHz is −50 dBm. Resolution bandwidth was 100 kHz.

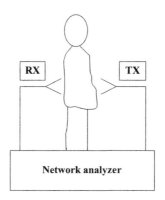

FIGURE 6.10 A test set up for loss measurement of the human body.

Using again the Friis equation:

$$Pr/Pt = Gr\,Gt\,\lambda^2/(4\pi)^2\,R^2L. \tag{6.19}$$

We now get:

$$L = Gr\,Gt\,\lambda^2/(4\pi)^2\,R^2\,(Pr/Pt) = Lc\,Lb \tag{6.20}$$

FIGURE 6.11   A printed inverted-F antenna with circular shape, Laird D-Puck [18].

FIGURE 6.12   A printed inverted-F antenna with a rectangular shape. Overall size is 30 × 30 mm. The antenna has a ground plane with a short at the low part of the print.

where
  Gr is the receive antenna gain −10 to −6 dBi
  Gt is the transmit antenna gain −10 to −6 dBi
  λ^2 is the wavelength = 0.12² m² = −18 dB
  Pr/Pt is the reading of the network analyzers
  $(4\pi)^2$ = 157 = 22 dB
  R^2 = 0.25 m² = −6 dB
  Lc is the cable loss = 2 dB
  Lb is the body loss

Typical readings in different antenna rotations at range of 0.5 m are shown in Figure 6.13 (free space) and Figure 6.14 (through the body) and are listed in Table 6.2.

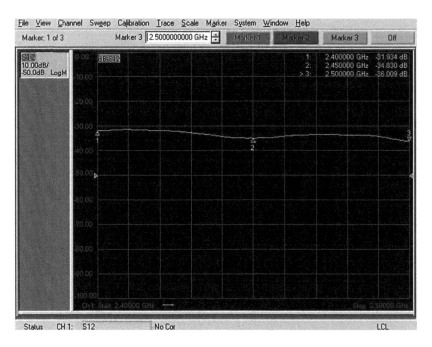

FIGURE 6.13 Network analyzer measurement number 1 **in free space link** at frequencies 2400–2500 MHz. The ripple is caused by the antennas.

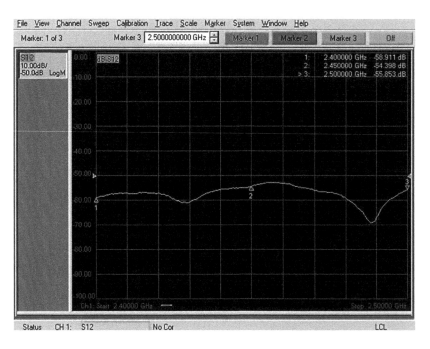

FIGURE 6.14 Network analyzer measurement number 4 **via the body** at frequencies 2400–2500 MHz. The ripple is caused by the body.

TABLE 6.2
Six measurements of a short link through the body

Number	Space	S12 at 2400 MHz (-dB)	S12 at 2450 MHz (-dB)	S12 at 2500 MHz (-dB)
1	Free	32	35	36
2	Free	38	35	38
3	Free	31	34	38
4	With body	59	54	56
5	With body	67	75	57
6	With body	62	61	58

For a nominal antenna gain of −8 dBi we find that the expected loss of the body is:

$$\text{Lb} = -8 - 8 + 18 - 22 + 6 + 35 - \text{Lc} \approx 20 - 30\,\text{dB}. \tag{6.21}$$

From Table 6.2 we see that the body loss (attenuation with body minus attenuation in free space) is within 20–30 dB in most of the cases. The major part of the loss is caused by reflection and absorption of the body and the minor part is caused by antenna polarization mismatch. One can see that the details of the body may create deep resonances, hence working at different frequencies is always recommended. For a transmitter with Pt = 10 dBm (as permitted by ECC ERC Recommendation 70-03 [21]) and antenna gain of −10 dBi, a receiver separated at a distance of 0.5 m with the body in between will get a sensitivity of −55 to −70 dBm.

6.4 MEASUREMENTS OF BODY PARAMETERS WITH EKG PADS

Wireless measurement of the human body has become one of the most challenging biomedical applications in recent years. Various sensors for detecting vital signs like breathing, fluid content and other biological information were proposed [22–26]. The relevant antenna, attached closely to the body, should be small, comfortable for the patient and wideband. An example of such an antenna, produced on an EKG pad, is presented here (Figures 6.15 and 6.16).

The total length of the monopole can vary from 2 to 10 cm, resulting in different ranges of frequencies. The matching of the antenna, while attached to the body, is totally different from free space conditions. Figure 6.17 shows three cases of monopoles under study which cover the frequency bands 1–4 GHz, 2–6 GHz and 5–9 GHz.

The return losses of the three models in free space and while attached to the chest of a patient (the author in this case) are shown in Figures 6.18–6.23.

The specific choice of the frequencies should obey regulatory aspects. The industrial, scientific and medical (ISM) frequencies of 2.4–2.5 GHz and 5.4–5.9 GHz seem a natural choice, but for power less than 0 dBm, radiated into the body, other frequencies can be considered as well. Examples for transmission coefficients between antennas 2 and 3, while mounted on the chest, at three distances of 5, 10 and 15 cm are shown in Figures 6.24–6.26.

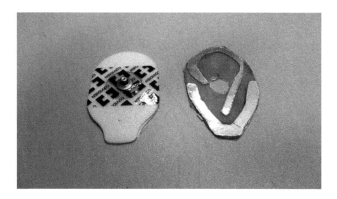

FIGURE 6.15   An electrocardiogram pad as a carrier for biomedical antenna attached to the body.

FIGURE 6.16   A folded monopole attached to an electrocardiogram pad.

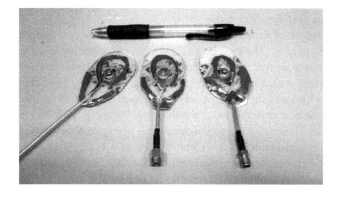

FIGURE 6.17   Three folded monopoles for different frequency ranges.

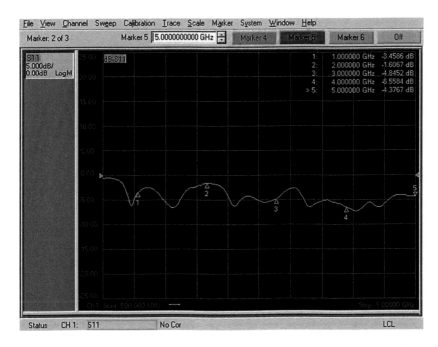

FIGURE 6.18 Return loss of **antenna 1 in free space**. Reasonable matching is achieved at 1–5 GHz.

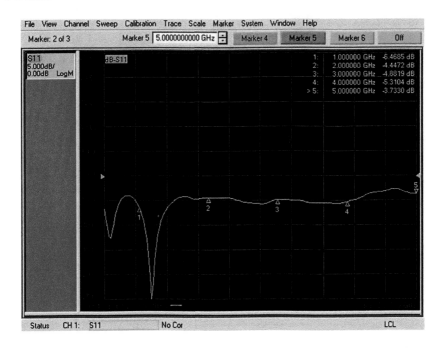

FIGURE 6.19 Return loss of **antenna 1 on the body**. Reasonable matching is achieved at 1–4 GHz.

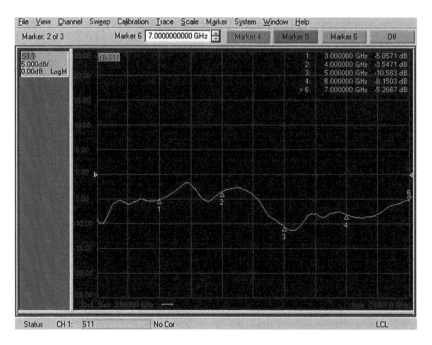

FIGURE 6.20   Return loss of **antenna 2 in free space**. Reasonable matching is achieved at 2–7 GHz.

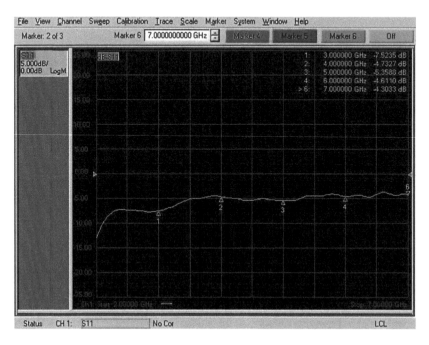

FIGURE 6.21   Return loss of **antenna 2 on the body**. Reasonable matching is achieved at 2–6 GHz.

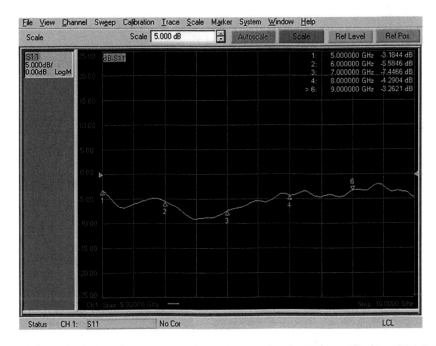

FIGURE 6.22 Return loss of **antenna 3 in free space**. Reasonable matching is achieved at 5–10 GHz.

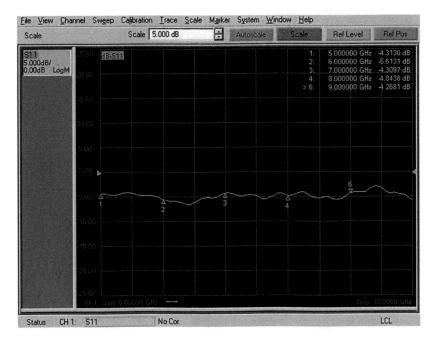

FIGURE 6.23 Return loss of **antenna 3 on the body**. Reasonable matching is achieved at 5–9 GHz.

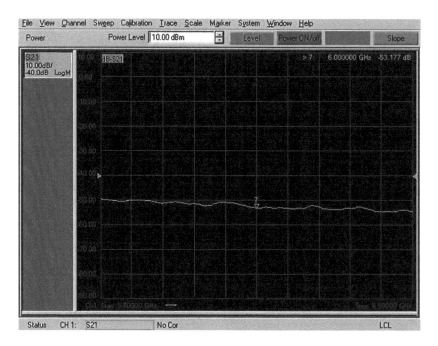

FIGURE 6.24 Transmission loss between two antennas, mounted on the body, **at distance 5 cm** at 5.5–6.5 GHz.

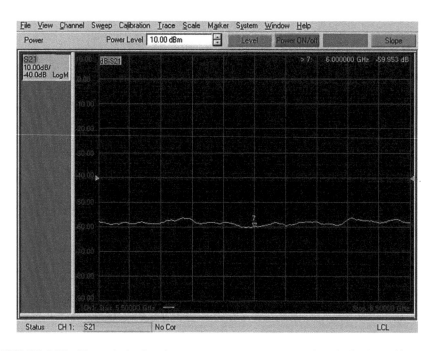

FIGURE 6.25 Transmission loss between two antennas, mounted on the body, **at distance 10 cm** at 5.5–6.5 GHz.

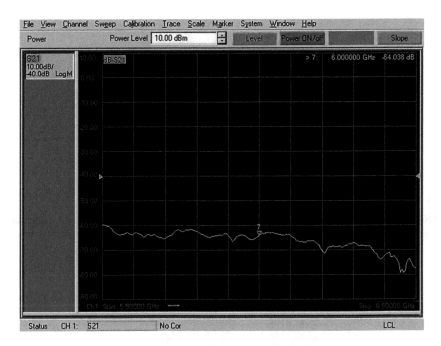

FIGURE 6.26 Transmission loss between two antennas, mounted on the body, **at distance 15 cm** at 5.5–6.5 GHz.

We see a significant change in the transmission loss as a function of the distance. For example, at a center frequency of 6 GHz the recorded values were −53 dB, −60 dB and −64 dB for distances 5 cm, 10 cm and 15 cm, respectively. The body does not create distinct resonances at this range of frequencies and the dynamic range is well above the noise floor of the network analyzer. In other experiments we used coaxial cable RF-316 of 25 cm length with absorbing ECCOSORB FGM-40 material [27]. Figure 6.27 shows a set of six antennas with cables and absorbing materials, which provide cleaner and reproducible results. The antennas were developed in collaboration with Dr Z. Reznik and Dr A. DeRowe [28].

6.5 CELLULAR ANTENNAS ON A PHANTOM

A commercial chip antenna for cellular terminals was tested as a wearable device on a phantom (half a watermelon with diameter of 250 mm). The antenna was soldered onto a 30 × 40 mm printed circuit board (PCB) with clearance area as recommended by the manufacturer. Figures 6.28 and 6.29 show examples of such antennas: Pulse *Larsen* 3530 [29] and Taoglas PA-700A [30]. Many products of this kind can be found in the industry [31–33]. The specific gain of these antennas is largely dependent on the size of the ground plane and the presence of metallic objects in close vicinity to the antenna.

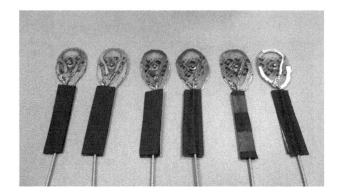

FIGURE 6.27   Set of six antennas on electrocardiogram pads, with coaxial cables and absorbing covers [26].

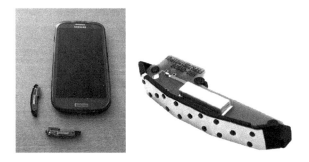

FIGURE 6.28   Cellular chip antenna Pulse *Larsen* 3530 [29].

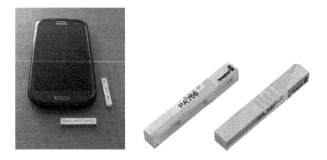

FIGURE 6.29   Cellular chip antenna Taoglas PA-700A [30].

The antenna radiation patterns in azimuth and in elevation of Taoglas PA-710 [30] were taken without the phantom and in the presence of the phantom, while the typical distance between the PCB and the watermelon was 2 mm. It is important to notice that the chip itself had no ground plane under the chip so the interaction with the phantom is significant. Measurements were taken in an anechoic chamber at a distance of 1.5 m (far field) as shown in Figure 6.30. A transmit antenna was mounted

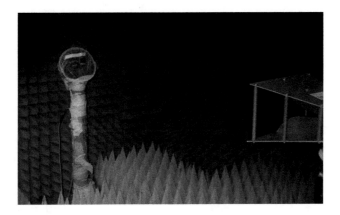

FIGURE 6.30   Cellular chip antennas mounted on a phantom in far-field range.

in two linear polarizations. The gain was found in comparison with a calibrated log-periodic dipole array (LPDA) with gain of 5–6 dBi.

Firstly, we show in Figures 6.31–6.34 the azimuth and elevation patterns of the antenna without and with the watermelon.

In wireless links in a multipath environment, the polarization in the receiver is a random variable. The performance of the antenna cannot be quantified in terms of the

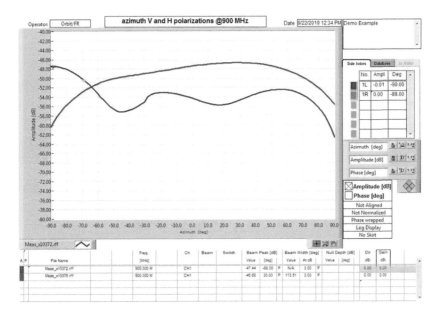

FIGURE 6.31   Azimuth cut, without watermelon, at two polarizations at 900 MHz. Red = vertical polarization. Blue = horizontal polarization.

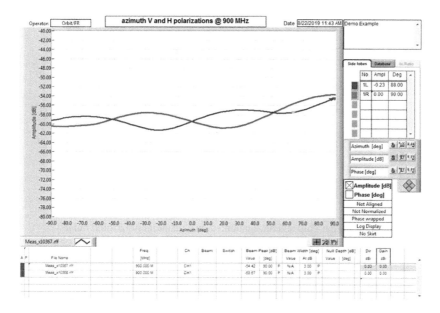

FIGURE 6.32   Azimuth cut, with watermelon, at two polarizations at 900 MHz. Red = vertical polarization. Blue = horizontal polarization.

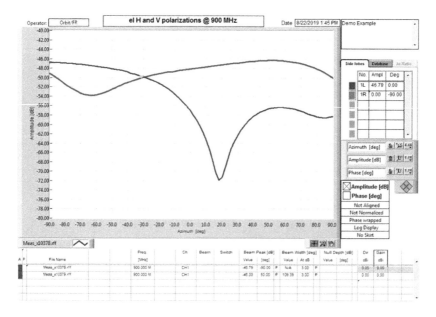

FIGURE 6.33   Elevation cut, without watermelon, at two polarizations at 900 MHz. Red = vertical polarization. Blue = horizontal polarization.

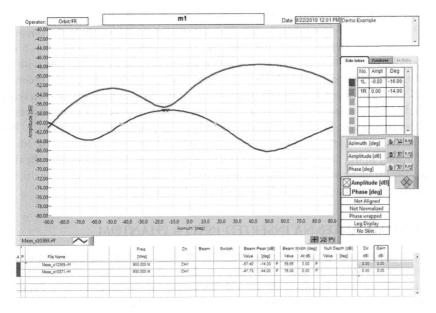

FIGURE 6.34 Elevation cut, with watermelon, at two polarizations at 900 MHz. Red = vertical polarization. Blue = horizontal polarization.

free space gain (radiation pattern) and it is often suggested to use the concept of MEG (mean effective gain) [34]:

$$\text{MEG} = \frac{1}{(1+X_P)} \iint\limits_{\theta,\varphi} \left[G_\theta(\theta,\varphi) p_\theta(\theta,\varphi) + X_P G_\varphi(\theta,\varphi) p_\varphi(\theta,\varphi) \right] d\theta d\varphi \qquad (6.22)$$

where

G_θ and G_φ are the antenna power gains in the θ and φ polarizations
X_p is the cross polarization p_φ/p_θ in any angle of observation
p_θ and p_φ are the angular density functions of the polarized waves.

We can see that the concept of MEG includes all radiation-dependent parameters including cross-polarization and the proximity of objects that rotate the polarization. For simplicity we can define the absolute gain of the antenna Gabs as:

$$G \text{ abs} = G_\theta + G_\varphi = G_v + G_h \qquad (6.23)$$

where G_v and G_h are the gains in vertical and in horizontal polarizations, respectively (note that the addition is not done in decibels but in scalar values).

The antenna readings in free space, with added polarizations shown in Figure 6.33, vary between −46 and −52 dB. The calibrated gain was found to be −4 to −10 dBi. The antenna readings while attached to the phantom are −51 to −56 dB (Figure 6.34) and the calibrated gain is −9 to −14 dBi. We see that the phantom adds loss of 6–8 dB. The phantom makes the pattern flatter and the polarizations more alike.

We can now see that the antenna in free space shows readings between −46 and −47 dB (amazingly omnidirectional). The antenna in close vicinity to the phantom shows readings between −48 and −56 dB. The phantom adds a loss of 4–6 dB. Secondly, Figures 6.35 and 6.36 compare the radiation patterns without and with the watermelon at frequency 850 MHz at the same scale. Again we see that the phantom adds a typical loss of 4–8 dB and it diverts the beam. In summary, it is shown that a cellular chip antenna, mounted near a phantom of the belly/chest of an adult person, suffers from a gain loss of 4–8 dB and possibly some beam shifting in elevation. Part of the loss can be attributed to frequency detuning (return loss) and part of the loss is absorption in the body. The absorption can be minimized by introducing a flexible metallic ground between the antenna and the body, which makes the antenna thicker. This case study was done in collaboration with FYC Ltd [35].

6.6 SMALL ANTENNA INSERTED INTO A PHANTOM

Communications from and into the body can serve many biomedical monitoring and device controlling applications. The implanted antennas must be small and well protected against body liquids [8, 9, 11, 14]. They must be smooth and soft and comply with safety standards of non-ionized radiation. We describe here two small antennas, inserted into a watermelon, in order to study the possible effects of antenna behavior inside the human body. The first antenna was a small coil in the lower UHF band shown in Figure 6.37 and the second antenna was a straight monopole in L band shown in Figure 6.38.

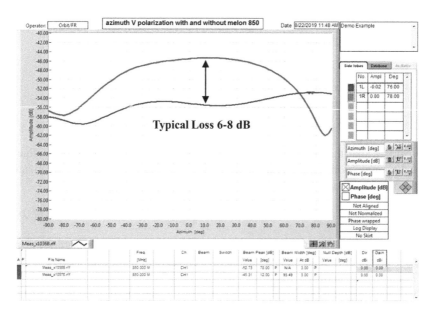

FIGURE 6.35   Azimuth cut, without and with watermelon, at V polarization at 850 MHz. Red without watermelon. Blue with watermelon.

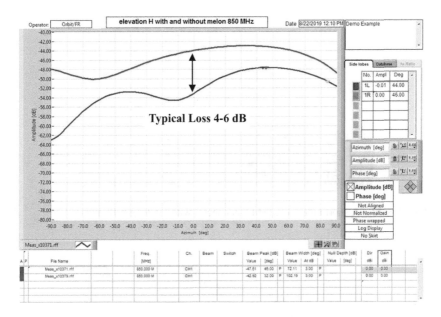

FIGURE 6.36 Elevation cut, without and with watermelon, at H polarization at 850 MHz. Red without watermelon. Blue with watermelon.

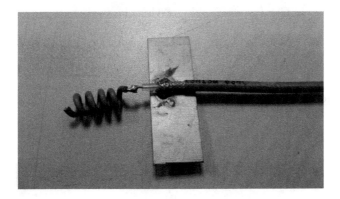

FIGURE 6.37 Miniature coil with six turns with a ground plane.

6.6.1 COIL

Figures 6.39–6.41 show the watermelon prior to the measurement, while the antenna is inserted to a nominal depth of 5 cm and while the watermelon is closed.

The resonant frequency of the coil in air is 915 ± 15 MHz. The resonant frequency of the coil inside the watermelon is 320 ± 60 MHz. We can conclude that the effective dielectric constant of the substance is close to 10. The available gain of the antenna was found to be −25 dBi as measured in a range of 1 m shown in Figure 6.42. This value is composed of original gain of the antenna −4 ± 1 dBi (measured in free space)

FIGURE 6.38 A short monopole (SMA connector with pin length 18 mm).

FIGURE 6.39 A watermelon as a phantom of the human belly.

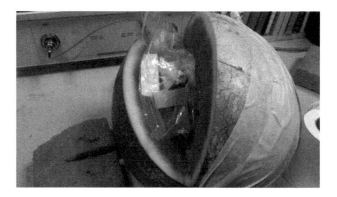

FIGURE 6.40 A coil antenna is inserted into the watermelon. The cable is isolated by a cylindrical ferrite (acts as a BALUN, balanced to unbalanced, transformer).

FIGURE 6.41 The watermelon closed with the coil antenna inside.

FIGURE 6.42 Transmission measurement at a distance of 1 m.

plus the loss of the watermelon at 320 MHz (−20 dB). Such a gain enables the operation of a narrow-band data link within the range of a few meters.

6.6.2 MONOPOLE

The resonant frequency of the monopole in air is 4 GHz ($\lambda_o/4 = 19$ mm), hence this antenna is very close to being a typical quarter-wave wire with a small ground. The resonant frequency of the monopole inside the watermelon (Figures 6.43 and 6.44) is 1.4 GHz. We can conclude that the effective dielectric constant of the substance in this band is also close to 10. The available gain of the antenna was found to be −22 dBi. This value is composed of the original gain of the antenna 3 ± 1 dBi (measured in free space) plus the loss of the watermelon at 1.15 GHz (−25 dB).

The return loss of the antenna while inserted in the watermelon is shown in Figure 6.45. The good matching at wide bandwidth indicates that a large part of the power is absorbed by the substance. The measured transmission loss is shown in Figure 6.46.

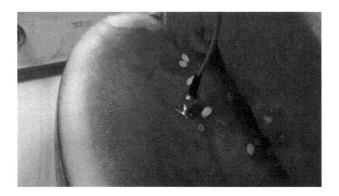

FIGURE 6.43   Monopole inserted into the watermelon.

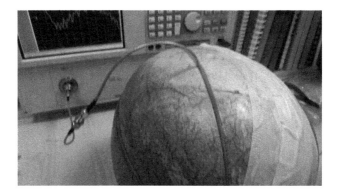

FIGURE 6.44   A closed watermelon with the monopole inside.

The link budget from the monopole to an external antenna (LPDA AARONIA HyperLOG 4060 Gt = 5–6 dBi) is as follows:

$$Pr/Pt = Gr\,Gt\,\lambda^2/(4\pi)^2\,R^2\,L \approx -55\,dB \qquad (6.24)$$

where
 Gr = 5–6 dBi
 $\lambda^2 = -12$ dB
 $(4\pi)^2 = 22$ dB
 $R^2 = 0$ dB (distance 1 m)
 L = 4 dB (cables).

Hence,

$$Gt = -22\,dBi.$$

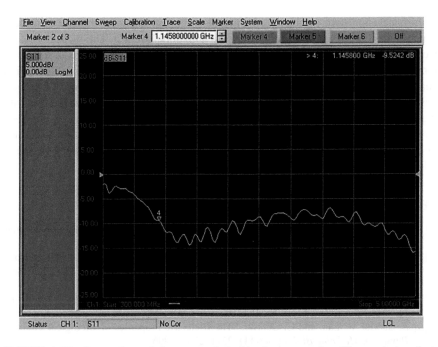

FIGURE 6.45 Return loss of the monopole inside the watermelon. The vertical scale is 5 dB/div. The bandwidth for voltage standing wave ratio = 2 is 1.15–2.55 GHz (75%).

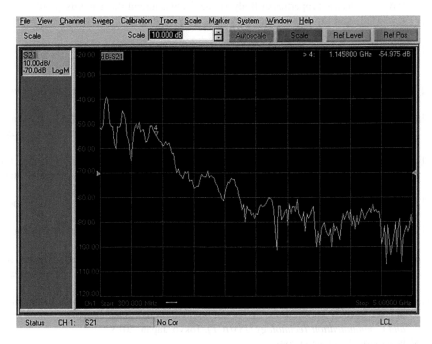

FIGURE 6.46 Transmission loss of the monopole inside the watermelon. The vertical scale is 10 dB/div. Pr/Pt ≈ −55 dB at 1150 MHz.

In conclusion, two miniaturized antennas were tested while implanted inside a watermelon that served as a phantom of the human belly/chest. The measured gains were −25 dBi at 300 MHz and −22 dBi at 1150 MHz (including the loss of the phantom). In this chapter it has been demonstrated that antenna matching and radiation characteristics depend on the interaction with human tissues. The electrical properties of human tissues (dielectric constant, loss tangent and conductivity) vary with the type of the tissue, the water content and the frequency. Human tissues have dielectric constants in the range of 5–60 and loss tangent in the range of 0.1–2; thus, in addition to frequency detuning, we generally observe reduction in the radiation efficiency, gain, radiation pattern and polarization.

REFERENCES

[1] R.G. Martin, "Electromagnetic Field Theory for Physicists and Engineers: Fundamentals and Applications," http://maxwell.ugr.es/rgomez/electrodinamicaprimeraparte.pdf.
[2] M. Okoniewski and M. A. Stuchly, A Study of the Handset Antenna and Human Body Interaction, *IEEE Transactions on Microwave Theory and Techniques* 44, 1855–1864, October 1996.
[3] R. Pethig, "Dielectric Properties of Biological Materials," *IEEE Transactions on Electrical Insulation*, EI-19(5), 453–466, 1984.
[4] S. Gabriel, R. Lau, and C. Gabriel, "The Dielectric Properties of Biological Tissues," *Physics in Medicine Biology*, 41(11), 2271–2293, 1996.
[5] D. Werber, A. Schwentner, and E.M. Biebl, "Investigation of RF Transmission Properties of Human Tissues," *Advances of Radio Sciences*, 4, 357–360, 2006.
[6] IFAC, "Dielectric Properties of Body Tissues," http://niremf.ifac.cnr.it/tissprop/.
[7] P.S. Hall and Y. Hao, *Antennas and Propagation for Body-Centric Wireless Communications*, Artech House, 2006.
[8] B. Gupta, S. Sankaralingam, and S. Dhar, "Development of Wearable and Implanted Antennas in the Last Decade," in *10th Mediterranean Microwave Conference*, Guzelyurt, Northern Cyprus, 2010.
[9] F. Merli, "Implantable Antennas for Biomedical Applications," PhD thesis, Ecole Polytechnique Federale de Lausanne, Lausanne, 2011.
[10] N. Chalhat et al., "Characterization of the Interactions between a 60-GHz Antenna and the Human Body in an Off-Body Scenario," *IEEE Transactions on Antennas and Propagation*, 60, 5958–5964, 2012.
[11] A.A.Y. Ibraheem, "Implanted Antennas and Intra-Body Propagation Channel for Wireless Body Area Network," PhD thesis, Virginia Polytechnic Institute and State University, Blacksburg, VA, 2014.
[12] A. Sabban, *Low Visibility Antennas for Communication Systems*, Taylor & Francis Group, 2015.
[13] K. Ito, M. Takahashi, and K. Saito, "Small Antennas Used in the Vicinity of Human Body," *IEICE Transactionson Communications*, E-99-3(1), 9–18, 2016.
[14] P. Loktongbam and L.S. Solanki, "A Brief Review of Implantable Antennas for Biomedical Applications," *International Journal of Advance Research in Science and Engineering*, 6(5), 821–851, 2017.
[15] T. Tuovinen et al., "Effect of Antenna-Human Body Distance on the Antenna Matching in UWB WBAN Applications," in *7th International Symposium on Medical Information and Communication Technology (ISMICT)*, Tokyo, Japan, 2013.
[16] http://virtual-extension.com.

[17] S. Saini, S. Singh, and N. Kumar, "A Review of Various Planar Inverted F Antenna (PIFA) Structures for Wireless Applications," *International Journal of Electrical & Electronics Engineering*, 2(1), 2015.

[18] www.lairdconnect.com.

[19] D. Wang, Y.J. Zhang, and M.S. Tong, "A Wearable UHF RFID Tag Antennna with Archimedean Spiral Strips," *Progress in Electromagnetics Research Symposium*, 2017, 1849–1852, 2017.

[20] S. Gao, Q. Luo, and F. Zhu, *Circularly Polarized Antennas*, Wiley and IEEE Press, Chichester, UK, 2014.

[21] ERC Recommendation 70-03, ed. 2019,www.ecodocdb.dk.

[22] H. Sato, Y. Li, and Q. Chen, "Experimental Study of Transmission Factor through Conducting Human Body Equivalent Liquid," in *ISAP 2016*, 2016, Okinawa, Japan.

[23] A.W. Basan, F.Y. Zullkifli, and E.T. Rahardjo, "Dual Band Magnetic Textile Antenna for Body Area Network Application," in *ISAP 2016*, 2016, Okinawa, Japan.

[24] Y. Nakatami and M. Takahashi, "Textile Antenna for Biological Information Monitoring," in *ISAP 2016*, 2016, Okinawa, Japan.

[25] T.W. Hsu and C.H. Tseng, "Compact 24-GHz Doppler Radar Module for Non Contact Human Vital-Sign Detection," in *ISAP 2016*, 2016, Okinawa, Japan.

[26] S.H. Eom and S. Lim, "RF Strechable Sensor Using Flexible Substrate and Eutectic Gallium-Indium," in *ISAP 2016*, 2016, Okinawa, Japan.

[27] www.eccosorb.com.

[28] zvi.reznic@gmail.com.

[29] Pulse *Larsen* 3530 or 3070.

[30] Taoglas PA-700A or PA-710A or PA-740A.

[31] Yageo ANT2112A or ANT01080918A.

[32] Linx ANT-LTE-CER.

[33] Molex 105263 or 146200.

[34] Y. Huang and K. Boyle, *Antennas: From Theory to Practice*, John Wiley & Sons, Chichester, UK, 2008.

[35] http://fyc.co.il.

7 Wideband RF Technologies for Wearable Communication Systems

Albert Sabban

CONTENTS

7.1 INTRODUCTION

The popular microwave technologies for wearable communication systems are:

– microwave integrated circuits (MICs)
– monolithic MICs (MMICs)
– micro-electro-mechanical systems (MEMS)
– low-temperature co-fired ceramic (LTCC).

MIC technology was the first step in the process of miniaturization of microwave modules. The invention of the microstrip transmission line, in 1950, led to the development of MIC technology. Active and passive elements can be assembled on the same substrate in MICs. The next step in the development of microwave technology was the invention of MMICS technology. MMICS are usually fabricated on gallium arsenide (GaAs) substrate. Microwave MMIC devices are compact. For example, the dimensions of a 30 GHz 1 W power amplifier are around $4 \times 3 \times 0.1$ mm. Modern power amplifier MMICs are fabricated on gallium nitride (GaN). GaN transistors can operate at much higher temperatures and work at much higher voltages than GaAs transistors. Indium phosphide (InP) offers better electrical performance than GaAs

MMICs in terms of gain, higher cutoff frequency and low noise. However, they also are more expensive due to smaller wafer size.

Compact wearable communication systems may be designed only by employing modern microwave technologies such as MMIC, MEMS and LTCC technologies. The communication industry in microwave and mm-wave frequencies is currently in continuous growth. Radio frequency (RF) integrated circuit design is presented in [1–5]. RF modules such as front end, filters, power amplifiers, printed antennas, passive components and limiters are important modules in wearable communication devices, see [1–21]. The electrical performance of the modules determines if the system will meet the required specifications. Moreover, in several cases the modules' performance limits the system's performance. Minimization of the size and weight of the RF modules is achieved by employing MIC, MMIC, MEMS and LTCC technology. However, the integration of MIC and MMIC components and modules raises several technical challenges. Design parameters that may be neglected in modular communication systems cannot be ignored in the design of wide band integrated RF modules. Powerful RF design software is required to achieve accurate design of RF modules and antennas in microwave frequencies. Accurate design of mm-wave antennas and RF modules is crucial in the development of low-cost communication. It is an impossible mission to tune mm-wave RF modules in the fabrication process.

7.2 MICS FOR 5G AND INTERNET OF THINGS APPLICATIONS

Traditional microwave systems consist of connectorized components (such as amplifier, filters and mixers) connected by coaxial cables. These modules have big dimensions and suffer from high losses and weight. Dimension and losses may be minimized by using MIC technology. There are three types of MICs: hybrid MIC (HMIC), standard MIC and miniature HMIC. Solid state and passive elements are bonded to the dielectric substrate. The passive elements are fabricated by using thick or thin film technology. Standard MICs use a single level metallization for conductors and transmission lines. Miniature HMICs use multilevel process in which passive elements, such as capacitors and resistors, are batch deposited on the substrate. Semiconductor devices, such as amplifiers and diodes, are bonded on the substrate. Figure 7.1 presents a MIC transceiver. Figure 7.2 presents the layout of the MIC receiving link. The receiving channel consists of a low noise amplifier (LNA), filters, dielectric resonant oscillators and a diode mixer.

7.3 K BAND COMPACT RECEIVING CHANNEL

In Section 7.3 an example of a MIC integrated receiving channel is presented.

7.3.1 INTRODUCTION

An increasing demand for wide bandwidth in communication links makes the K/Ka band attractive for future commercial systems. The frequency allocations for the Ka/K band Very Small Aperture Terminal (VSAT) system are 17.7–21.2 GHz for the receiving channel and 27.5–31 GHz for the transmitting channel. The communication industry is currently in continuous growth. In particular, the VSAT networks have

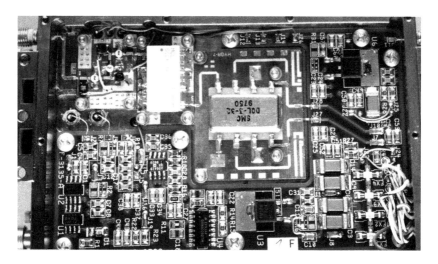

FIGURE 7.1 A microwave integrated circuit transceiver.

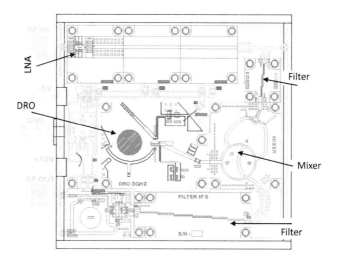

FIGURE 7.2 Layout of a microwave integrated circuit receiving link.

gained wide use for business and private applications. Private organizations and banks use VSAT networks to communicate between their various sites. VSAT applications cover a wide range, such us telephony, message distribution, lottery, credit card approval and inventory management. Commercial VSAT systems operate in C band and Ku band; however, there are many advantages to developing wide band K/Ka band communication systems. However, only some commercial low-cost power amplifiers, LNAs, mixers and dielectric resonators (DROs) are published in commercial catalogs. Moreover, development of low-cost RF components is crucial for K/Ka band satellite communication industry. This section describes the design and performance of a compact and low-cost K band receiving channel.

TABLE 7.1
Receiving channel specification

Parameter	Specification
RF frequency range	18.8–19.3 GHz, 19.7–20.2 GHz
IF frequency range	0.95–1.45 GHz
Gain	50 dB
Noise figure	2 dB
Input voltage standing wave ratio	2:1
Output voltage standing wave ratio	2:1
Spurious level	−40 dBc
Frequency stability versus temperature	±2 MHz
Supply voltage	±5 V
Connectors	K connectors
Operating temperature	−40°C–60°C
Storage temperature	−50°C–80°C
Humidity	95%

7.3.2 RECEIVING CHANNEL DESIGN

The major objectives in the design of the receiving channel were electrical specifications and cost.

7.3.2.1 Receiving Channel Specifications

The receiving channel specification is listed in Table 7.1.

7.3.3 DESCRIPTION OF THE RECEIVING CHANNEL

A block diagram of the receiving channel is shown in Figure 7.3.

The receiving channel consists of a RF side coupled band pass filter, LNA, mixer, DRO, intermediate frequency (IF) filter, MMIC down-converter block. Noise figure (N.F), gain and third intercept point (IP3) budget are given in Table 7.2.

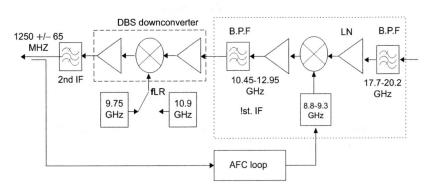

FIGURE 7.3   Block diagram of the receiving channel.

TABLE 7.2
Noise figure, gain, third intercept point budget

	LNA		RF Low-Pass Filter	1st Mixer	1st IF Amp	Band-Pas Filter	2nd Mixer IF	Cascade Total
NF (dB)	0.8	0.8	0.5	10	2.5	6.5	6.5	1.66
Gain (dB)	10	10	−0.5	−10	16	−6.5	37	56.00
OIP3 (dBm)	14	14	90	8	18	100	3	8.00

A receiving channel with improved N.F 0.95 dB and gain (77 dB) is obtained by adding a gain block after the LNA. However, because of cost consideration we decided to realize the first configuration shown in Table 7.2. The major objectives in the design of the receiving channel were specifications and cost.

7.3.4 DEVELOPMENT OF THE RECEIVING CHANNEL

A MIC and MMIC LNA were developed for K band receiving links. The dimensions of the MMIC LNA are much smaller than the MIC LNA. However, the N.F of the MIC LNA is around 1.2–1.5 dB and the N.F of the MMIC LNA is around 1.7–2 dB.

Components of the receiving channel were printed on 10 mil thick RT-5880 Duroid substrate. Drawings of the MMIC LNA, and of the receiving channel are shown in Figures 7.4 and 7.5. A photograph of the receiving channel with a MMIC LNA is shown in Figure 7.6.

7.3.5 MEASURED TEST RESULTS OF THE RECEIVING CHANNEL

The measured test results of the receiving channel are summarized in Table 7.3.

7.4 MMICs

MMICs are circuits in which active and passive elements are formed on the same dielectric substrate, as presented in Figure 7.7, by using a deposition scheme such as epitaxy, ion implantation, sputtering, evaporation and diffusion. The MMIC chip in Figure 7.7 consists of passive elements such as resistors, capacitors, inductors and a field-effect transistor (FET).

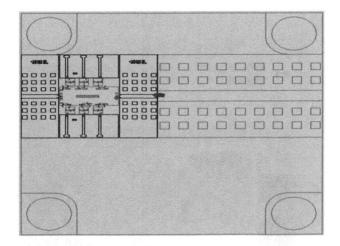

FIGURE 7.4   Monolithic microwave integrated circuit low noise amplifier on carrier.

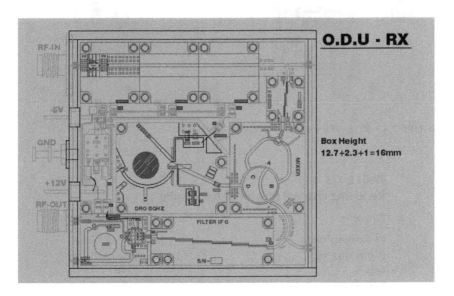

FIGURE 7.5   Outdoor unit receiving channel.

7.4.1 FEATURES OF MMIC TECHNOLOGIES

Accurate design is crucial in the design of MMICs. Accurate design may be achieved by using 3D electromagnetic software.

Materials employed in the design of MMICs are GaAs, InP, GaN and silicon germanium (SiGe).

Large statistic scattering of all electrical parameters cause sensitivity of the design.

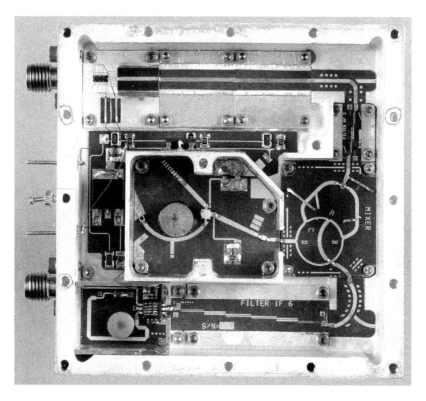

FIGURE 7.6 Photo of the outdoor receiving channel.

TABLE 7.3
Receiving channel measured test results

Parameter	Measured results
RF frequency range	18.8–19.3 GHz
IF frequency range	0.95–2. GHz
Gain	50 dB
Noise figure	2 dB
Input voltage standing wave ratio	2:1
Output voltage standing wave ratio	2:1
Spurious level	−40 dBc
Frequency stability versus temperature	±2 MHz

Fabrication facility (FAB) runs are expensive, around US$200,000 per run. Miniaturization of components yields lower cost of the MMICs. Figure 7.8 presents MMIC design flow.

A designer's goal is to comply with customer specifications in one design iteration. MMIC components cannot be tuned.

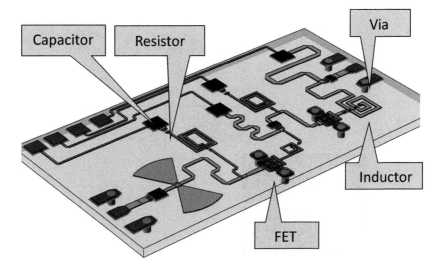

FIGURE 7.7   Monolithic microwave integrated circuit basic components.

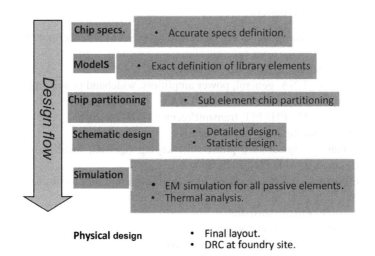

FIGURE 7.8   Monolithic microwave integrated circuit design flow.

- 0.25 µm GaAs pseudo-morphic high-electron-mobility transistor (PHEMT) for power applications to Ku band.
- 0.15 µm GaAs PHEMT for applications to high Ka band.
- GaAs P-N intrinsic semiconductor (PIN) process for low loss power switching applications.
- Future new process – InP hetero-junction bipolar transistor (HBT), SiGe, GaN, RF/complementary metal-oxide semiconductor (RF/CMOS), RF/MEMS.

Figure 7.9 presents a GaAs wafer layout. Wafer size may be "3, 5" or 6"

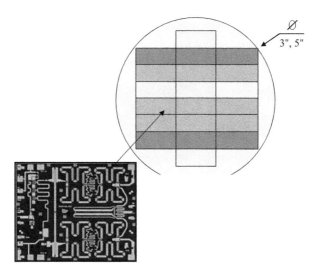

FIGURE 7.9   GaAs wafer layout.

7.4.2 MMIC COMPONENTS

Several microwave components are fabricated by using MMIC technology.

- Mixers – balanced, Star, sub-harmonic.
- Amplifiers – LNA, general, power amplifiers, wideband power amplifiers, distributed traveling wave amplifier (TWA).
- Switches – PIN, PHEMT, transmit/receive (T/R) matrix.
- Frequency multipliers – active, passive.
- Modulators – quadrature phase-shift keying (QPSK), quadrature amplitude modulation (QAM; PIN, PHEMT).
- Multifunction – RX chip, TX chip, switched amp chip, local oscillator (LO) chain.
  FET – field-effect transistor
  BJT – bipolar junction transistor
  HEMT – high-electron-mobility transistor
  PHEMT – pseudo-morphic HEMT
  MHEMT – metamorphic HEMT
  D-HBT – double hetero-junction bipolar transistor
  CMOS – complementary metal-oxide semiconductor.

Table 7.4 presents types of devices fabricated by using MMIC technology.

7.4.3 ADVANTAGES OF GaAs VERSUS SILICON

MMICs were originally fabricated by using GaAs, a III-V compound semiconductor. MMICs are dimensionally small (from around 1 mm^2 to 10 mm^2) and can be mass

TABLE 7.4
Monolithic microwave integrated circuit technology

Material	FET	BJT	Diode
III-V-based	PHEMT GaAs	HBT GaAs	Schottky GaAs
	HEMT InP	D-HBT InP	
	MHEMT GaAs		
	HEMT GaN		
Silicon	CMOS	HBT SiGe	

produced. GaAs has some electronic properties which are better than those of silicon. It has a higher saturated electron velocity and higher electron mobility, allowing transistors made from GaAs to function at frequencies higher than 250 GHz. Unlike silicon junctions, GaAs devices are relatively insensitive to heat due to their higher band gap. Also, GaAs devices tend to have less noise than silicon devices, especially at high frequencies, which is a result of higher carrier mobility and lower resistive device parasitic. These properties recommend GaAs circuitry in mobile phones, satellite communications, microwave point-to-point links and higher frequency radar systems. It is used in the fabrication of Gunn diodes to generate microwave. GaAs has a direct band gap, which means that it can be used to emit light efficiently. Silicon has an indirect band gap and so it is very poor at emitting light. Nonetheless, recent advances may make silicon light-emitting diodes (LEDs) and lasers possible. Due to its lower band gap though, silicon LEDs cannot emit visible light and rather work in infrared (IR) range while GaAs LEDs function in visible red light. As a wide direct band gap material and resulting resistance to radiation damage, GaAs is an excellent material for space electronics and optical windows in high power applications.

Silicon has three major advantages over GaAs for an integrated circuit manufacturer. First, silicon is a cheap material. In addition, a Si crystal has an extremely stable structure mechanically and it can be grown to very large diameter boules and can be processed with very high yields. It is also a decent thermal conductor; thus, it enables very dense packing of transistors, which is very attractive for the design and manufacturing of very large integrated circuits (ICs). The second major advantage of Si is the existence of silicon dioxide—one of the best insulators. Silicon dioxide can easily be incorporated onto silicon circuits, and such layers are adherent to the underlying Si. GaAs does not easily form such a stable adherent insulating layer and does not have stable oxide either. The third, and perhaps most important, advantage of silicon is that it possesses a much higher hole mobility. This high mobility allows the fabrication of higher-speed P-channel FETs, which are required for CMOS logic. Because they lack a fast CMOS structure, GaAs logic circuits have much higher power consumption, which has made them unable to compete with silicon logic circuits. The primary advantage of Si technology is its lower fabrication cost compared with GaAs. Silicon wafer diameters are larger. Typically, 8" or 12" compared with 4" or 6" for GaAs. Si wafer costs are much lower than GaAs wafer costs, contributing to a less expensive Si IC.

TABLE 7.5
Comparison of material properties

Property	Si	Si or sapphire	GaAs	InP
Dielectric constant	11.7	11.6	12.9	14
Resistivity (Ω/cm)	10^3–10^5	$>10^{14}$	10^7–10^9	10^7
Mobility (cm^2/v-s)	700	700	4300	3000
Density (gr/cm^3)	2.3	3.9	5.3	4.8
Saturation velocity (cm/s)	9×10^6	9×10^6	1.3×10^7	1.9×10^7

Other III-V technologies, such as InP, offer better performance than GaAs in terms of gain, higher cutoff frequency, and low noise. However, they are more expensive due to smaller wafer sizes and increased material fragility.

SiGe is a Si-based compound semiconductor technology offering higher speed transistors than conventional Si devices but with similar cost advantages.

GaN is also an option for MMICs. Because GaN transistors can operate at much higher temperatures and work at much higher voltages than GaAs transistors, they make ideal power amplifiers at microwave frequencies. In Table 7.5, the properties of material used in MMIC technology are compared.

7.4.4 SEMICONDUCTOR TECHNOLOGY

The cutoff frequency of Si CMOS MMIC devices is lower than 200 GHz. Si CMOS MMIC devices are usually low power and low-cost devices. The cutoff frequency of SiGe MMIC devices is lower than 200 GHz. SiGe MMIC devices are used as medium power high gain devices. The cutoff frequency of InP HBT devices is lower than 400 GHz. InP HBT devices are used as medium power high gain devices. The cutoff frequency of InP HEMT devices is lower than 600 GHz. InP HEMT devices are used as medium power high gain devices. In Table 7.6 the properties of MMIC technologies are compared. Figure 7.10 presents a 0.15-μm PHEMT on GaAs substrate.

TABLE 7.6
Summary of semiconductor technology

	Si CMOS	SiGe HBT	InP HBT	InP HEMT	GaN HEMT
Cutoff frequency	>200 GHz	>200 GHz	>400 GHz	>600 GHz	>200 GHz
Published MMICs	170 GHz	245 GHz	325 GHz	670 GHz	200 GHz
Output power	Low	Medium	Medium	Medium	High
Gain	Low	High	High	Low	Low
RF noise	High	High	High	Low	Low
Yield	High	High	Medium	Low	Low
Mixed signal	Yes	Yes	Yes	No	No
1/Frequency noise	High	Low	Low	High	High
Breakdown voltage	−1 V	−2 V	−4 V	−2 V	>20 V

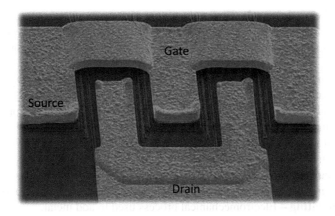

FIGURE 7.10 0.15 μm pseudo-morphic high-electron-mobility transistor on GaAs substrate.

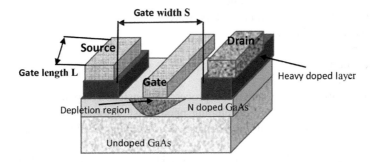

FIGURE 7.11 Metal–semiconductor field-effect transistor cross section on GaAs substrate.

7.4.5 MMIC FABRICATION PROCESS

The MMIC fabrication process consists of several controlled processes in a semiconductor FAB. The processes are listed in Section 7.4.6. In Figure 7.11 a metal–semiconductor field-effect transistor (MESFET) cross section on GaAs substrate is shown.

7.4.5.1 MMIC Fabrication Process List

Wafer fabrication – Preparing the wafer for fabrication.

Wet cleans – Wafer cleaning by wet process.

Ion implantation – Dopants are embedded to create regions of increased or decreased conductivity. Selectively implant impurities. Create p or n type semiconductor regions.

Dry etching – Selectively remove materials.

Wet Etching – Selectively remove materials chemical process.

Plasma etching – Selectively remove materials.

Thermal treatment – High temperature process to remove stress.

Rapid thermal anneal – High temperature process to remove stress.

Furnace anneal – After ion implantation, thermal annealing is required. Furnace annealing may take minutes and causes too much diffusion of dopants for some applications.

Oxidation – Substrate oxidation, for example: dry oxidation $Si + O_2 \rightarrow SiO_2$; wet oxidation $Si + 2H_2O \rightarrow SiO_2 + 2H_2$.

Chemical vapor deposition (CVD) – Chemical vapor deposited on the wafer. Pattern defined by photoresist.

Physical vapor deposition (PVD) – Vapor produced by evaporation or sputtering deposited on the wafer. Pattern defined by photoresist.

Molecular beam epitaxy (MBE) – A beam of atoms or molecules produced in high vacuum. Selectively grow layers of materials. Pattern defined by photoresist.

Electroplating – Electromechanical process used to add metal.

Chemical mechanical polish (CMP). Chemical mechanical polishing (**CMP**) is a fabrication technique that uses chemical oxidation and mechanical abrasion to remove material and achieve a near-perfect flat and smooth surface upon which layers of integrated circuitry are built.

Wafer testing – Electrical test of the wafer.

Wafer back-grinding

Die preparation

Wafer mounting

Die cutting

Lithography – Lithography is the process of transferring a pattern onto the wafer by selectively exposing and developing photoresist. Photolithography consists of four steps; the order depends on whether we are etching or lifting off the unwanted material.

Contact lithography – A glass plate is used that contains the pattern for the entire wafer. The glass plate is led against the wafer during exposure of the photoresist. In this case the entire wafer is patterned in one shot.

Electron-beam lithography – Electron-beam lithography is a form of direct-write lithography. Using E-beam lithography you can write directly to the wafer without a mask. Because an electron beam is used, rather than light, much smaller features can be resolved.

Exposure can be done with light, ultraviolet (UV) light, or electron beam, depending on the accuracy needed. E beam provides much higher resolution than light, because the particles are bigger (greater momentum), the wavelength is shorter.

7.4.5.2 Etching versus Lift-off Removal Processes

There are two principal means of removing material, etching and lift-off.

The steps for an etch-off process are:

1. Deposit material.
2. Deposit photoresist.
3. Pattern (expose and develop).
4. Remove material where it is not wanted by etching.

Etching can be isotropic (etching wherever we can find the material we like to etch) or anisotropic (directional, etching only where the mask allows). Etches can be dry (reactive ion etching; RIE) or wet (chemical). Etches can be very selective (only etching what we intend to etch) or non-selective (attaching a mask to the substrate).

In a lift-off process, the photoresist *forms a mold* into which the desired material is deposited. The desired features are completed when photo-resist B under unwanted areas is dissolved, and unwanted material is "lifted off".

Lift-off processes are:

– Deposit photoresist.
– Pattern.
– Deposit material conductor or insulator.
– Remove material where it is not wanted by *lifting off.*

In Figure 7.12 Metal–semiconductor field-effect transistor cross section on GaAs substrate is shown. In Figure 7.13 a MMIC resistor cross section is shown.

In Figure 7.14 a MMIC capacitor cross section is shown.

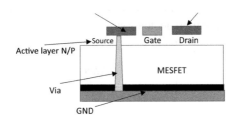

FIGURE 7.12   Metal–semiconductor field-effect transistor cross section.

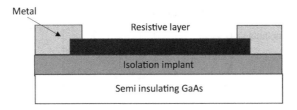

FIGURE 7.13   Resistor cross section.

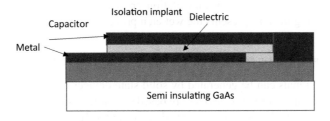

FIGURE 7.14   Capacitor cross section.

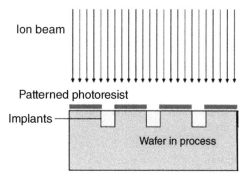

FIGURE 7.15   Ion implantation.

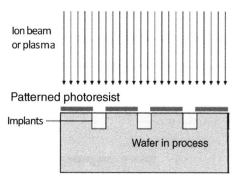

FIGURE 7.16   Ion etch.

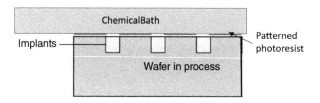

FIGURE 7.17   Wet etch.

Figure 7.15 presents the ion implantation process. Figure 7.16 presents the ion etch process. Figure 7.17 presents the wet etch process.

7.4.6 GENERATION OF MICROWAVE SIGNALS IN MICROWAVE AND MM WAVE

Microwaves signals can be generated by solid state devices and vacuum tube-based device. Solid state microwave devices are based on semiconductors such as silicon or GaAs, and include FETs, BJTs, Gunn diodes, and IMPact ionization Avalanche

Transit-Time (IMPATT) diodes. Microwave variations of BJTs include the HBT; microwave variants of FETs include the MESFET, the HEMT (also known as HFET), and laterally diffused metal-oxide semiconductor (LDMOS) transistor. Microwaves can be generated and processed using ICs and MMICs. They are usually manufactured using GaAs wafers, though SiGe and heavy-doped silicon are increasingly used. Vacuum tube-based devices operate on the ballistic motion of electrons in a vacuum under the influence of controlling electric or magnetic fields, and include the magnetron, klystron, traveling wave tube (TWT) and gyrotron. These devices work in the density modulated mode, rather than the current modulated mode. This means that they work on the basis of clumps of electrons flying ballistically through them, rather than using a continuous stream.

7.4.7 MMIC Circuit Examples and Applications

Figure 7.18 presents a wide band mm-wave power amplifier. The input power is divided by using a power divider. The RF signal is amplified by power amplifiers and combined by a power combiner to get the desired power at the device output.

Figure 7.19 presents a wide band mm-wave up-converter. MMIC process cost is listed in Table 7.7.

7.4.7.1 MMIC Applications
- Ka band satellite communication.
- 60 GHz wireless communication.
- Automotive radars.
- Imaging in security.
- Gbit wireless local area network (WLAN).

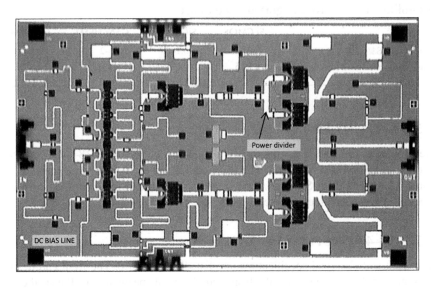

FIGURE 7.18 Wide band power amplifier.

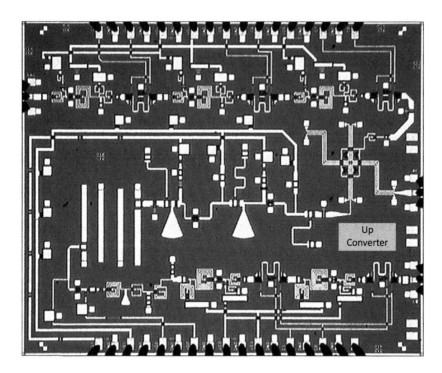

FIGURE 7.19   Ka band up-converter.

TABLE 7.7
Monolithic microwave integrated circuit cost

	Si CMOS	SiGe HBT	GaAs HEMT	InP HEMT
Chip cost (US$/mm2)	0.01	0.1–0.5	1–2	10
Mask cost (million US$/mask set)	1.35	0.135	0.0135	0.0135

7.5 18–40 GHZ FRONT END

Development and design considerations of a compact wideband 18–40 GHz front end are described in this section. The RF modules and the system were designed by using Advanced Design System (ADS) system software and Momentum RF software. There is a good agreement between computed and measured results.

7.5.1 18–40 GHz FRONT END REQUIREMENTS

The front end electrical specifications are listed in Table 7.8. The front end design presented in this section meets the front end electrical specifications. Physical characteristics, interfaces and connectors are listed in Table 7.9.

TABLE 7.8
Front end electrical specifications

Parameter	Requirements	Performance
Frequency range	18–40 GHz	Comply
Gain	24/3 dB typical, switched by external control. (−40 dB or lower for off state)	Comply
Gain flatness	±0.5 dB max for any 0.5 GHz bandwidth in 18–40 GHz. ±2 dB max for any 4 GHz bandwidth in 18–40 GHz. ±3 dB max for the whole range 18–40 GHz	Comply
Noise figure (high gain)	10 dB max for 40°C baseplate temperature 11 dB over temperature	Comply
Inputs power range	−60 dBm to 10 dBm	Comply
Output power range	−39 dBm to 11 dBm not saturated, 13 dBm saturated	Comply
Linearity	Output 1 dB compression point at 12 dBm min. Third intercept point (IP3) at 21 dBm. Single tone. Second harmonic power −25 dBc maximum for 10 dBm output	Comply
Voltage standing wave ratio	2:1	Comply
Power input protection	No damage at +30 dBm continuous wave and +47 dBm pulses (for average power higher than 30 dBm) input power at 0.1–40 GHz Test for pulses: pulse width (PW) = 1 usec, pulse repetition frequency (PRF) = 1 KHz	Comply
Power supply voltages	±5 V, ±15 V	Comply
Control logic	Low voltage transistor–transistor logic (LVTTL) standard "0" = 0–0.8 V; "1" = 2.0–3.3 V	Comply
Switching time	Less than 100 ns	Comply
Non-harmonic spurious (output)	−50 dBm maximum (when it is not correlative with the input signals)	Comply
Video leakage	Video leakage signals will be below the RF output level for terminated input	Comply
Dimension	60 × 40 × 20 mm	Comply

TABLE 7.9
Physical characteristics – interfaces' connectors

Interface	Type
Radio frequency input	Wave guide WRD180 (double ridge)
Radio frequency output	K connector
Direct current supply	D type
Dimensions	60 × 40 × 20 mm
Control	D type

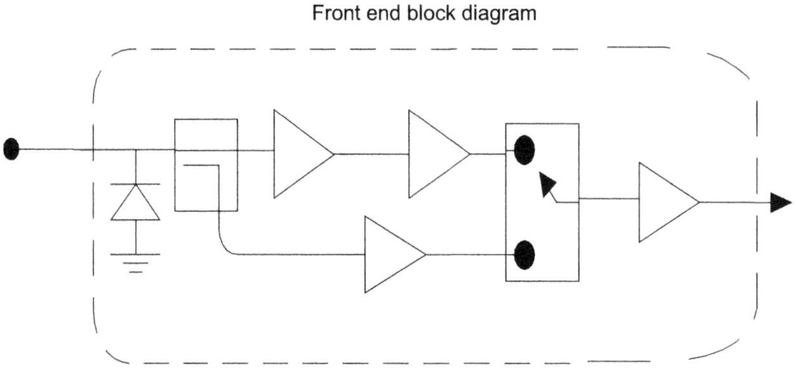

Front end block diagram

FIGURE 7.20   Front end block diagram.

7.5.2 FRONT END DESIGN

A front end block diagram is shown in Figure 7.20.

The front end module consists of a limiter and a wideband 18–40 GHz Filtronic, LNA LMA406. The LMA406 gain is around 12 Db with 4.5 dB N.F and 14 dBm saturated output power. The LNA dimensions are 1.44 × 1.1 mm. We used a wide band PHEMT MMIC single pole double throw (SPDT) manufactured by Keysight, AMMC-2008. The SPDT insertion loss is lower than 2 dB. The isolation between the SPDT input port to the output ports is better than 25 dB. The SPDT 1 dBc compression point is around 14 dBm. The SPDT dimensions are 1 × 0.7 × 0.1 mm. The front end electrical characteristics were evaluated by using ADS Keysight software and by using SYSCAL software. Figure 7.21 presents the front end module N.F and gain for LNA N.F of 6 dB. The overall computed module N.F is 9.46 dB. The module gain is 21 dB. Figure 7.22 presents the front end module N.F and gain for LNA N.F of 5.5 dB. The overall computed module N.F is 9.25 dB. The module gain is 21 dB.

System1

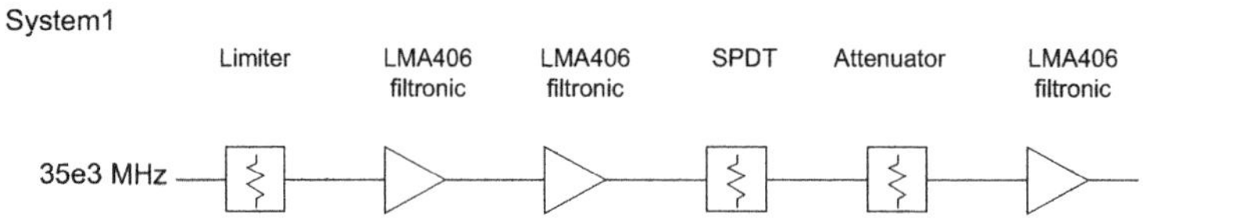

	Limiter	LMA406 filtronic	LMA406 filtronic	SPDT	Attenuator	LMA406 filtronic	

							Total
NF (dB)	3.00	6.00	6.00	3.00	3.00	6.00	9.46
Gain (dB)	−3.00	10.00	10.00	−3.00	−3.00	10.00	21.00
OIP3 (dBm)	30.00	25.00	25.00	30.00	30.00	25.00	23.14

Input Pwr (dBm)	−60.00	System temp (K)	290.00	IM offset (MHz)	.025
OIP2 (dBm)	23.14	OIP3 (dBm)	23.14	Output P1dB (dBm)	11.37
IIP2 (dBm)	2.14	IIP3 (dBm)	2.14	Input P1dB (dBm)	−8.63
OIM2 (dBm)	−101.14	OIM3 (dBm)	−163.28	Compressed (dB)	0.00
ORR2 (dB)	62.14	ORR3 (dB)	124.28	Gain, actual (dB)	21.00
IRR2 (dB)	31.07	IRR3 (dB)	41.43		
SFDR2 (dB)	61.34	SFDR3 (dB)	81.79	Gain, linear (dB)	21.00
AGC Controlled range:		Min input (dBm)	N/A	Max input (dBm)	N/A

FIGURE 7.21   Front end module design for low noise amplifier noise figure = 6 dB.

System1	Limiter	LMA406 filtronic	Attenuator	LMA406 filtronic	SPDT	Attenuator	LMA406 filtronic	
								Total
NF (dB)	3.00	5.50	3.00	5.50	3.00	1.00	5.50	9.25
Gain (dB)	−3.00	10.50	−3.00	10.50	−3.00	−1.00	10.00	21.00
OIP3 (dBm)	30.00	25.00	30.00	25.00	30.00	30.00	25.00	23.61

Input Pwr (dBm)	−60.00	System temp (K)	290.00	IM offset (MHz)	.025
OIP2 (dBm)	24.04	OIP3 (dBm)	23.61	Output P1dB (dBm)	11.74
IIP2 (dBm)	3.04	IIP3 (dBm)	2.61	Input P1dB (dBm)	−8.26
OIM2 (dBm)	−102.04	OIM3 (dBm)	−164.23	Compressed (dB)	0.00
ORR2 (dB)	63.04	ORR3 (dB)	125.23	Gain, actual (dB)	21.00
IRR2 (dB)	31.52	IRR3 (dB)	41.74		
SFDR2 (dB)	61.90	SFDR3 (dB)	82.24	Gain, linear (dB)	21.00
AGC Controlled range:		Min input (dBm)	N/A	Max input (dBm)	N/A

FIGURE 7.22 Front end module design for low noise amplifier noise figure = 5.5 dB.

TABLE 7.10
Front end module voltage and current consumption

Voltage (v)	3	5	−12	−5	5 Digital
Current (A)	0.25	0.15	0.1	0.1	0.1

The MMIC amplifiers and the SPDT are glued to the surface of the mechanical box. The MMIC chips are assembled on covar carriers. During development it was found that the spacing between the front end carriers should be less than 0.03 mm in order to achieve flatness requirements and voltage standing wave ratio (VSWR) better than 2:1.

The front end voltage and current consumption are listed in Table 7.10. The front end module has a high gain and low gain channels. The gain difference between high gain and low gain channels is around 15–20 dB. Measured front end gain is presented in Figure 7.23. The front end gain is around 20 ± 4 dB for the 18–40 GHz frequency range.

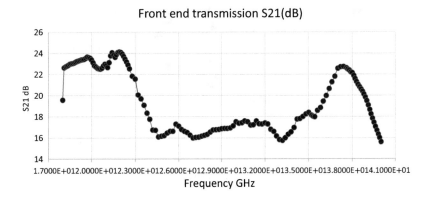

FIGURE 7.23 Measured front end gain.

7.5.3 High Gain Front End Module

To achieve a high gain front end module a medium power Hittite MMIC amplifier, HMC283, was added to the front end module presented in Figure 7.24. The HMC283 gain is around 21 dB with 10 dB N.F and 21 dBm saturated output power. The amplifier dimensions are 1.72 × 0.9 mm. The high gain front end module block diagram is shown in Figures 7.24 and 7.25. The front end module has a high gain and low gain channels. The gain difference between high gain and low gain channels presented in

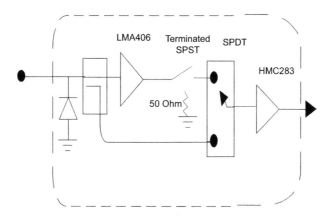

FIGURE 7.24   High gain front end block diagram.

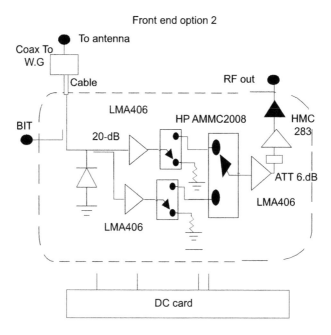

FIGURE 7.25   High gain front end block diagram with amplifier in the low gain channel.

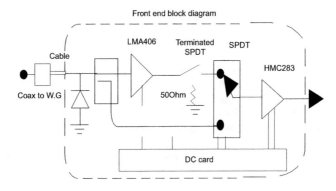

FIGURE 7.26 Detailed block diagram high gain front end.

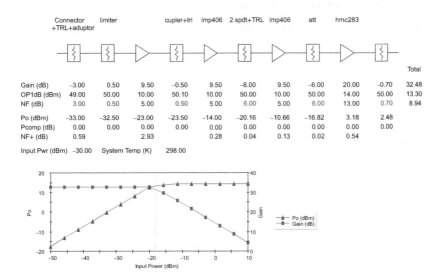

	Connector +TRL+aduptor	limiter		cupler+trl	lmp406	2 spdt+TRL	lmp406	att	hmc283		Total
Gain (dB)	-3.00	0.50	9.50	-0.50	9.50	-6.00	9.50	-6.00	20.00	-0.70	32.48
OP1dB (dBm)	49.00	50.00	10.00	50.10	10.00	50.00	10.00	50.00	14.00	50.00	13.30
NF (dB)	3.00	0.50	5.00	0.50	5.00	6.00	5.00	6.00	13.00	0.70	8.94
Po (dBm)	-33.00	-32.50	-23.00	-23.50	-14.00	-20.16	-10.66	-16.82	3.18	2.48	
Pcomp (dB)	0.00	0.00	0.00	0.00	0.00	0.00	0.00	0.00	0.00	0.00	
NF+ (dB)	0.59		2.93		0.28		0.04	0.13	0.02	0.54	

Input Pwr (dBm) -30.00   System Temp (K) 298.00

FIGURE 7.27 Front end module design for low noise amplifier noise figure = 9.5 dB.

Figure 7.25 is around 15–20 dB. The gain difference between high gain and low gain channels presented in Figure 7.26 is around 10–15 dB. A detailed block diagram of high gain module is shown in Figure 7.27.

7.5.4 HIGH GAIN FRONT END DESIGN

The high front end electrical characteristics were evaluated by using ADS Keysight software and by using SYSCAL software. Figure 7.27 presents the front end module N.F and gain for LNA N.F of 9.5 dB. The overall computed module N.F is 13.3 dB. The module gain is 32.48 dB. Figure 7.28 presents the front end module N.F and gain for LNA N.F of 5 dB. The overall computed module N.F is 10 dB. The module gain is 29.5 dB.

						Total
NF (dB)	3.50	5.00	5.00	5.00	6.00	9.00	10.02
Gain (dB)	−3.50	11.00	−5.00	11.00	−6.00	22.00	29.50
OIP3 (dBm)		20.00		20.00		20.00	19.87
NF+ (dB)		2.85		0.56		0.57	
Po (dBm)	−63.50	−52.50	−57.50	−46.50	−52.50	−30.50	
Input Pwr (dBm)	−60.00	System Temp (K)		290.00			

FIGURE 7.28 Front end module design for low noise amplifier noise figure = 5 dB.

TABLE 7.11
Measured results of front end modules

Parameter	DF6	DF4	DF3	DF2	DF1	OMNI02	OMNI01
High gain max	31	32	32.5	32.5	31	31.5	32
High gain min	26	26	27.5	27	26	28.5	28
High gain avg	29	29	29	29	29	30	30
Amp. bal.	5	6	5	5	4	3	4
S_{11} (dB)	4.5	5	5	5	5	5	4
S_{22} dB)	7.5	6	5	6	5	7	6
Isolation (dB)	9	9	10	10	6.5	21.5	22.5
Low gain max	19	18	17	17	17	16.5	18
Low gain min	13	10	7.5	12	12	10.5	11
Low gain avg	15	14	12	14	14	13.5	14.5
Amp. bal.	6	8	9.5	5	5	6	7
P1dB 30 GHz	11.6	11.93	11.7	11.4	10.9	14	15.96
P1 dB 40 GHz	13.96	14.5	15.58	15.28	14	14.48	16.8
NF 30 GHz	8.68	9.48	8.65	8.45	10.5	8.14	8.75
NF 40 GHz	9.28	10.1	8.64	9.17	10.24		8.75

Measured results of front end modules are listed in Table 7.11. HMC283 assembly is shown in Figure 7.29. A photo of the front end is shown in Figure 7.30. There is a good agreement between computed and measured results.

A photo of the compact wideband 18 to 40 GHz RF modules is shown in Figure 7.31.

7.6 MEMS TECHNOLOGY

MEMS is the integration of mechanical elements, sensors, actuators and electronics on a common silicon substrate through micro-fabrication technology. These devices replace bulky actuators and sensors with micron-scale equivalent that can be produced in large quantities by fabrication process used in ICs in photolithography. They reduce cost, bulk, weight and power consumption while increasing performance, production volume and functionality by orders of magnitude. The physical

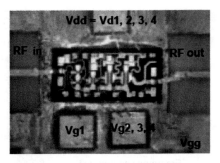

FIGURE 7.29   HMC283 assembly.

FIGURE 7.30   18–40 GHz front end module.

dimensions of MEMS devices can vary from around one micron on the lower end and all the way to several millimeters. The types of MEMS devices can vary from relatively simple structures having no moving elements, to extremely complex electromechanical systems with multiple moving elements under the control of integrated microelectronics.

Over the past 20 years MEMS researchers and developers have fabricated an extremely large number of microsensors for almost every possible sensing function. Such as temperature sensing, pressure, inertial forces, chemical species, magnetic fields and radiation detection. Many of these micromachined sensors have demonstrated performances exceeding those of conventional sensors and actuators.

The electronic modules are fabricated using IC process sequences, for example, CMOS, bipolar or bipolar CMOS (BiCMOS) processes; the micromechanical

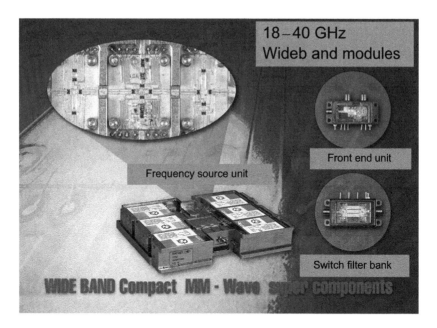

FIGURE 7.31 Photo of 18–40 GHz compact modules.

components are fabricated using compatible "micromachining" processes that selectively etch away parts of the silicon wafer or add new structural layers to form the mechanical and electromechanical devices.

The real potential of MEMS may be fulfilled when these miniaturized sensors, actuators and other components can all be merged onto a common silicon substrate along with ICs and microelectronic ICs. The electronic components are fabricated using IC process sequences (such as CMOS, bipolar, or BiCMOS processes). Usually, micromechanical components are fabricated using compatible "micromachining" processes that selectively etch away parts of the semiconductor wafer or add new structural layers and components to form the mechanical and electromechanical, MEMS modules.

7.6.1 MEMS Technology Advantages

- Low insertion loss <0.1 dB.
- High isolation >50 dB.
- Low distortion.
- High linearity.
- Very high Q.
- Size reduction, system-on-a-chip.
- High power handling ~40 dBm.
- Low power consumption (~mW and no LNA).
- Low-cost high-volume fabrication.

7.6.2 MEMS TECHNOLOGY PROCESS

Bulk micromachining fabricates mechanical structures in the substrate by using orientation-dependent etching. A bulk micromachined substrate is shown in Figure 7.32.

  Surface micromachining fabricates mechanical structures above the substrate surface by using sacrificial layer. A surface micromachined substrate is presented in Figure 7.33.

  In bulk micromachining process silicon is machined using various etching processes. Surface micromachining uses layers deposited on the surface of a substrate as the structural materials rather than using the substrate itself. The surface micromachining technique is relatively independent of the substrate used, and therefore can be easily mixed with other fabrication techniques which modify the substrate first. An example is the fabrication of MEMS on a substrate with embedded control

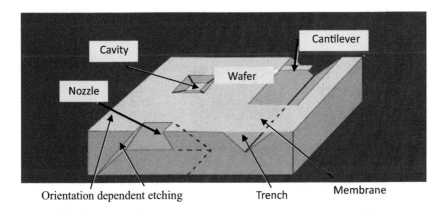

FIGURE 7.32 Bulk micromachining.

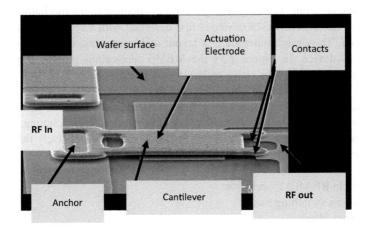

FIGURE 7.33 Surface micromachining.

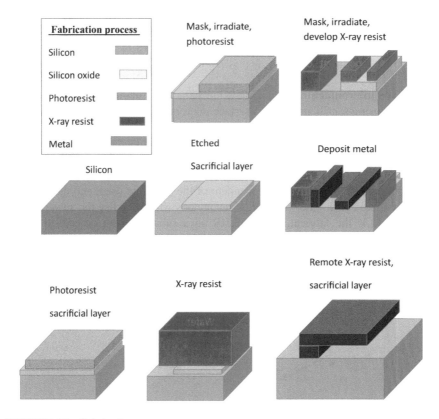

FIGURE 7.34   Fabrication process.

circuitry, in which MEMS technology is integrated with IC technology. This is being used to produce a wide variety of MEMS devices for many different applications. On the other hand, bulk micromachining is a subtractive fabrication technique, which converts the substrate, typically a single-crystal silicon, into the mechanical parts of the MEMS device. A MEMS device is first designed with a computer-aided design (CAD) tool. The design outcome is a layout and masks that are used to fabricate the MEMS device. In Figure 7.34 the MEMS fabrication process is presented. A summary of MEMS fabrication technology is listed in Table 7.12. In Figure 7.35 the block diagram of a MEMS bolometer coupled antenna array is presented.

Packaging of the device tends to be more difficult, but structures with increased heights are easier to fabricate when compared to surface micromachining.

This is because the substrates can be thicker resulting in relatively thick unsupported devices. Applications of RF MEMS technology are:

- tunable RF MEMS inductor
- low loss switching matrix
- tunable filters
- bolometer coupled antenna array
- low-cost W band detection array.

TABLE 7.12
Fabrication technology

Fabrication technology	Process
Surface micromachining	Release and drying systems to realize free-standing microstructures
Bulk micromachining	Dry etching systems to produce deep 2D free-form geometries with vertical sidewalls in substrates. Anisotropic wet etching systems with protection for wafer front sides during etching. Bonding and aligning systems to join wafers and perform photolithography on the stacked substrates.

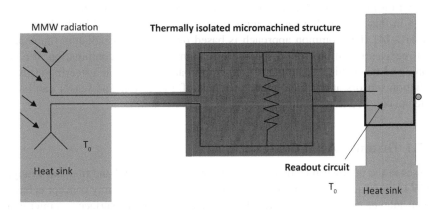

FIGURE 7.35   Bolometer coupled antenna array.

7.6.3 MEMS COMPONENTS

MEMS components are categorized in one of several applications. Such as:

1. **Sensors** are a class of MEMS that are designed to sense changes and interact with their environments. These classes of MEMS include chemical, motion, inertia, thermal, RF sensors and optical sensors. Microsensors are useful because of their small physical size, which allows them to be less invasive.
2. **Actuators** are a group of devices designed to provide power or stimulus to other components or MEMS devices. MEMS actuators are either electrostatically or thermally driven.
3. **RF MEMS** are a class of devices used to switch or transmit high frequency, RF signals. Typical devices include metal contact switches, shunt switches, tunable capacitors and antennas.
4. **Optical MEMS** are devices designed to direct, reflect, filter and/or amplify light. These components include optical switches and reflectors.
5. **Microfluidic MEMS** are devices designed to interact with fluid-based environments. Devices such as pumps and valves have been designed to move, eject and mix small volumes of fluid.

6. **Bio MEMS** are devices that, much like microfluidic MEMS, are designed to interact specifically with biological samples. Devices such as these are designed to interact with proteins, biological cells, medical reagents and so on and can be used for drug delivery or other in situ medical analysis.

7.7 W BAND MEMS DETECTION ARRAY

In this section we present the development of millimeter wave radiation detection array. The detection array may employ around 256–1024 patch antennas. These patches are coupled to a resistor. Optimization of the antenna structure, feed network dimensions and resistor structure allow us to maximize the power rate dissipated on the resistor. Design considerations of the detection antenna array are given in this section. Several imaging approaches are presented in the literature, see [10–14]. The common approach is based on an array of radiators (antennas) that receives radiation from a specific direction by using a combination of electronic and mechanical scanning. Another approach is based on a steering array of radiation sensors at the focal plane of a lens of a reflector. The sensor can be an antenna coupled to a resistor.

7.7.1 DETECTION ARRAY CONCEPT

Losses in the microstrip feed network are very high in the W band frequency range. In W band frequencies we may design a detection array. The array concept is based on an antenna coupled to a resistor. A direct antenna-coupling surface to a micromachined micro bridge resistor is used for heating and sensing. The feed network determines the antenna efficiency. The insertion loss of a gold microstrip line with width of 1 μm and 188 μm length is 4.4 dB at 95 GHz. The insertion loss of a gold microstrip line with width of 10 μm and 188 μm length is 3.6 dB at 95 GHz. The insertion loss of a gold microstrip line with width of 20 μm and 188 μm length is 3.2 dB at 95 GHz. To minimize losses the feed line dimensions were selected as 60 × 10 × 1 μm. An analog CMOS readout circuit may be employed as a sensing channel per pixel. Figure 7.36 presents a pixel block diagram.

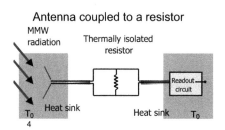

FIGURE 7.36   Antenna coupled to a resistor.

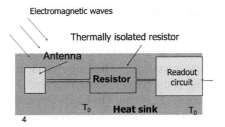

FIGURE 7.37   A single array pixel.

7.7.2 THE ARRAY PRINCIPLE OF OPERATION

The antenna receives effective mm-wave radiation. The radiation power is transmitted to a thermally isolated resistor coupled to a titanium resistor. The electrical power raises the structure temperature with a short response time. The same resistor changes its temperature and therefore its electrical resistance. Figure 7.37 shows a single array pixel. The pixel consists of a patch antenna, a matching network, printed resistor and DC pads. The printed resistor consists of titanium lines and a titanium resistor coupled to an isolated resistor.

The operating frequency range of 92 to 100 GHz is the best choice. In the frequency range of 30–150 GHz there is a proven contrast between land, sky and high transmittance of clothes. Size and resolution considerations promote higher frequencies above 100 GHz. Typical penetration of clothing at 100 GHz is 1 dB and 5–10 dB at 1 THz. Characterization and measurement considerations promote lower frequencies. The frequency range of 100 GHz allows sufficient bandwidth when working with illumination. The frequency range of 100 GHz is the best compromise. Figure 7.38 presents the array concept. Several types of printed antennas may be used as the array element such as bowtie dipole, patch antenna and ring resonant slot.

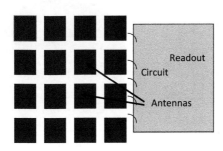

FIGURE 7.38   Array concept.

7.7.3 W Band Antenna Design

The bowtie dipole and a patch antenna have been considered the array element. Computed results show that the directivity of the bowtie dipole is around 5.3 dBi and the directivity of a patch antenna is around 4.8 dBi. However, the length of the bowtie dipole is around 1.5 mm; the size of the patch antenna is around 700 ×700 µm. We used a quartz substrate with thickness of 250 µm. The bandwidth of the bowtie dipole is wider than that of a patch antenna. However, the patch antenna bandwidth meets the detection array's electrical specifications. We chose the patch antenna as the array element since the patch size is significantly smaller than that of the bowtie dipole. This feature allows us to design an array with a higher number of radiating elements. The resolution of a detection array with a higher number of radiating elements is improved. We also realized that the matching network between the antenna and the resistor has a smaller size for a patch antenna than that for a bowtie dipole. The matching network between the antenna and the resistor consists of microstrip open stubs. Figure 7.39 shows the 3D radiation pattern of the bowtie dipole. Figure 7.40 presents S11 parameter of the patch antenna. The electrical performance of the bowtie dipole and the patch antenna was compared. The VSWR of the patch antenna is better than 2:1 for 10% bandwidth. Figure 7.41 presents the 3D radiation pattern of the patch antenna at 95 GHz.

7.7.4 Resistor Design

As described in [6], the resistor is thermally isolated from the patch antenna by using a sacrificial layer. Optimizations of the resistor structure maximize the power rate dissipated on the resistor. Ansys High Frequency Structure Simulator (HFSS)

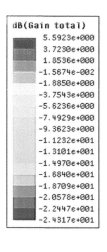

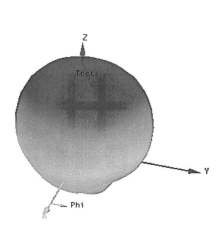

FIGURE 7.39   Dipole three-dimensional radiation pattern.

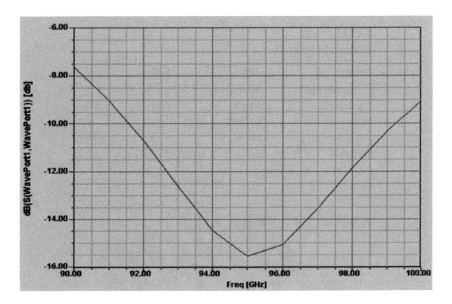

FIGURE 7.40   Patch S_{11} computed results.

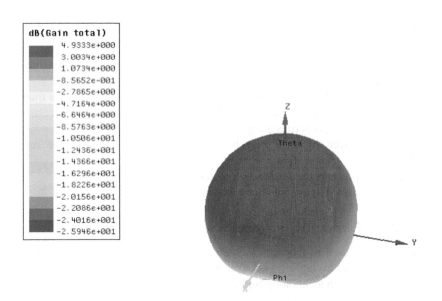

FIGURE 7.41   Patch three-dimensional radiation pattern.

TABLE 7.13
Material properties

Property	Units	SiNi	Ti
Conductivity [K]	W/m/K	1.6	7
Capacity [C]	J/kg/K	770	520
Density [ρ]	Gr/cm³	2.85	4.5
Resistance	Ω/cm²	>1e8	90
Thickness	μm	0.1	0.1

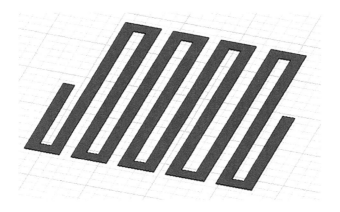

FIGURE 7.42   Resistor configuration.

software is employed to optimize the height of the sacrificial layer, the transmission line width and length. Dissipated power on a titanium resistor is higher than the dissipated power on a platinum resistor. The rate of the dissipated power on a titanium resistor is around 25%. The rate of the dissipated power on a platinum resistor is around 4%. Material properties are given in Table 7.13.

The sacrificial layer thickness may be 2–3 μm. Figure 7.42 shows the resistor configuration.

Figure 7.43 presents the MEMS bolometer layout.

The patch coupled to a bolometer is shown in Figure 7.44.

7.7.5 Array Fabrication and Measurement

Nine masks are used to fabricate the detection array. Masks' processes and layer thickness are listed in Table 7.14. Layer thickness has been determined as the best compromise between technology limits and design considerations.

The dimensions of detection array elements were measured in several array pixels as part of a visual test of the array after fabrication of the array; some of the measured results are listed in Table 7.15. From results listed in Table 7.15 we may conclude that the fabrication process is very accurate. There is a good agreement between computed and measured results of the array's electrical and mechanical parameters.

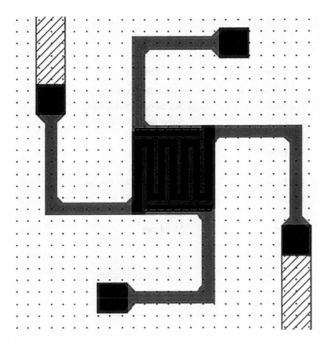

FIGURE 7.43   Micro-electro-mechanical systems bolometer layout.

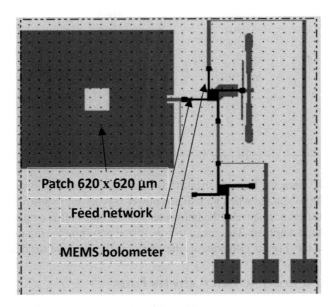

FIGURE 7.44   Patch antenna coupled to bolometer.

TABLE 7.14
Masks' processes

Masks	Layer	Process	Layer thickness (μm)
1	L 1 Lift or Etch	Gold reflector Au	1
2	L 2 Etch	Streets open Street layer	3
3	L 3 Etch	Street layer contacts	
4	L 4 Etch	SiN + Contacts	0.1
5	L 5 Etch	Ti_1	0.1
		SiN	0.15
6	L 6 Etch	Vanadium oxide (Vox)	0.1
7	L 7 Etch	Contacts for Ti_2	
		Ti_2	0.1
	L 3 Lift	Metal cap	0.1–0.5
8	L 8 Etch	Ti_2	
		SiN	0.1
9	L 9 Etch	Membrane definition	

TABLE 7.15
Comparison of design and fabricated array dimensions

Element	Design (μm)	Pixel 1 (μm)	Pixel 2 (μm)
Patch width	600	599.5	600.5
Patch length	600	600.3	600.5
Hole width	100	99.8	100
Hole length	100	100	99.8
Feed line	10	10	10
Feed line	10	9.8	10
Stub width	2	2	1.8
Tapered line	15	15.2	14.8
Stub width	2	1.8	2
Tapered line	25	25.3	25.2

Figure 7.45 presents how we measure the bolometer output voltage. Vref is the bolometer output voltage when no radiated power is received by the detection array. The voltage difference between the bolometer voltage and Vref is amplified by a low noise differential amplifier. The rate of the dissipated power on the titanium resistor is around 25%–30%. Figure 7.46 presents the operational concept of the detection array.

7.7.6 Mutual Coupling Effects between Pixels

HFSS software has been used to compute mutual coupling effects between pixels in the detection array as shown in Figure 7.47. Computation results indicate that the power dissipated on the centered pixels in the array is higher by 1% to 2% than the pixels located at the corners of the array.

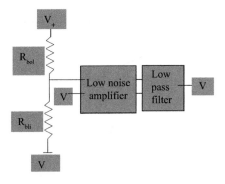

FIGURE 7.45   Measurements of bolometer voltage.

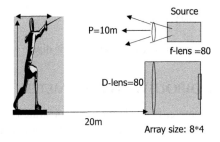

FIGURE 7.46   Detection array.

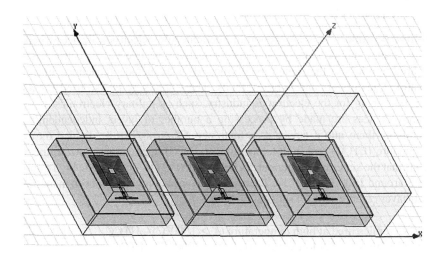

FIGURE 7.47   Computation of mutual coupling between pixels.

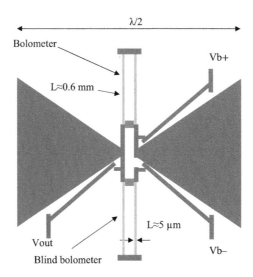

FIGURE 7.48 Micro-electro-mechanical systems bowtie dipole with bolometer.

7.8 MEMS BOWTIE DIPOLE WITH BOLOMETER

A bowtie dipole with bolometer printed on quartz substrate is shown in Figure 7.48. The length of the bowtie dipole is around 1.5 mm. The bolometer length is 0.6 mm. Fig. 7.49 presents the MEMS bowtie dipole S_{11} computed results. The bowtie dipole 3D radiation pattern is shown in Fig. 7.50.

7.9 LTCC AND HIGH-TEMPERATURE CO-FIRED CERAMIC (HTCC) TECHNOLOGY

Co-fired ceramic devices are monolithic, ceramic microelectronic devices where the entire ceramic support structure and any conductive, resistive and dielectric materials are fired in a kiln at the same time. Typical devices include capacitors, inductors, resistors, transformers and hybrid circuits. The technology is also used for multi-layer packaging for the electronics industry, such as military electronics. Co-fired ceramic devices are made by processing a number of layers independently and assembling them into a device as a final step. Co-firing can be divided into low-temperature (LTCC) and high-temperature (HTCC) applications: low temperature means that the sintering temperature is below 1000°C (1830°F), while high tempera-ture is around 1600°C (2910°F). There are two types of raw ceramics to manufacture multi-layer ceramic (MLC) substrate:

- Ceramics fired at high temperature (T ≥ 1500°C) – HTCC,
- Ceramics fired at low temperature (T ≤ 1000°C) – LTCC.

The base material of HTCC is usually Al_2O_3. HTCC substrates are row ceramic sheets. Because of the high firing temperature of Al_2O_3 the material of the embedded

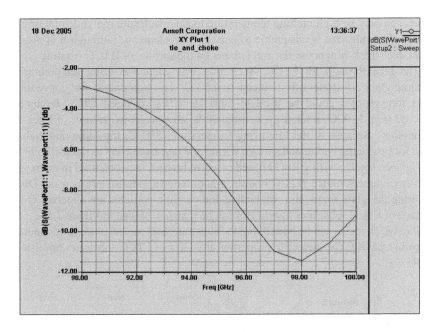

FIGURE 7.49 Micro-electro-mechanical systems bowtie dipole S_{11} computed results.

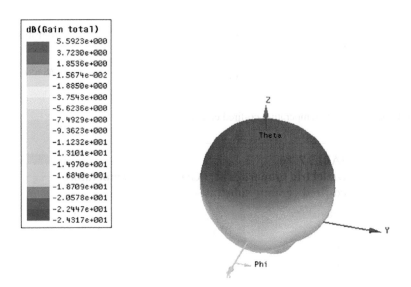

FIGURE 7.50 Bowtie dipole three-dimensional radiation pattern.

layers can only be high melting temperature metals: wolfram, molybdenum or manganese. The substrate is unsuitable to bury passive elements, although it is possible to produce thick-film networks and circuits on the surface of HTCC ceramic.

The breakthrough for LTCC fabrication was when the firing temperature of ceramic-glass substrate was reduced to 850°C. The equipment for conventional thick-film process could be used to fabricate LTCC devices. LTCC technology evolved from HTCC technology, combined the advantageous features of thick-film technology. Because of the low firing temperature (850°C) the same materials are used for producing buried and surface wiring and resistive layers as thick-film hybrid IC (i.e. Au, Ag, Cu wiring RuO2 based resistive layers). It can be fired in an oxygen-rich environment unlike HTCC boards, where reduced atmosphere is used. During co-firing the glass melts, the conductive and ceramic particles are sintered. On the surface of LTCC substrates hybrid ICs can be realized, shown in Figure 7.51.

Passive elements can be buried into the substrate, and we can place semiconductor chips in a cavity. The dielectric properties at 9 GHz of LTCC substrates are listed in Table 7.16.

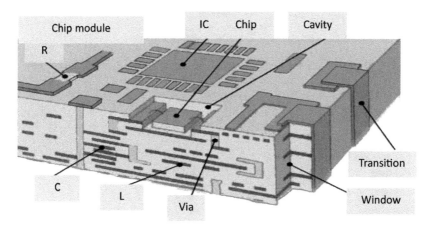

FIGURE 7.51   Low-temperature co-fired ceramic module.

TABLE 7.16
Dielectric properties at 9 GHz of low-temperature co-fired ceramic substrates

Material	ε_r	Tan δ × 10^{-3}
99.5% AL	9.98	0.1
LTCC1	7.33	3.0
LTCC2	6.27	0.4
LTCC3	7.2	0.6
LTCC4	7.44	1.2
LTCC5	6.84	1.3
LTCC6	8.89	1.4

7.9.1 LTCC AND HTCC TECHNOLOGY PROCESS

- Low temperature   LTCC 875°C.
- High temperature   HTCC 1400–1600°C.
- Co-fired   Co-firing of (di)electric pastes
  LTCC – precious metals (Au, Ag, Pd, Cu)
  HTCC – refractory metals (W, Mo, MoMn).
- Ceramic   Mix of alumina Al_2O_3
  Glasses SiO_2 - B_2O_3 - CaO - MgO
  Organic binders.
  HTTC – essentially Al_2O_3.

7.9.2 ADVANTAGES OF LTCC

- low permittivity tolerance
- good thermal conductivity
- low thermal coefficient of expansion (TCE) (adapted to silicon and GaAs)
- excellently suited for multi-layer modules
- integration of cavities and passive elements such as R, L, and C-components
- very robust against mechanical and thermal stress (hermetically sealed)
- composable with fluidic, chemical, thermal and mechanical functionalities
- low material costs for silver conductor paths
- low production costs for medium and large quantities.

Advantages for high frequency applications:

- parallel processing (high yield, fast turnaround, lower cost)
- precisely defined parameters
- high performance conductors
- potential for multi-layer structures
- high interconnect density.

In Table 7.17 LTCC process steps are listed. LTCC raw material comes as sheets or rolls. Material manufacturers are DuPont, ESL, Ferro and Heraeus. In Table 7.18, several electrical, thermal and mechanical characteristics of several LTCC materials are listed. In Figure 7.52 a LTCC process block diagram is presented.

In Table 7.19 LTCC line losses at 2 GHz are listed for several LTCC materials. For LTTC1 material losses are 0.004 dB/mm.

7.9.3 DESIGN OF HIGH PASS LTCC FILTERS

The trend in the wireless industry toward miniaturization, cost reduction and improved performance drives microwave designers to develop microwave components in LTCC technology. A significant reduction in the size and the cost of microwave components may be achieved by using LTCC technology. In LTCC technology, discrete surface-mounted components such as capacitors and inductors are replaced by integrated printed components. LTCC technology allows the designer to use multi-layer design

TABLE 7.17
Low-temperature co-fired ceramic process list

Process Number	LTCC process
1	Tape casting
2	Sheet cutting
3	Laser punching
4	Printing
5	Cavity punching
6	Stacking
7	Bottom side printing
8	Pressing
9	Side hole formation
10	Side hole printing
11	Snap line formation
12	Pallet firing
13	Plating Ni-Au

TABLE 7.18
Low-temperature co-fired ceramic material characteristics

Material	LTCC DP951	Al2O3 96%	BeO	AIN 98%
Electrical characteristics at 10 MHz				
Dielectric constant, εr	7.8	9.6	6.5	8.6
Dissipation factor, $\tan\delta$	0.00015	0.0003	0.0002	0.0005
Thermal characteristics				
Thermal expansion $10^{-6}/°C$	5.8	7.1	7.5	4.6
Thermal conductivity W/mk 25°C–300°C	3	20.9	251	180
Mechanical characteristics				
Density	3.1	3.8	2.8	3.3
Flexural strength, MPa	320	274	241	340
Young's modulus, GPa	120	314	343	340

if needed to reduce the size and cost of the circuit. However, multi-layer design results in more losses due to via connections and due to parasitic coupling between different parts of the circuit. To improve the filter performance, all the filter parameters have been optimized. Package effects were taken into account in the design.

7.9.3.1 High Pass Filter Specification

Frequency 1.5–2.5 GHz.
Insertion loss 1.1 Fo – 1 dB.
Rejection 0.9 Fo – 3 dB.
Rejection 0.75 Fo – 20 dB.

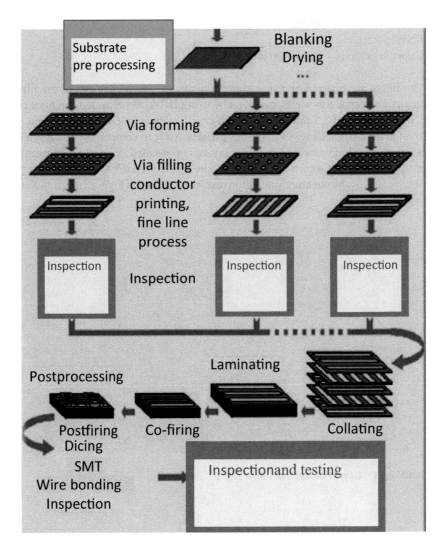

FIGURE 7.52 Low-temperature co-fired ceramic process.

TABLE 7.19
Low-temperature co-fired ceramic line loss

Material	Dissipation factor, tanδ × 10^{-3}	Line loss dB/mm, at 2 GHz
LTTC1	3.8	0.004
LTTC2	2.0	0.0035
LTTC6-CT2000	1.7	0.0033
Alumina 99.5%	0.65	0.003

Rejection 0.5 Fo – 40 dB.
VSWR – 2:1.
Case dimensions – 700 × 300 × 25.5 mil inch.

The filters are realized by using lumped elements. The parameters of the filter's inductors and capacitors were optimized by using HP ADS software. The filter consists of five layers of 5.1 mil substrate with εr = 7.8. Package effects were taken into account in the design. Changes in the design were made to compensate and minimize package effects. In Figure 7.53 the filter layout is presented. S_{11} and S_{12} Momentum simulation results are shown in Figure 7.54. In Figure 7.55 the filter 2 layout is presented. S_{11} and S_{12} Momentum simulation results are shown in Figure 7.56. Simulation

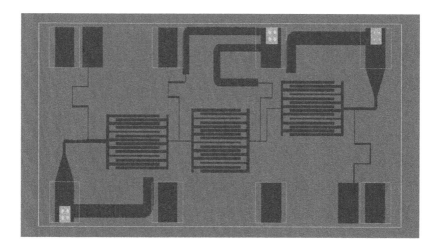

FIGURE 7.53   Layout of high pass filter No. 1.

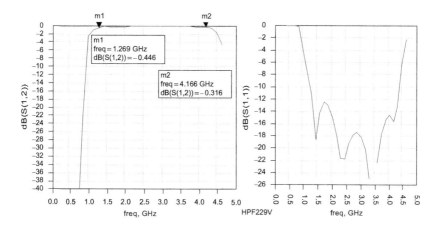

FIGURE 7.54 S_{12} and S_{11} results of high pass filter No. 1.

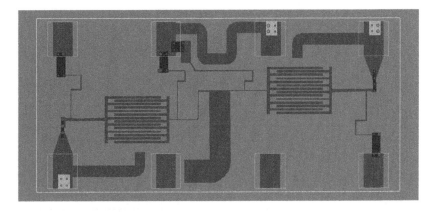

FIGURE 7.55   Layout of high pass filter No. 2.

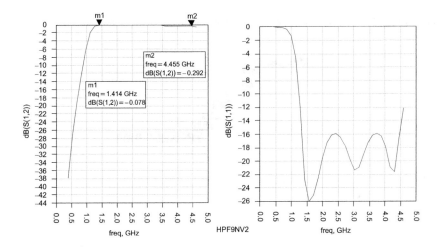

FIGURE 7.56 S_{12} and S_{11} results of high pass filter No. 2.

results of tolerance check are shown in Figure 7.57. The parameters that were tested in the tolerance check are: inductor and capacitors line width and length and spacing between capacitor fingers.

7.10 COMPARISON OF SINGLE-LAYER AND MULTI-LAYER PRINTED CIRCUITS

In a single-layer microstrip circuit all conductors are in a single layer. Coupling between conductors is achieved through edge or end proximity (across narrow gaps). Single-layer microstrip circuits are cheap in production. In Figure 7.58 a single-layer microstrip edge coupled filter is shown.

Figure 7.59 presents the layout of single-layer microstrip directional coupler. Figure 7.60 presents the structure of a multi-layer microstrip coupler.

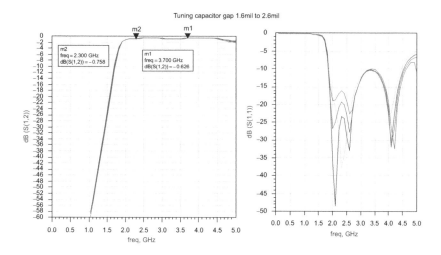

FIGURE 7.57   Tolerance simulation for spacing between capacitor fingers.

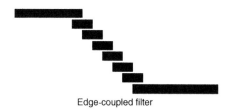

Edge-coupled filter

FIGURE 7.58   Edge coupled filter.

FIGURE 7.59   Single-layer microstrip directional coupler.

In multi-layer microwave circuits, conductors are separated by dielectric layers and stacked on different layers. This structure allows for (strong) broadside coupling. Registration between layers is not difficult to achieve as there are narrow gaps between strips in single-layer circuits. Multi-layer structure technique is well-suited to thick-film print technology and is suitable for LTCC technology.

7.11 A COMPACT INTEGRATED TRANSCEIVER

This section describes the design, performance and fabrication of a compact and low-cost integrated transceiver. Surface-mount MIC technology is employed to fabricate a RF-Head.

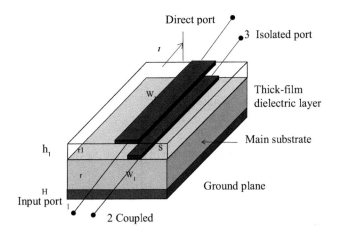

FIGURE 7.60   Multi-layer microstrip coupler.

7.11.1 INTRODUCTION

The mobile telecommunication industry is currently in continuous growth. Moreover, the great public demand for cellular and cordless telephones has stimulated a wide interest in new mobile services, such as portable satellite communication terminals.

Communication links to and from mobile installations are established via geostationary satellite and the associated ground station.

The RF-Head includes receiving and transmitting channels, RF controller, synthesizers, modem and a DC supply unit. The RF-Head size is $30 \times 20 \times 2.5$ cm and weighs 1 kg. The transmitting channel may be operated in high power mode to transmit 10 W or in low power mode to transmit 4 W. The transmitted power level is controlled by an automatic leveling control unit to ensure low power consumption over all the frequency and temperature ranges. The RF-Head vent is set to On and Off automatically by the RF controller. The gain of the receiving channel is 76 dB and is temperature compensated by using a temperature sensor and a voltage-controlled attenuator. Surface-mount technology is employed to fabricate the RF-Head.

7.11.2 DESCRIPTION OF THE RECEIVING CHANNEL

A block diagram of the receiving channel is shown in Figure 7.61.

The receiving channel consists of LNAs, filters, active mixer, saw filter, temperature sensor, voltage-controlled attenuator and IF amplifiers.

The LNA has 10 dB gain and 0.9 dB N.F for frequencies ranging from 1.525 to 1.559 GHz. The total gain of the receiving channel is 76 dB.

The LNA employs a $1.8 GaAs FET. The receiving channel N.F and power budget calculation are given in Table 7.20. The channel gain is temperature compensated by connecting the output port of a temperature sensor to the reference voltage port of a voltage-controlled attenuator. The variation of the sensor output voltage as a

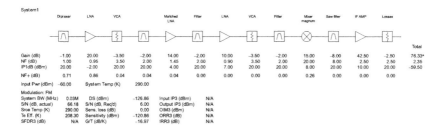

FIGURE 7.61 Block diagram and gain budget of the receiving channel.

TABLE 7.20
Noise filter and gain calculation

Component	Noise figure (dB)	Gain (dB)	Pout (dBm)
Diplexer	1	−1	−131
LNA	0.95	20	−111
Voltage-controlled amplifier	3.5	−3.5	−114.5
Filter	2	−2	−116.5
Matched LNA unit	1.45	14	−102.5
Filter	2	−2	−104.5
LNA	0.9	10	−94.5
Voltage-controlled amplifier	3.5	−3.5	−98
Filter	2	−2	−100
Mixer	20	15	−85
Saw filter	8	−8	−93
IF amplifiers	2.5	42.5	−50.5
Tr. line losses	2.5	−2.5	−53
Total	2.35	77	−53

function of temperature varies the attenuation level of the attenuator. Gain stability as a function of temperature is less than 1 dB. The receiving channel N.F is less than 2.2 dB, including 1 dB diplexer losses. The receiving channel rejects out-of-band signals with power level lower than −20 dBm.

7.11.2.1 Receiving Channel Specifications
Frequency range is 1525–1559 MHz.
LO frequency is 1355–1389 MHz.
IF frequency is 170 MHz.
Channel 1 dB bandwidth is 50 KHz.
N.F is 3 dB.
Input signals are −105 to 135 dBm.
Output signals are −30 to 60 dBm.
Receiving sensitivity – carrier/noise (C/N) = 41 dB/Hz for −130 dBm input signal.

7.11.3 Receiving Channel Design and Fabrication

A major parameter in the receiving channel design was to achieve a very low target price in manufacturing hundreds of units. The components were selected to meet the electrical requirements for a given low target price assigned to each component. Low production cost of the RF-Head is achieved by using surface mount (SMT) technology to manufacture the RF-Head. Trimming is not required in the fabrication procedure of the receiving channel. The gain and N.F values of the receiving channel are measured in each RF-Head.

7.11.4 Description of the Transmitting Channel

A block diagram of the transmitting channel is shown in Figure 7.62. The transmitting channel consists of low power amplifiers, pass band filters, voltage-controlled attenuator, active mixer, medium power and high-power amplifiers, high power isolator, coupler, power detector and DC supply unit.

The power budget of the transmitting channel is given in Table 7.21.

Five stages of power amplifiers amplify the input signal from 0 dBm to 40 dBm. The fifth stage is a 10 W power amplifier with high efficiency. The amplifier may transmit 10 W in high power mode or 4 W in low power mode. The DC bias voltage of the power amplifier is automatically controlled by the RF controller to set the power amplifier to the required mode and power level. A −30 dB coupler and a power detector are used to measure the output power level. The measured power level is transferred via an A/D converter to the RF controller to monitor the output power level of the transmitting channel by varying the attenuation of the voltage-controlled attenuator. This feature ensures low DC power consumption and high efficiency of the RF-Head. The RF controller sets the transmitting channel to On and Off. A temperature sensor is used to measure the RF-Head temperature. The RF controller sets the RF-Head vent to On and Off according to the measured RF-Head temperature.

7.11.4.1 Transmitting Channel Specifications

Frequency range is 1626.5–1660.5 MHz.
IF frequency range is 99.5–133.5 MHz.
LO frequency is 1760 MHz.

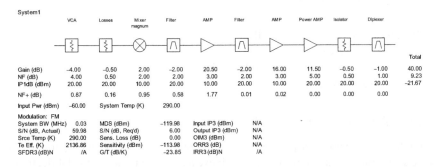

FIGURE 7.62   Block diagram and gain budget of the transmitting channel.

TABLE 7.21
Transmitting channel power budget

Component	Gain/Loss (dB)	Pout (dBm)
Input	0	0
Voltage-controlled amplifier	−4	−4
Tr. line loss	−0.5	−4.5
Mixer	2	−2.5
Filter	−2	−4.5
Low power amplifiers	20.5	16
Filter	−2	14
Medium power amplifier	16	30
Power amplifier	11.5	41.5
Isolator	−0.5	41
Diplexer	−1	40
Total	40.0	40

Input power is 0 dBm.
Output power (high mode) is 38–40 dBm.
Output power (low mode) is 34–36 dBm.
Power consumption is 42 W.

7.11.4.2 Diplexer Specifications

A very compact and light weight diplexer connects the receiving and transmitting channels to the antenna. The diplexer simple structure and easy manufacturability ensure lower costs in production than similar diplexers.

7.11.4.2.1 Transmit Filter

Pass band frequency range is 1626.5–1660.5 MHz.
Pass band insertion loss is 0.7 dB.
Pass band VSWR <1.3:1.
Rejection >54 dB at 1525–1559 MHz.

7.11.4.2.2 Receive Filter

Pass band frequency range is 1525–1559 MHz.
Pass band insertion loss < 1.3 dB.
Pass band VSWR <1.3:1.
Rejection >65 dB at 1626.5–1660.5 MHz.
Size is 86 × 36 × 25 mm.

7.11.5 TRANSMITTING CHANNEL FABRICATION

A photo of the transceiver prototype is shown in Figure 7.63. A major parameter in the transmitting channel design was to achieve a low target price in the fabrication of hundreds of units. The components were selected to meet the target price given to

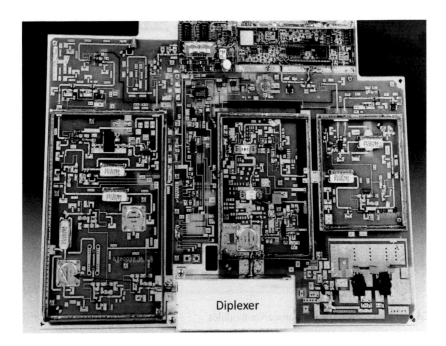

FIGURE 7.63   The transceiver prototype.

each component and the electrical requirements. A low production cost is achieved by using SMT technology to manufacture the transmitting channel. A quick trimming procedure is required in the fabrication of the transmitting channel to achieve the required output power and efficiency. The output power and spurious level of the transmitting channel are tested in the fabrication procedure of each RF-Head. Around 300 transmitting channels were manufactured during the first production cycle.

A photo of the transceiver modules is shown in Figure 7.64. The RF-Head is separated into five sections: receiving and transmitting channels, diplexer, synthesizers, RF controller and a DC supply unit. A metallic fence and cover separate the transmitting and receiving channels.

The RF-Head size is $30 \times 20 \times 2.5$ cm and it weighs less than 1 Kg.

7.11.6 RF Controller

The RF controller is based on an 87C51 microcontroller. The RF controller communicates with the system controller via a full duplex serial bus. The communication is based on message transfer. The RF controller sets the transmitting channel to On and Off by controlling the DC voltage switching unit. The RF controller monitors the output power level of the transmitting channel by varying the attenuation of the voltage-controlled attenuator in the transmitting channel. The RF controller sets the transmitting channel to burst or single channel per carrier (SCPC) modes with high or low power level. The RF controller produces the clock data and enable signals for the RX and TX synthesizers.

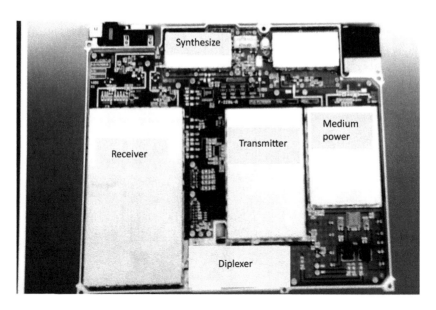

FIGURE 7.64   Photo of the transceiver modules.

7.12 CONCLUSIONS

Dimension and losses of microwave systems are minimized by using MICs, MMIC technology, MEMS and LTCC technology. Dimension and losses of microwave systems are minimized by using multi-layer structure technique. Multi-layer structure technique is well-suited to thick-film print technology and for LTCC technology. LTCC technology allows integration of cavities and passive elements such as R, L, and C-components as part of the LTCC circuits. Sensors, actuators and RF switches may be manufactured by using MEMS technology. Losses of MEMS components are considerably lower than MIC and MMIC RF components. MMICs are circuits in which active and passive elements are formed on the same dielectric substrate. MMICs are dimensionally small (from around 1–10 mm^2) and can be mass produced. MMIC components cannot be tuned. Accurate design is crucial in the design of MMICs. The goal of designers of MMICs, MEMS and LTCC is to comply with customer specifications in one design iteration.

A compact and low-cost transceiver was presented in this chapter. The transceiver supplies phone and fax services to the customer. The RF controller automatically monitors the output power level to ensure low DC power consumption. A DC to DC converter supplies to the power amplifier a controlled DC bias voltage to set the power amplifier to high power level mode, 10 W or 4 W.

The receiving channel N.F is less than 2.2 dB. The total gain of the receiving channel is 76 dB with gain stability of 1 dB as function of temperature.

REFERENCES

[1] Sabban, A., *Low Visibility Antennas for Communication Systems*, Taylor & Francis Group, New York, 2015.

[2] Rogers, J. and Plett, C., *Radio frequency Integrated Circuit Design*, Artech House, Boston USA, 2003.

[3] Sabban, A., *Wideband RF Technologies and Antenna in Microwave Frequencies*, Taylor & Francis Group, New York, 2016.

[4] Sabban, A., Applications of MM Wave Microstrip Antenna Arrays, ISSSE 2007 Conference, Montreal, Canada, August 2007.

[5] Sabban, A., Wideband RF Modules and Antennas at Microwave and MM Wave Frequencies for Communication Applications, *Journal of Modern Communication Technologies & Research*, 3, 89–97, March 2015.

[6] Milkov, M.M., Millimeter-Wave Imaging System Based on Antenna-Coupled Bolometer, MSc. Thesis, *UCLA*, 2000.

[7] Sabban, A., W Band MEMS Detection Arrays, *Journal of Modern Communication Technologies & Research*, 2, 9–13, December 2014.

[8] Sabban, A., Ultra-Wideband RF Modules for Communication Systems, *PARIPEX, Indian Journal of Research*, 5(1), 91–95, January 2016.

[9] Mass, S.A., *Nonlinear Microwave and RF Circuits*, Artech House, Norwood, MA, 1997.

[10] Sabban, A., Microstrip Antenna Arrays, *Microstrip Antennas*, Nasimuddin Nasimuddin (Ed.), InTech, Croatia, 361–384, 2011,

[11] Sabban, A., A New Wideband Stacked Microstrip Antenna, IEEE Antenna and Propagation Symp., Houston, Texas, USA, June 1983.

[12] Sabban, A., Wideband Microstrip Antenna Arrays, IEEE Antenna and Propagation Symposium MELCOM, Tel-Aviv, 1981.

[13] Sabban, A., New Wideband printed Antennas for Medical Applications, *IEEE Journal, Transaction on Antennas and Propag.*, 61(1), 84-91, January 2013.

[14] Sabban, A., *Wideband RF Technologies and Antenna in Microwave Frequencies*, Taylor & Francis Group, New York, 2016

[15] Sabban, A., Small Wearable Meta Materials Antennas for Medical Systems, *The Applied Computational Electromagnetics Society Journal*, 31(4), 434–443, April 2016.

[16] Rahman, A. et al., Micro-Machined Room Temperature Micro Bolometers for MM-Wave Detection, *Applied Physical Letters* 68(14), 2020–2022, 1996.

[17] Sabban, A., New Compact Wearable Meta-Material Antennas, *Global Journal for Research and Analysis*, IV, 268–271, August 2015.

[18] Sabban, A., New Wideband Meta Materials Printed Antennas for Medical Applications, *Journal of Advance in Medical Science (AMS)*, 3, 1–10, April 2015. Invited paper.

[19] Maluf, N. and Williams, K., *An Introduction to Microelectromechanical System Engineering*, Artech House, 2004.

[20] Gauthier, G.P., Raskin, G.P., Rebiez, G.M., and Kathei, P.B.A 94 GHz Micro-machined Aperture- Coupled Microstrip Antenna, *IEEE Transactions on Antenna and Propagation*, 47(12), 1761–1766, December 1999.

[21] de Lange, G. et al., A 3*3 mm-Wave Micro Machined Imaging Array with Sis Mixers, *Applied Physical Letters*, 75(6), 868–870, 1999.

8 Wearable Metamaterial Antennas for Communication, IOT and Medical Systems

Herwansyah Lago, Ping Jack Soh and Guy A. E. Vandenbosch

CONTENTS

8.1 WIRELESS BODY AREA NETWORK (WBAN)

A wireless body area network (WBAN) is a network of wearable short-range, low power and highly reliable wireless communication devices which operates in, on and around the human body [1]. Equipped with wireless communication capabilities, WBAN is a natural progression from the concept of the wireless personal area network (WPAN), a network for interconnecting short range wireless devices in an ad hoc fashion (with little or no infrastructure), located around the workspace of an individual [2]. For WBAN to be accepted by the majority of consumers, the radio system component, including the antenna, needs to be seamless, compact and lightweight. This requires a possible integration of these systems within daily clothes or garment as presented in [3]. By integrating wearable antennas and radio frequency (RF) systems into clothes, the wearer will be comfortable and uninterrupted by those devices.

8.2 WEARABLE ANTENNAS

Wearable antennas are antennas operated on the human body, typically employed in sensing, health monitoring, sports, tracking and navigation, communication, mobile computing, public safety and military applications. Their application is illustrated in Figure 8.1. To implement wearable antennas for body worn scenarios, flexible materials such as textiles and thin dielectrics are preferred, as they can be easily integrated with clothing or worn directly on the human body without restraining body movements. Moreover, the implementation of such materials in low profile antennas, such as patch antennas, is ideal for wearable applications as these antennas will be conformal and comfortable. Textile-based planar antennas typically use textile substrates such as cotton, felt and denim, thanks to their cost, comfort and easy off-the-shelf access. Conductive textiles which have been used for electromagnetic shielding purposes in the past are now being adapted as materials for textile antennas. These textiles are conventionally made from polymer threads or conductive metals which are combined with ordinary fabric threads. Their structure is very similar to the structure of conventional fabrics that can be sewn for daily clothing.

Various implementations of textile antennas have been presented in recent years. One of them is the wearable antenna for application into protective clothing of fire fighters in [5]. Further, an ultra-wideband (UWB)-WBAN wearable antenna was developed based on microstrip technology with full ground plane using textiles in [6], as shown in Figure 8.2. The proposed full ground plane minimizes coupling to the body. The intrinsically narrowband behavior of the microstrip topology is overcome by implementing a combination of broad banding concepts while maintaining the full ground plane. This ensures the reduction of on-body performance degradation.

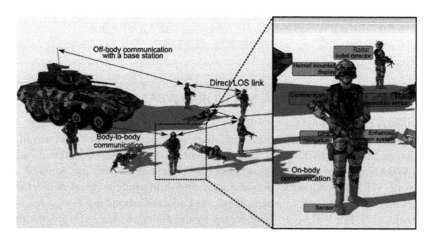

FIGURE 8.1   Wearable antenna technology used for military application [4].

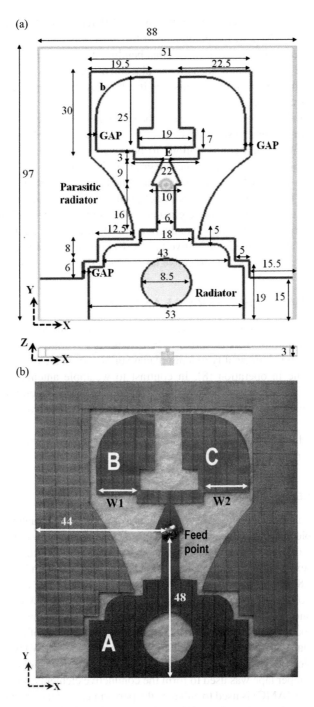

FIGURE 8.2 Ultra-wideband wireless body area network antenna. (a) Simulated antenna design (dimensions in mm) and (b) fabricated antenna in [6].

Reproduced courtesy of The Electromagnetics Academy.

Wearable antennas can be designed for use on different on-body locations, such as the chest, arm and back of the body. One of the most popular locations is the upper human torso due to the low risk of antenna bending due to body movement and the relatively large area available for antenna placement. Several works have promoted the use of the human chest due to this, see [6], by providing measurements in comparison with free space.

Wearable antennas for WBANs need to be small in size and flexible. For example, the body worn-antenna system developed in [7], with a flexible modular design, allows different antenna configurations to be selected for different platforms. An inflatable life vest was chosen as the integration platform, and the antenna system operates in the 121.5 MHz (using a meander line topology) and 406 MHz (using a shorted patch) bands. The antenna is inkjet-printed to form a low-profile structure. Both antenna structures use lightweight and low-loss foam as their substrate to avoid moisture absorption, while their conductive elements are thin flexible foil implemented using inkjet-printed conductive layers. To provide mechanical-abrasion protection, the researchers used a protective layer which shields the cover fabric.

8.3 MATERIALS FOR WEARABLE ANTENNAS

The properties of flexibility and stretchability in wearable antennas potentially open up new opportunities. This is particularly beneficial for large area devices and in systems requiring conformality to a nonplanar surface, or in the case of bending and stretching while in operation [8]. In contrast to wearable antennas built on non-stretchable textiles, such mechanical compliance increases user comfort in wearable and implantable devices, besides simplifying its integration for a range of new applications when subjected to flexing functionality. To meet these requirements, the materials used to fabricate wearable antennas need to be chosen with care to result in wearable antennas with satisfactory quality.

8.3.1 TEXTILE

Textile materials are widely used by researchers to design wearable antennas as they are easily available and easily integrated with clothing. Due to their low dielectric constant and relatively high loss tangent, the impedance bandwidths produced by such antennas are good. An example is the dual-band diamond antenna proposed in [9] for application at 2.45 and 5.8 GHz. Denim material is used as its substrate, while conductive copper tape is used for the conductive elements. The antenna shows 500 MHz of bandwidth with a gain and efficiency of 2 dB and 99.4% at 2.45 GHz. Meanwhile, at 5.8 GHz, 700 MHz of bandwidth is reached with 3.22 dB of gain and 96% of efficiency.

Next, a dual-band antenna fed by a coplanar waveguide (CPW) was designed in [10] for wireless local area network (WLAN) applications using cotton jeans as substrate. Pure copper tape was used to form the conductive elements. An artificial magnetic conductor (AMC) is used to enhance the performance of the antenna. The result shows a gain of 5.12 dB at 2.45 GHz and a gain of 3.97 dB at 5.8 GHz. Wearable antennas for wearable protective clothing operating in the industrial, scientific and medical (ISM) bands have been presented in [5]. Aramid-based fibers are used as

TABLE 8.1
Several types of textile materials used in [11]

| Textile substrate | Material properties | | Thickness (mm) |
	Permittivity	Loss tangent	
Wash cotton	1.51	0.02	3.0
Curtain cotton	1.47	0.02	3.0
Polyester	1.44	0.01	2.85
Polycot (polyester combined cotton, 65:35)	1.48	0.02	3.0

substrate in order to ensure its robustness against high temperatures. Flectron is used to form the patch and ground plane. The antenna shows an acceptable performance with a bandwidth of 100 MHz, a gain of 4.4 dB and 47% of efficiency. Finally, three textile patch antennas were designed for WLAN application in [11] for various cotton and polyester clothing. Copper is applied as patch and ground plane. The first antenna uses wash cotton as substrate. Its impedance bandwidth is 148 MHz with a gain and efficiency of 7.22 dBi and 63%, respectively. Next is a curtain cotton as substrate, resulting in a gain of 7.52 dBi with 61% of efficiency and 128 MHz of impedance bandwidth. Finally, a polycot as substrate yields 152 MHz of bandwidth with a gain of 9.623 dBi and an efficiency of 70%. Table 8.1 shows the comparison of several textile materials used as substrate in [11].

8.3.2 POLYMER

An interesting material is a transparent polymer, which enables antennas to be hidden from sight. Besides their suitable electrical properties, the mechanical characteristics of polymer-based substrates may result in waterproofness, increase the ease of fabrication, have high flexibility, and thermal stability, making it advantageous over other materials [12]. Furthermore, the dielectric properties of such materials can also be easily modified by mixing with other materials, such as carbon, to produce new dielectric properties suited for a required application. Kapton® polyimide, liquid crystal polymer (LCP) and poly-di-methyl-siloxane (PDMS) are the most frequently used [13, 14]. Polymer-based antennas such as the one proposed in [15] can be transparent and flexible.

8.4 METAMATERIALS

Metamaterials are a new class of electromagnetic materials which have attracted significant interest among material and electromagnetic engineers. While there are numerous definitions of metamaterials, they are generally defined as a class of engineered (artificial) structures exhibiting extraordinary electromagnetic properties that do not exist in nature. Their unique structure offers the possibility to manipulate electromagnetic wave propagation in one, two or three dimensions. Metamaterials are subdivided into several categories based on their permittivity (ε)

TABLE 8.2
Categories of metamaterials

Metamaterial category	Metamaterial properties	
	$\mathcal{E}$	μ
DPS	>0	>0
ENG	<0	>0
MNG	>0	<0
DNG	<0	<0

and permeability (μ). Normal materials are classified in a first category, known as double positive (DPS) materials, with both $\mathcal{E}$ and μ greater than zero. Metamaterials that have a negative permittivity while keeping a positive permeability are classified in a second category, called epsilon negative (ENG) materials. The third category contains the mu-negative (MNG) materials with a positive $\mathcal{E}$ and a negative μ. The final category concerns double negative (DNG) materials. This type of metamaterials has both $\mathcal{E}$ and μ lower than zero. The categories of metamaterials are tabulated in Table 8.2.

Several classes of metamaterials have been proposed in [17–19]. However, an upgraded metamaterial classification, including a discussion on their characteristics, is given further below in this section. The chart shown in Figure 8.3 illustrates different types of already studied metamaterials. It concerns artificial dielectrics, electromagnetic bandgap (EBG) materials, negative index materials, frequency selective surfaces (FSSs) and AMCs.

8.4.1 Artificial Dielectric

An artificial dielectric consists of numerous subwavelength metal elements embedded in a medium. The conducting properties of the metal elements result in a material with the properties of a dielectric. Such a material is typically characterized by a high permittivity which is capable of boosting the antenna performance [20–22].

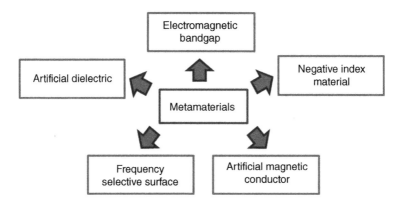

FIGURE 8.3 Metamaterial classification.

An artificial dielectric can be designed as an MNG material. In microwave engineering, the artificial dielectric is widely used for beam shaping, as reviewed in [23, 24]. This class of metamaterials is also suitable to be integrated with an antenna to reach a high gain and a compact size. This material can also be used as a substrate shield for a transmission line thanks to the high permittivity [25–27].

8.4.2 FSS

A FSS is a periodically arranged structure which contains metallic planar elements (resonators). Normally, an FSS acts as a band pass or band stop filter in different frequency ranges. The resonators are constructed with normal shapes: rectangle, ring, slot and so on. These conducting surfaces will determine the equivalent circuit, as such defining the filter type [28]. FSSs are not only applied in filters [29], but also in sensors [30], polarizers [28, 31] and absorbers [32]. FSSs are also widely implemented in antennas to enhance their performance [33].

8.4.3 EBG

An EBG material is based on a periodic structure that affects electromagnetic wave propagation. An EBG can be formed by dielectric or metallic elements or a combination of these. EBG structures have been implemented in various electromagnetic devices [34–36]. One common feature of EBG structures is the occurrence of a stopband. A comprehensive band-gap is exhibited when the stopband occurs in all wave propagation directions [37]. EBG structures are widely applied in polarizers and filters [38, 39]. In antennas, EBGs are used to enhance impedance matching and the efficiency, and to diminish the mutual coupling between array elements [40–43].

8.4.4 NEGATIVE INDEX MATERIAL

Negative index materials are a class of metamaterials with a negative refractive index in a specific frequency range [44]. They are also known as left-handed materials. They are essentially DNG materials, see Table 8.1. This type of metamaterials has been widely used in electromagnetic engineering [45–48]. A DNG material has the capability to improve an antenna's performance, just as an EBG, an FSS or an artificial dielectric does.

8.4.5 AMC

AMCs are an interesting class of metamaterials in electromagnetics. They are also known as high impedance surfaces (HIS). An AMC also involves a periodical arrangement, similar to the one in an FSS or an EBG structure [46, 49–51]. However, a unique property of an AMC is that it is designed to completely reflect electromagnetic waves in a certain frequency range. Moreover, it provides a reflection phase that is zero or almost zero degrees, which matches the theoretical boundary condition of a perfect magnetic conductor (PMC). Although this boundary condition is reached in principle only at a single frequency, in many practical applications it is sufficient that the AMC's reflection phase is in the range of ± 90° in a sufficiently wide band [49]. Since this property does not occur in nature, AMCs can be classified as metamaterials. The AMC is often applied to reduce the effect of a ground plane on an antenna's performance. This can be explained as follows. The spacing between a normal ground plane and a radiating element can strongly affect the performance of the antenna. The gain and efficiency of the antenna may reduce significantly when the ground plane is placed too close to the radiating element. However, an AMC is allowed to be placed very close to the radiating element without affecting the radiation characteristics too much, due to the PMC-like properties [49].

There are various potential applications where AMCs can be used. RF energy harvesting is very popular and was presented in [34, 52, 53]. In [54], an AMC plane is used in combination with a loop antenna operating at 500 and 875 MHz. The AMC acts as a reflector and is placed very near to the loop antenna, in this way enhancing the RF energy reception efficiency.

In [55], a printed dipole antenna with linear polarization at 5 GHz is placed over an AMC mushroom structure. A normal mushroom-type AMC has a centered grounded via and will reflect the same polarization as the incident one. However, when a dual grounded via is used, the AMC has the potential to change incident linear polarization into reflected circular polarization. The generated right-handed circular polarization has an extensive axial ratio bandwidth and a huge beam width.

AMCs are not limited to rigid applications, they can also be used in wearable applications [56, 57]. In a wearable application, the back radiation is critical since it determines the Specific Absorption Rate (SAR) level. An AMC can be applied in a wearable antenna to act as a reflector, in this way reducing the back radiation [57], besides improving the antenna gain and the impedance bandwidth [51, 58]. In [57], both the AMC structure and dual-band patch antenna are fabricated using textile materials. The antenna shows a high front-to-back ratio (FBR) and a low SAR level in the human body.

8.5 WEARABLE METAMATERIAL-BASED ANTENNAS

8.5.1 MULTIBAND TEXTILE ANTENNAS WITH METASURFACE

In this section, previous studies on wearable antennas integrated with metasurface structures, more specifically the available designs involving dual bands, are reviewed. Integration of metasurface structures is mainly used to reduce backward radiation toward human body tissue and minimize SAR.

The work in [57] proposed a coaxially fed rectangular patch antenna which operates at 2.45 GHz. A CPW line is used to connect the two slot dipoles, resonating at 5.5 GHz. The substrate material used is felt, while the radiating elements are made from ShieldIt Super. The overall size with the AMC integrated is 100 mm × 100 mm. The simulated reflection coefficient in the lower band at 2.45 GHz is −16.2 dB, while in the upper band at 5.5 GHz it is −11.7 dB. The measured reflection coefficient in the lower band degrades slightly to −8.3 dB, whereas the reflection coefficient is improved to −12.5 dB in the upper frequency band. The slot dipole enables a wider bandwidth compared to a basic rectangular microstrip patch. The FBR in both bands is higher than 12 dB. The realized gain in the lower band is about 2.5 dB while in the upper band it goes up to 4 dB. When the antenna is measured near a human user, the S_{11} is only slightly affected by the coupling between the human body and the antenna due to the effectiveness of the full ground plane, with maximum S_{11} of −9.8 and −8.4 dB in the lower and upper band, respectively. The ground plane of the antenna is made from conducting textile material to enable the antenna's conformability for users' comfort. The antenna was simulated when bent over cylinders with different radii (r) of 20, 30, 50 mm along the two main axes. The S_{11} is almost unaffected when the antenna is bent along the x-axis, due to the fact that the bending is parallel to the current distribution direction on the patch. When the radius is decreased, the resonant frequency shifts upward. The total efficiency throughout the whole operating frequency band is above 40%. A series of SAR simulations were performed using CST software. Averaged over 10 g of body tissue, and with the antenna placed about 10 mm away from the ground of the proposed antenna, the large ground is capable of reducing SAR values. When fed using an input power of 0.5 W (root mean square, rms), the SAR value for the antenna with larger ground is reduced from 0.0464 to 0.0042 W/kg in the lower band and from 0.03 to 0.0016 W/kg in the upper band. Besides that, the use of this larger ground plane resulted in the reduction of the back radiation by between 5 and 10 dB.

A dual-band textile-based planar fork-shaped dipole is proposed in [59]. It is integrated with a 3×1 EBG ground plane which operates in the ISM band (2.4 and 5.2 GHz). The overall antenna dimension is 30 mm × 90 mm, with a thickness of 8 mm. The substrate of the proposed antenna is felt, while the conducting element is made from Zelt textile. The performance of the EBG-integrated antenna is compared with a conventional antenna. The conventional antenna's bandwidth in flat conditions is about 12.7% and 4.45% in the upper and lower band, respectively. Meanwhile, the EBG antenna produces 5.4% and 11.7% of bandwidth in the lower and upper frequency bands, respectively. For the conventional antenna, the radiation pattern is omnidirectional in the azimuth plane with a maximum gain of 2.0 dBi at 2.45 GHz and 1.74 dBi at 5.2 GHz. Meanwhile, the proposed EBG antenna provides a directional radiation pattern with reduced backward radiation and improved maximum gain of 4.9 dBi and 6.15 dBi at 2.45 and 5.2 GHz, respectively. The reflection coefficient is slightly worse in the lower band but improved up to −31 dB in the upper band. Where the conventional dipole produces very high SAR values, these are reduced by the use of the EBG. The SAR value (using the European standard <2.0 W/kg over 10 g tissue) is reduced by 83.3% and 92.8% in the 2.4 and 5.2 GHz band, respectively. Both the conventional and the metamaterial antenna were bent over an arm with a radius of 40 mm along the two main axes. When bent, the conventional antenna shows a lowering of the resonant frequency in the lower and upper band from 2.4 to 2.1 GHz and from 5.2 to 5 GHz, respectively. This is due to the close proximity of the human body. Conversely, for the EBG antenna the resonant frequencies are retained, both in the planar condition and when bent along the x- and y-axes. The EBG ground plane acts as a shield, which is beneficiary when the antenna is operating near the human body.

A dual-band coplanar textile antenna integrated with an EBG plane which operates at 2.45 and 5.5 GHz was proposed in [60]. It is formed using Zelt for its conducting elements, and nylon as its substrate. The EBG plane consists of 3×3-unit elements to act as a shield to reduce the interaction with the human body. The overall antenna size is 102 mm $\times$ 102 mm, with a thickness of 7 mm. The reflection coefficient is below -10 dB and the fractional bandwidth is 2.3% and 12.4% in the lower and upper bands, respectively. The proposed EBG antenna has a low back radiation, with an FBR of about 15 dB. When the antenna is operated without EBG, the radiation of the antenna is dipole-like, radiating significantly toward the human body. However, placing the EBG structure behind the coplanar antenna enabled this to be reduced. As a result, the realized gain of the EBG antenna is 5.2 dB (improved by about 83%) and 3 dB (improved by about 86%) in the lower and upper bands, respectively. Furthermore, the EBG antenna demonstrates reduced SAR values (average over 1 g of tissue) of about 1.7 W/kg (a reduction of 96.5%) at 2.45 GHz and 1.0 W/kg (a reduction of 98.5%) at 5.5 GHz.

A dual-band metamaterial loaded antenna with composite right or left-handed transmission line (CRLH TL) metamaterial operating at 2.45 and 5.2 GHz is proposed in [61] for WLAN. The antenna is fully made of textile, using felt as a substrate (with ε_r is 1.3) and ShieldIt Super as conductor. The antenna was measured in three locations: planar on chest, bent on upper arm, and bent on forearm. When bent slightly, the resonant frequencies are consistent with the resonant frequencies of the planar case. However, when the radius of bending is decreased, both the lower and the upper band shift upward. The bandwidth of the antenna in free space is 135 and 583 MHz in the lower and upper band, respectively. The SAR values are calculated with CST software for the antenna located 5 mm away from different parts of the HUGO human body model—forearm, upper arm, thigh and chest. The SAR values in the upper band (5.2 GHz) are higher than in the lower band (2.45 GHz) for the same bending radius. This is due to the intrinsic properties of the body phantom, since the conductivity of human tissue increases with frequency. However, the SAR values are below the European safety limit of 2 W/kg over 10 g of tissue.

Next, the performances of the two antennas with and without EBG plane are compared in [62]. Both are designed based on a CPW-fed antenna which operates at 2.45 and 5 GHz. The antennas use felt as a substrate and Zelt as a conductor. The overall size of the EBG antenna is 120 mm $\times$ 120 mm, and the thickness is 4.48 mm. The bandwidth in the lower band is 4% in both simulations and measurements. The bandwidth in the upper band is about 12% in measurements compared to 8% in simulations. When the CPW antenna is integrated with the EBG plane, the back lobe in the lower band is reduced by 12 dB and the gain is improved by about 3 dB. The EBG antenna shows a large 96.5% SAR reduction when averaged over 1 g of tissue, and 97.3% when averaged over 10 g of tissue. Both antennas were tested in two ways: in free space and on-body. The diameters of bending used are 80 mm (critical case) and 140 mm. When the diameter is increased, the resonant frequency of the EBG antenna is shifted downward by about 10 MHz in the 2.45 GHz band, and upward by about 65 MHz in the 5.8 GHz band. However, there are no changes in terms of bandwidth.

8.5.2 Broad/Wideband Textile Antennas with Metasurface

In [63], a novel compact uniplanar AMC design without via holes is proposed and designed on a flexible dielectric substrate, Rogers 3003 with a thickness $h = 0.762$ mm, relative dielectric permittivity $\varepsilon_r = 3.0$ and loss tangent $\tan\delta = 0.0013$. The simulated phase of the reflection coefficient for the designed AMC indicates that it is resonating at 6 GHz with an AMC operation bandwidth of 500 MHz, which is considered broadband for a low-profile AMC. This is equivalent to about 8.33% of fractional bandwidth with respect to the central frequency. Upon the optimization of the AMC performance, a 12×12 cell planar AMC prototype was manufactured using laser micromachining. The flat prototype resonates at 6.178 GHz, which is a deviation of 2.9% with respect to the simulated resonance at 6.0 GHz. Next, the AMC performance was evaluated under bending. Testing indicated that there is almost no frequency shift when the prototype is bent in a creeping configuration, relative to the resonance of the flat prototype. Meanwhile, smooth bending conditions yield a resonance at 6.208 GHz, yielding a small deviation of 1.69% with respect to the flat prototype.

Next, the AMC performance under different polarizations of the incident field for normal and oblique incidence was investigated. Incident angles from 0° to 60° for transverse-electric (TE)-polarized waves were simulated. It is noticed that the AMC operation bandwidth is slightly reduced above an angle of 40°. Overall, this proposed AMC operates with a broad bandwidth, shows a polarization angle independency under normal incidence, and a high angular stability under oblique incidence. Besides, its uniplanar low-profile design without via holes, its implementation with flexible characteristic, low cost, simple fabrication and integration, makes it very attractive for applications involving antennas in radio-frequency identification (RFID) tags, wearable systems and radar cross section (RCS) reduction.

Next, a wideband AMC was designed in [64] for operation in the WBAN UWB mandatory channel 6. The proposed AMC is incorporated onto a rectangular-ring patch antenna for operation centered at 8 GHz with 2 GHz of bandwidth, as shown in Figure 8.4. A rectangular patch structure is chosen as the AMC unit element notched on its four corners, duplicated and placed periodically in the complete AMC plane. A metallic ground is placed at the bottom-most layer. The patch is printed on a 1.5-mm-thick felt substrate with a dielectric constant and loss tangent of 1.2 and 0.044, respectively. Meanwhile, ShieldIt Super conductive textile from LessEMF Inc. is used to form the metallic layers. The operating bandwidth of the proposed AMC plane, defined as the band within ±90° from the zero-phase reflection (i.e. PMC) point, is from 6.78 to 8.77 GHz.

Two antenna designs are investigated in this chapter. The first design is the proposed antenna integrated with an AMC plane, and the second antenna is a conventional antenna as a reference. The incorporation of the AMC structure onto the antenna improved the matching to achieve a minimum value of −24.03 dB compared to the reference antenna resonating at 8 GHz. Besides that, an additional band operating in the 5.5 GHz WLAN band is also noticed. This is due to the rectangular ring which allows the excitation of two modes. It is observed that the fabricated antenna

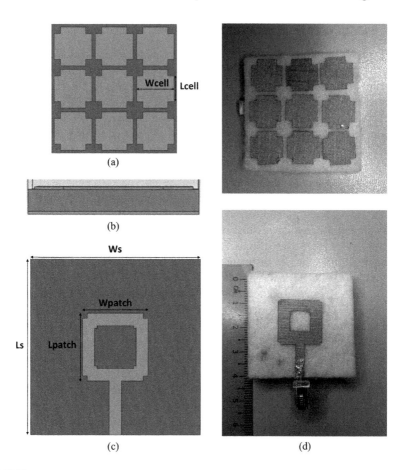

FIGURE 8.4 Antenna from [64]. (a) Design of the artificial magnetic conductor unit cell, (b) fabricated prototype of the artificial magnetic conductor plane, (c) simulated antenna where the artificial magnetic conductor is implemented, and (d) fabricated prototype.

operates from 5.2 to 5.69 and 6.77 to 8.48 GHz, while simulations indicate an operation from 5.27 to 5.68 and 7.73 to 8.27 GHz.

Finally, the body's influence on the operation of the antenna is studied using a three-layered rectangular biological tissue model consisting of layers of skin, fat and muscle. The stacked tissue layers model is defined behind the antenna at 10 and 20 mm distances from the antenna ground layer, with an average thickness of 3, 7 and 60 mm for the skin, fat and muscle layer, respectively. It is observed that there is no significant frequency detuning due to the effect of the AMC plane. Meanwhile, the calculated SAR values demonstrate that the AMC has decreased the power absorption. The obtained SAR values are 0.305 and 0.624 W/kg at 5.5 and 8 GHz, respectively.

Next, an AMC operating between 8 and 12 GHz is designed based on square conductive elements with annularly shaped slots in [1]. The unit cell is designed based on [36]. It consists of a square patch with an annular slot engraved on a felt substrate.

The results indicate that the phase of the unit cell reflection coefficient exhibits a behavior with a phase between $-90°$ and $90°$ that spreads over a wide frequency band (between 8 and 12 GHz). This proposed unit cell is then designed into a 3×4 array to form the AMC plane. It is prototyped and integrated for use under a monopole antenna. A 2-mm-thick foam layer with a permittivity of 1.05 is combined to avoid the coupling between the AMC and the antenna.

Next, a flexible wideband metamaterial-based reflector and EBG structure is proposed in [65] to be integrated with an antenna array designed and embedded into a multi-layer, low cost, low loss, transparent and robust polymer PDMS substrate. The average permittivity and loss tangent of PDMS are 2.8 and 0.0013, respectively. The metamaterial unit cells are located behind the antenna array in the form of a 4×4 array to enable unidirectional radiation. The antenna is designed onto a 2.5-mm-thick PDMS substrate, which is then covered by a 0.8 mm PDMS layer. Its unit cell is designed using a square patch with four extended open-ended U-slots. The patch is connected to a three-turn meander line from each corner. To enable double negative metamaterial (DNM) behavior, the size of the metamaterial unit cell should be much smaller than the wavelength. Therefore, the total length and width of the proposed unit cell is optimized to be smaller than a tenth of the guided wavelength ($\lambda g/10$) at 1.45 GHz. As a result, the unit cell operates as a reflector over a wide frequency range, between 0.94 and 2.7 GHz.

On the other hand, an EBG intended to reduce surface waves around the feeding network is designed in the form of a mushroom-like unit cell. Similar to the metamaterial unit cell, the unit cell for the EBG is also designed based on a square patch (8 mm $\times$ 8 mm) integrated with four symmetrically extended open-ended U-slots originating from each corner. The U-slots on the EBG patch are 0.4 mm in width. They are introduced to extend the electrical path of the current, which is flowing along the adjacent edges of the U-slots, in this way creating additional LC resonant circuits. Each EBG patch is connected to the ground plane through a via. This results in a compact EBG unit cell size and a wide band stop behavior at the lower microwave frequencies. The side lengths of the EBG unit cell are optimized based on the $\lambda g/16$ criterion, calculated at 1.45 GHz. The transmission coefficient (S_{21}) of the proposed EBG unit cell shows an efficiently suppressed transmission within the band between 0.85 and 4.3 GHz, which covers the targeted operation band between 1 and 2 GHz. Finally, the overall structure is assessed via simulations and measurements. It operates from 1.16 GHz to 1.94 GHz, with a fractional bandwidth of 53.8%, and at least 80% of radiation efficiency.

8.6 CONCLUSION

This chapter presents an overview of wearable metamaterial antennas intended for communication applications, the Internet of Things, and use in the medical domain. The overview includes the underlying principles of designing this type of antenna. The benefits of implementing metamaterials in the design of wearable antennas are vast and are very relevant to overcome limitations when operating in a WBAN environment. Metamaterials are capable of enhancing the gain of wearable antennas, which is in principle intrinsically limited. This is due to the use of more lossy flexible conductive materials, such as conductive textiles, in place of conventional highly conducting materials, such as copper. Besides that, in using metamaterials, the bandwidth of the designed wearable antennas can be broadened. This can be used to compensate the shift in operating frequency due to the dielectric coupling with the body. This coupling is different for different parts of the human body, or when worn by different wearers with varying morphologies. The replacement of perfect electrical conductor (PEC) planes needed as antenna ground planes or reflectors by PMC-like metamaterial-based surfaces is able to provide an effective mechanism in controlling/redirecting antenna beams, for example by transforming omnidirectional patterns into unidirectional patterns. In a wearable design, this will enable the antennas to radiate outward from the body without compromising the low profile. Finally, the use of metamaterial-based techniques in antennas also enables an increased compactness of the finally designed structures. This then enables wearable antennas to be implemented in space-limited on-body locations, such as hats, caps or within a logo, which adds flexibility in locating antennas intended to be pointed to a certain direction. Good examples are antennas for Global Navigation Satellite System (GNSS) applications, which are intended to be pointed toward nadir when being used on body.

REFERENCES

[1] A. Mersani, L. Osman, and J. M. Ribero, "Flexible UWB AMC antenna for early stage skin cancer identification," *Prog. Electromagn. Res. M*, 80(March), 71–81, 2019.
[2] B. C. Jeyhan Karaoguz, "High-rate wireless personal area networks," *IEEE Commun. Mag.*, 39(12), 96–102, 2001.
[3] J. G. Santas, A. Alomainy, and Y. Hao, "Textile antennas for on-body communications: Techniques and properties," in *2007 2nd European Conference on Antennas Propagation, EuCAP 2007*, Edinburgh, UK, pp. 537–537, 2007.
[4] R. S. Nacer Chahat, M. Zhadobov, and R. Sauleau, *Antennas for Body Centric Wireless Communications at Millimeter Wave Frequencies*, London, UK, Intech, vol. i, pp. 23–55, 2014.
[5] C. Hertleer, H. Rogier, L. Vallozzi, and F. Declercq, "A textile antenna based on high-performance fabrics," *IET Semin. Dig*, 2007(11961), 1–5, 2007.
[6] P. B. Samal, P. J. Soh, and G. A. E. Vandenbosch, "A Systematic Design Procedure for Microstrip-based Unidirectional UWB Antennas," *Prog. Electromagn. Res.*, 143, 105–130, 2013.
[7] J. Lilja et al., "Body-worn antennas making a splash: Lifejacket-integrated antennas for global search and rescue satellite system," *IEEE Antennas Propag. Mag.*, 55(2), 324–341, 2013.
[8] A. A. Nawaz, X. Mao, Z. S. Stratton, and T. J. Huang, "Unconventional microfluidics: Expanding the discipline," *Lab Chip*, 13(8), 1457–1463, 2013.
[9] M. E. Jalil, M. K. A. Rahim, N. A. Samsuri, N. A. Murad, N. Othman, and H. A. Majid, "On-body investigation of dual band diamond textile antenna for wearable applications at 2.45 GHz and 5.8 GHz," in *2013 7th European Conference on Antennas Propagation, EuCAP 2013*, Gothenburg, Sweden, pp. 414–417, 2013.
[10] A. Mersani, L. Osman, and I. Sfar, "Dual-band textile antenna on AMC substrate for wearable applications," *Mediterr. Microw. Symp.*, 2015, 1–3, 2015.
[11] S. Sankaralingam and B. Gupta, "Development of textile antennas for body wearable applications and investigations on their performance under bent conditions," *Prog. Electromagn. Res. B*, 22(22), 53–71, 2010.

[12] R. B. V. B. Simorangkir, Y. Yang, L. Matekovits, and K. P. Esselle, "Dual-band dual-mode textile antenna on PDMS substrate for body-centric communications," *IEEE Antennas Wirel. Propag. Lett.*, 16, 677–680, 2017.

[13] H. A. Elmobarak Elobaid, S. K. Abdul Rahim, M. Himdi, X. Castel, and M. Abedian Kasgari, "A transparent and flexible polymer-fabric tissue UWB antenna for future wireless networks," *IEEE Antennas Wirel. Propag. Lett.*, 16, 1333–1336, 2017.

[14] G. DeJean, R. Bairavasubramanian, D. Thompson, G. E. Ponchak, M. M. Tentzeris, and J. Papapolymerou, "Liquid Crystal Polymer (LCP): A new organic material for the development of multilayer dual-frequency/dual-polarization flexible antenna arrays," *IEEE Antennas Wirel. Propag. Lett.*, 4(1), 22–26, 2005.

[15] A. S. M. Alqadami, M. F. Jamlos, P. J. Soh, and G. A. E. Vandenbosch, "Assessment of PDMS technology in a MIMO antenna array," *IEEE Antennas Wirel. Propag. Lett.*, 15, 1939–1942, 2016.

[16] N. J. Kirsch, N. A. Vacirca, T. P. Kurzweg, A. K. Fontecchio, and K. R. Dandekar, "Performance of transparent conductive polymer antennas in a MIMO ad-hoc network," *2010 IEEE 6th International Conference on Wireless and Mobile Computing, Networking and Communication (WiMob'2010)*, Niagara Falls, Canada, pp. 9–14, 2010.

[17] I. A. Buriak and V. O. Zhurba, "A review of microwave metamaterial structures classifications and applications," in *IEEE 9th International Kharkiv Symposium on Physics and Engineering of Microwave, Milimeter and Submilimeter Waves (MSMW)*, Kharkiv, Ukraine, pp. 1–3, 2016.

[18] A. Erentok, P. L. Luljak, and R. W. Ziolkowski, "Characterization of a volumetric metamaterial realization of an artificial magnetic conductor for antenna applications," *IEEE Trans. Antennas Propag.*, 53(1 part I), 160–172, 2005.

[19] H. Mosallaei and Y. Rahmat-Samii, "Electromagnetic band-gap structures: Classification, characterization and applications," in *11th International Conference on Antennas and Propagation*, Manchester, UK, pp. 17–20, 2001.

[20] S. Barzegar-Parizi and B. Rejaei, "Calculation of effective parameters of high permittivity integrated artificial dielectrics," *IET Microw. Antennas Propag.*, 9(12), 1287–1296, 2015.

[21] A. Dadgarpour, B. Zarghooni, B. S. Virdee, and T. A. Denidni, "Enhancement of tilted beam in elevation plane for planar end-fire antennas using artificial dielectric medium," *IEEE Trans. Antennas Propag.*, 63(10), 4540–4545, 2015.

[22] W. H. Syed, D. Cavallo, H. T. Shivamurthy, and A. Neto, "Wideband, wide-scan planar array of connected slots loaded with artificial dielectric superstrates," *IEEE Trans. Antennas Propag.*, 64(2), 543–553, 2016.

[23] A. C. Brown, "Pattern shaping with a metal plate lens," *IEEE Trans. Antennas Propag.*, 28(4), 564–568, 1980.

[24] I. J. Bahl and K. C. Gupta, "Frequency scanning by leaky-wave antennas using artificial dielectrics," *IEEE Trans. Antennas Propag.*, 23(4), 57–62, 1975.

[25] I. Awai, H. Kubo, T. Iribe, D. Wakamiya, and A. Sanada, "An artificial dielectric material of huge permittivity with novel anisotropy and its application to a microwave BPF," in *IEEE MTTS International Microwave Symposium Digest 2003*, Philadelphia, PA, vol. 2, pp. 1085–1088, 2003.

[26] J. Machac, "Microstrip line on an artificial dielectric substrate," *IEEE Microw. Wirel. Components Lett.*, 16(7), 416–418, 2006.

[27] Y. Ma, B. Rejaei, and Yan Zhuang, "Artificial dielectric shields for integrated transmission lines," *IEEE Microw. Wirel. Components Lett.*, 18(7), 431–433, 2008.

[28] P. C. Zhao, Z. Y. Zong, W. Wu, and D. G. Fang, "A convoluted structure for miniaturized frequency selective surface and its equivalent circuit for optimization design," *IEEE Trans. Antennas Propag.*, 64(7), 2963–2970, 2016.

[29] M. Gao, S. M. A. Momeni Hasan Abadi, and N. Behdad, "A dual-band, inductively coupled miniaturized-element frequency selective surface with higher order bandpass response," *IEEE Trans. Antennas Propag.*, 64(8), 3729–3734, 2016.

[30] F. Dincer, K. Delihacioglu, M. Karaaslan, M. Bakir, and C. Sabah, "U-shaped frequency selective surfaces for single- and dual-band applications together with absorber and sensor configurations," *IET Microw. Antennas Propag.*, 10(3), 293–300, 2016.

[31] S. Sheikh, "Miniaturized-element frequency-selective surfaces based on the transparent element to a specific polarization," *IEEE Antennas Wirel. Propag. Lett.*, 15, 1661–1664, 2016.

[32] J. Li et al., "Design of a tunable low-frequency and broadband radar absorber based on active frequency selective surface," *IEEE Antennas Wirel. Propag. Lett.*, 15, 774–777, 2016.

[33] B. Döken and M. Kartal, "Triple band frequency selective surface design for global system for mobile communication systems," *IET Microw. Antennas Propag.*, 10, 1154–1158, 2016.

[34] B. Alavikia, T. S. Almoneef, O. M. Ramahi, B. Alavikia, T. S. Almoneef, and O. M. Ramahi, "Electromagnetic energy harvesting using complementary split-ring resonators," *Appl. Phys. Lett.*, 64(16), 163903, 2014.

[35] J. Ahn, K. Kim, Y. Yoon, and K. Hwang, "Pattern reconfigurable antenna for wireless sensor network system," *Electron. Lett.*, 48(16), 984–985, 2012.

[36] D. Sievenpiper, L. Zhang, R. F. Jimenez Broas, N. G. Alexöpolous, and E. Yablonovitch, "High-impedance electromagnetic surfaces with a forbidden frequency band," *IEEE Trans. Microw. Theory Tech.*, 47(11), 2059–2074, 1999.

[37] F. Elek, R. Abhari, and G. V. Eleftheriades, "A uni-directional ring-slot antenna achieved by using an electromagnetic band-gap surface," *IEEE Trans. Antennas Propag.*, 53(1 I), 181–190, 2005.

[38] A. Kanso, R. Chantalat, M. Thevenot, E. Arnaud, and T. Monediere, "Offset parabolic reflector antenna fed by EBG dual-band focal feed for space application," *IEEE Antennas Wirel. Propag. Lett.*, 9(6615), 854–858, 2010.

[39] M. S. Toubet, M. Hajj, R. Chantalat, E. Arnaud, B. Jecko, and T. Monediere, "Wide bandwidth, high-gain , and low-profile EBG prototype for high power applications," *IEEE Antennas Wirel. Propag. Lett.*, 10, 1362–1365, 2011.

[40] M. F. Abedin, M. Z. Azad, and M. Ali, "Wideband smaller unit-cell planar EBG structures and their application," *IEEE Trans. Antennas Propag.*, 56(3), 903–908, 2008.

[41] R. Coccioli, F. R. Yang, K. P. Ma, and T. Itoh, "Aperture-coupled patch antenna on UC-PBG substrate," *IEEE Trans. Microw. Theory Tech.*, 47(11), 2123–2130, 1999.

[42] K. Li, C. Zhu, L. Li, Y. M. Cai, and C. H. Liang, "Design of electrically small metamaterial antenna with ELC and EBG loading," *IEEE Antennas Wirel. Propag. Lett.*, 12, 678–681, 2013.

[43] F. Yang and Y. Rahmat-Samii, "Microstrip antennas integrated with electromagnetic band-gap (EBG) structures: A low mutual coupling design for array applications," *IEEE Trans. Antennas Propag.*, 51(10 part II), 2936–2946, 2003.

[44] R. A. Shelby, D. R. Smith, and S. Schultz, "Experimental verification of a negative index of refraction," *Science*, 292(5514), 77–79, 2001.

[45] C. L. Holloway, E. F. Kuester, J. Baker-Jarvis, and P. Kabos, "A double negative (DNG) composite medium composed of magnetodielectric spherical particles embedded in a matrix," *IEEE Trans. Antennas Propag.*, 51(10 I), 2596–2603, 2003.

[46] L. Ji and V. V. Varadan, "Negative refractive index and negative refraction of waves in lossy metamaterials," *Electron. Lett.*, 52(4), 260–262, 2016.

[47] G. J. Molina-Cuberos, Á. J. García-Collado, I. Barba, and J. Margineda, "Chiral metamaterials with negative refractive index composed by an eight-cranks molecule," *IEEE Antennas Wirel. Propag. Lett.*, 10(1), 1488–1490, 2011.

[48] N. Wongkasem, A. Akyurtlu, K. A. Marx, Q. Dong, J. Li, and W. D. Goodhue, "Development of chiral negative refractive index metamaterials for the terahertz frequency regime," *IEEE Trans. Antennas Propag.*, 55(11 part I), 3052–3062, 2007.

[49] Z. Bayraktar, J. P. Turpin, and D. H. Werner, "Nature-inspired optimization of high-impedance metasurfaces with ultrasmall interwoven unit cells," *IEEE Antennas Wirel. Propag. Lett.*, 10, 1563–1566, 2011.

[50] S. Clavijo, R. E. Díaz, and W. E. McKinzie, "Design methodology for sievenpiper high-impedance surfaces: An artificial magnetic conductor for positive gain electrically small antennas," *IEEE Trans. Antennas Propag.*, 51(10 I), 2678–2690, 2003.

[51] N. M. Mohamed-Hicho, E. Antonino-Daviu, M. Cabedo-Fabres, and M. Ferrando-Bataller, "A novel low-profile high-gain UHF antenna using high-impedance surfaces," *IEEE Antennas Wirel. Propag. Lett.*, 14, 1014–1017, 2015.

[52] J. A. Hagerty, F. B. Helmbrecht, W. H. McCalpin, R. Zane, and Z. B. Popović, "Recycling ambient microwave energy with broad-band rectenna arrays," *IEEE Trans. Microw. Theory Tech.*, 52(3), 1014–1024, 2004.

[53] M. Pinuela, P. D. Mitcheson, and S. Lucyszyn, "Ambient RF energy harvesting in urban and semi-urban environments," *IEEE Trans. Microw. Theory Tech.*, 61(7), 2715–2726, 2013.

[54] H. Kamoda, S. Kitazawa, N. Kukutsu, and K. Kobayashi, "Loop antenna over artificial magnetic conductor surface and its application to dual-band RF energy harvesting," *IEEE Trans. Antennas Propag.*, 63(10), 4408–4417, 2015.

[55] W. Yang, K. Tam, S. Member, W. Choi, and S. Member, "Novel polarization rotation technique based on an artificial magnetic conductor and its application in a low-profile circular polarization antenna," *IEEE Trans. Antennas Propag.*, 62(12), 6206–6216, 2014.

[56] S. Kim, Y. J. Ren, H. Lee, A. Rida, S. Nikolaou, and M. M. Tentzeris, "Monopole antenna with inkjet-printed EBG array on paper substrate for wearable applications," *IEEE Antennas Wirel. Propag. Lett.*, 11, 663–666, 2012.

[57] S. Yan, P. J. Soh, and G. A. E. Vandenbosch, "Low-profile dual-band textile antenna with artificial magnetic conductor plane," *IEEE Trans. Antennas Propag.*, 62(12), 6487–6490, 2014.

[58] N. A. Abbasi and R. J. Langley, "Multiband-integrated antenna/artificial magnetic conductor," *IET Microwaves, Antennas Propag.*, 5(6), 711, 2011.

[59] A. Afridi, S. Ullah, S. Khan, A. Ahmed, A. H. Khalil, and M. A. Tarar, "Design of dual band wearable antenna using metamaterials," *J. Microw. Power Electromagn. Energy*, 47(2), 126–137, 2013.

[60] N. Chahat, M. Zhadobov, R. Sauleau, and K. Mahdjoubi, "Improvement of the on-body performance of a dual-band textile antenna using an EBG structure," in *2010 Loughbrough Antennas and Propagation Conference, LAPC 2010*, Loughborough, pp. 465–468, November 2010.

[61] S. Yan, P. J. Soh, and G. A. E. Vandenbosch, "Compact all-textile dual-band antenna loaded with metamaterial-inspired structure," *IEEE Antennas Wirel. Propag. Lett.*, 14, 1486–1489, 2015.

[62] S. Zhu and R. Langley, "Dual-band wearable textile antenna on an EBG substrate," *IEEE Trans. Antennas Propag.*, 57(4 part 1), 926–935, 2009.

[63] M. E. De Cos and F. Las Heras, "Novel uniplanar flexible artificial magnetic conductor," *Appl. Phys. A: Mater. Sci. Process.*, 109(4), 1031–1035, 2012.

[64] N. F. M. Aun, P. J. Soh, M. F. Jamlos, H. Lago, and A. A. Al-Hadi, "A wideband rectangular-ring textile antenna integrated with corner-notched artificial magnetic conductor (AMC) plane," *Appl. Phys. A: Mater. Sci. Process.*, 123(19), 1–6, 2017.

[65] A. S. M. Alqadami, K. S. Bialkowski, A. T. Mobashsher, and A. M. Abbosh, "Wearable electromagnetic head imaging system using flexible wideband antenna array based on polymer technology for brain stroke diagnosis," *IEEE Trans. Biomed. Circuits Syst.*, 13(1), 124–134, 2019.

9 Wearable Technologies for 5G, Medical and Sport Applications

Albert Sabban

CONTENTS

9.1 INTRODUCTION

Mobile communication, radio-frequency identification (RFID) and Internet of Things (IOT) industries have been in continuous growth in the past decade. There is continuous increasing demand for wearable sensors, and RFID tags and devices. Wearable technology has several applications in personal wireless communication, IOT, RFID, sport and medical devices as presented in [1–8]. Several medical devices and systems have been developed to monitor patients' health as presented in several books and papers [1–44]. Wearable technology provides a powerful new tool for medical services, surgical rehabilitation services and IOT systems. Wireless body area networks (WBANs) can record electrocardiograms and can measure body temperature, blood pressure, heartbeat rate, arterial blood pressure, electro-dermal activity and other healthcare parameters. Wearable devices will be in the next decade an important part of individuals' daily lives. RFID technology and devices are employed as wearable devices. In this chapter, RFID technology and RFID antennas are presented. RFID technology and antennas were presented in several publications [47–53].

9.2 WEARABLE TECHNOLOGY

Devices, RFID tags and sensors that can be comfortably worn on the body are called wearable devices. Wearable technology is a multidisciplinary developing field. Knowledge in bioengineering, electrical engineering, software engineering and mechanical engineering is needed to design and to develop wearable communication and medical systems. Wearable medical systems and sensors are used to measure and monitor physiological parameters of the human body. Biomedical systems in the vicinity of the human body may be wired or wireless. Many physiological parameters may be analyzed by using wearable medical systems and sensors. Wearable medical systems and sensors can measure body temperature, heartbeat, blood pressure, sweat rate and other physiological parameters of the person wearing the medical device. Wearable technology may provide scanning and sensing features that are not offered by mobile phones and laptop computers. Wearable technology usually has communication capabilities and the user may have access to information in real time. Several wireless technologies are used to handle the data collection and processing in medical systems. The collected data may be stored or transmitted to a medical center so that the collected data can be analyzed. Wearable devices gather raw data which is fed to a database or to a software application for analysis. This analysis may result in a response that might alert a physician to contact a patient who is experiencing abnormal symptoms or a message may be sent to a person who achieves a fitness goal. Examples of wearable devices include belts, headbands, smart wristbands, watches, glasses, contact lenses, e-textiles and smart fabrics, jewelry, bracelets and hearing aid devices.

Usually, wearable communication devices consist of wearable antennas, transmitting unit, receiving unit and data processing unit.

Wearable technology may influence the fields of transportation, health and medicine, IOT, fitness, ageing, disabilities, education, finance, gaming, entertainment and music.

9.3 WEARABLE MEDICAL SYSTEMS

One of the main goals of wearable medical systems is to increase disease prevention. By using more wearable medical devices people can handle and be aware of their personal health. Sophisticated analysis of continuously measured medical data from a large number of patients at medical centers may result in better low-cost medical treatment.

9.3.1 APPLICATIONS OF WEARABLE MEDICAL SYSTEMS

- Wearable medical devices may help to monitor hospital activities.
- Wearable devices may help to operate and monitor home accessories.
- Wearable devices may help to operate and monitor IOT devices.
- Wearable medical devices may assist patients with diabetes.
- Wearable medical devices may assist patients with asthma.
- Wearable medical devices may help in solving sleep disorders.
- Wearable medical devices may assist in solving obesity problems.
- Wearable medical devices may assist in the treatment of cardiovascular diseases.
- Wearable medical devices may assist patients with epilepsy.
- Wearable medical devices may help in the treatment of patients with Alzheimer's disease.
- Wearable medical devices may help to gather data for clinical research trials and academic research studies.

9.4 PHYSIOLOGICAL PARAMETERS MEASURED BY WEARABLE MEDICAL SYSTEMS

Several physiological parameters may be measured by using wearable medical systems and sensors. Some of these physiological data are presented in this chapter.

9.4.1 MEASUREMENT OF BLOOD PRESSURE

Blood pressure indicates the arterial pressure of blood circulating in the human body.

Some of the causes of changes to blood pressure may be stress and being overweight. The blood pressure of a healthy person is around 80 by 120. Where the systole is 120 and the diastole is 80.

Changes that are 10% above or below these values are a matter of concern and should be examined. Usually, blood pressure and heartbeat are measured in the same set of measurements. The blood pressure and heartbeat may be transmitted to a medical center and, if needed, a doctor may contact the patient for further analysis and treatment.

9.4.2 MEASUREMENT OF HEART RATE

Measurement of heart rate is one of the most important tests needed to examine the health of a patient. A change in the heart rate will change the blood pressure and the amount of blood delivered to all parts of the body. The heart rate of a healthy person is 72 times per minute. Changes in heartbeat may cause several kinds of cardiovascular diseases. Traditionally, heart rate is measured by using a stethoscope. However, this is a manual test and is not very accurate. To measure and analyze the heart rate a wearable medical device may be connected to a patient's chest.

Medical devices that measure heartbeat may be wired or wireless.

9.4.3 Measurement of Respiration Rate

Measurement of respiration rate indicates if a person is breathing normally and if the patient is healthy. Elderly and overweight people have difficulties in breathing normally. Wearable medical devices are used to measure a person's respiration rate. A wired medical device used to measure respiration rate may cause uneasiness in a patient and cause an error in the measurement of their respiration rate. It is better to use a wireless medical device to measure respiration rate. The measured respiration rate may be transmitted to a medical center and, if needed, a doctor may contact the patient for further analysis and inspection.

9.4.4 Measurement of Human Body Temperature

The temperature of a healthy person ranges between 35°C and 38°C. Temperatures below or above this range may indicate that the person is sick. Temperatures above 40°C may cause death. A patient's body temperature may be transmitted to a medical center and, if needed, a doctor may contact the patient for further analysis.

9.4.5 Measurement of Sweat Rate

Glucose is the primary energy source of human beings. Glucose is usually supplied to the human body as sugar; sugar is a monosaccharide that provides energy to the human body. When a person does extensive physical activity, glucose comes out of their skin as sweat. A wearable medical device may be used to monitor and measure the sweat rate of a person when extensive physical activity is done. A wearable medical device may be attached to the person's clothes, which are in proximity to the skin, to monitor and measure their sweat rate. This device may also be used to measure the sweat PH, which is important in the diagnosis of diseases. The water vapor evaporated from the skin is absorbed in the medical device to determine the sweat PH. If the amount of sweat coming out of the body is too high, the body may dehydrate. Dehydration causes tiredness and fatigue. Measurements of sweat rate and PH may be used to monitor the physical activity of a person.

9.4.6 Measurement of Human Gait

Movements of human limbs is called human gait. Human gait is the way in which a human moves their limbs to go from one place to another place. Different gait patterns are characterized by differences in limb movement patterns, overall velocity, forces, kinetic and potential energy cycles, and changes in contact with the ground. Walking, jogging, skipping and sprinting are defined as natural human gait. Gait analysis is a helpful and fundamental research tool to characterize human locomotion. Wearable devices may be attached to different parts of the human body to measure and analyze human gait. The movement signals recorded by these devices can be used to analyze human gait. Temporal characteristics of gait are collected and estimated from wearable accelerometers and pressure sensors inside footwear.

In sports, gait analysis based on wearable sensors can be used for sport training and analysis and for the improvement of an athlete's performance. Results of ambulatory gait analysis may determine whether or not a particular treatment is appropriate for a patient. Motion analysis of human limbs during locomotion is applied in pre-operative planning for patients with cerebral palsy and can alter medical treatment decisions. Parkinson's disease is characterized by motor difficulties, such as gait difficulty, slowing of movement and limb rigidity. Gait analysis has been verified as one of the most reliable diagnostic signs of this disease. For patients with neurological problems, such as Parkinson's disease and stroke, ambulatory gait analysis is an important tool in their recovery process and can provide low-cost and convenient rehabilitation monitoring.

Gait analysis based on wearable devices may be applied in healthcare monitoring, such as in the detection of gait abnormalities, the assessment of recovery, fall risk estimation and in sport training. In healthcare centers, gait information is used to detect walking behavior abnormalities that may predict health problems or the progression of neurodegenerative diseases. Fall is the most common type of home accident among elderly people. Falls are a major threat to health and independence among elderly people. Gait analysis using wearable devices was used to analyze and predict fall among elderly patients [40–44].

9.4.7 WEARABLE DEVICES TRACKING AND MONITORING DOCTORS AND PATIENTS INSIDE HOSPITALS

Each patient may have a wearable device attached to their body. The wearable device is connected to several sensors; each sensor has its own specific task to perform. For example, one sensor node may be detecting the heart rate and body temperature while another is detecting the blood pressure. Doctors can also carry a wearable device, which allows other hospital personnel to locate them within the hospital.

9.5 WBANs

The main goal of WBANs is to provide continuous biofeedback data. WBANs can record electrocardiograms and can measure body temperature, blood pressure, heartbeat rate, arterial blood pressure, electro-dermal activity and other healthcare parameters in an efficient way. For example, accelerometers can be used to sense heartbeat rate, movement or even muscular activity. Body area networks (BANs) include applications and communication devices using wearable and implantable wireless networks. A sensor network that senses health parameters is called a body sensor network (BSN). A WBAN is a special purpose wireless-sensor network that incorporates different networks and wireless devices to enable remote monitoring in various environments.

An application of WBANs is in medical centers where the conditions of a large number of patients are constantly being monitored as presented in Figures 9.1 to 9.3. Wireless monitoring of the physiological signals of a large number of patients is needed in order to deploy a complete wireless sensor network (WSN) in healthcare centers. Human health monitoring is emerging as a significant application of

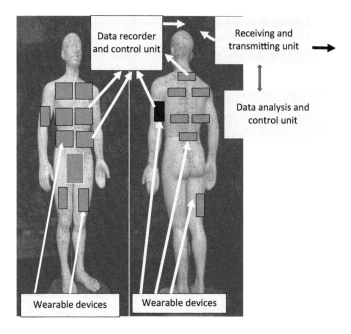

FIGURE 9.1 Wearable devices for various medical applications.

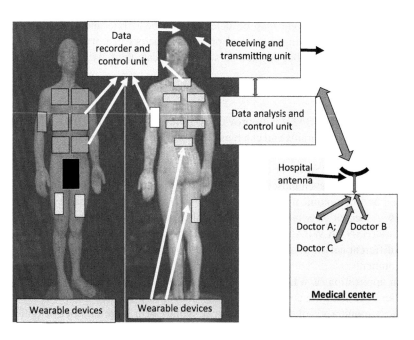

FIGURE 9.2 Wearable body area network for various medical applications.

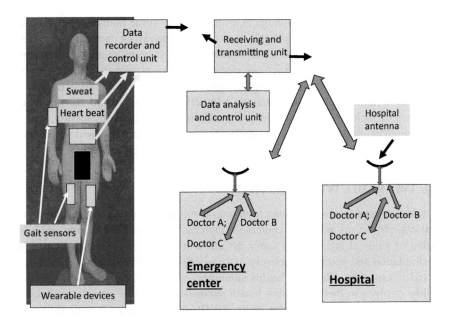

FIGURE 9.3   Wearable body area network health monitoring system.

embedded sensor networks. A WBAN can monitor vital signs, providing real-time feedback to allow many patient diagnostics procedures using continuous monitoring for chronic conditions or to monitor progress of recovery from an illness. Recent technological advances in wireless networking promise a new generation of WSNs suitable for human body wearable network systems.

Data acquisition in WBAN devices can be point-to-point or multipoint-to-point, depending on specific applications. Detection of an athlete's health condition would require point-to-point data sharing across various on-body sensors. Human body monitoring of vital signs will require data to be routed from several wearable sensors, multipoint-to-point, to a sink node, which in turn can relay the information wirelessly to an out-of-body computer. Data may be transferred in real-time mode or non-real time. Human body monitoring applications require real-time data transfer. The results of monitoring an athlete's physiological data can be collected offline for processing and analysis purposes.

A typical WBAN consists of a number of compact low-power sensing devices, control unit and wireless transceivers. The power supply for these components should be compact, lightweight and long lasting as well. WBANs consists of small devices, fewer opportunities for redundancy and cover less space. To improve the efficiency of a WBAN it is important to minimize the number of nodes in the network. Adding more devices and path redundancy to solve node failure and network problems cannot be a practical option in WBAN systems. WBANs receive and transmit a large amount of data constantly. Data processing must be hierarchical and efficient to deal with asymmetry of several resources, to maintain system efficiency and to ensure the availability of data. WBANs in a medical area consist of wearable and implantable sensor nodes that can sense biological information from the human body and transmit it over a short distance wirelessly to a control device worn on the body or placed in an accessible location. The sensor electronics must be miniaturized, low power and detect medical signals such as electrocardiograms, electroencephalography, pulse rate, pressure and temperature. The gathered data from the control devices are then transmitted to remote destinations in a WBAN for diagnostic and therapeutic purposes by including other wireless networks for long-range transmission.

A wireless control unit is used to collect information from wired and wireless sensors that transmit the information to a remote monitoring station.

9.6 WEARABLE WBAN (WWBAN)

Wireless communication systems offer a wide range of benefits to medical centers, patients, physicians and sport centers through the continuous measuring and monitoring of medical information, early detection of abnormal conditions, supervised rehabilitation and potential discovery of knowledge through data analysis of the collected information. Wearable health monitoring systems allow a person to closely follow changes in their important health parameters and they provide feedback for maintaining optimal health status. If the WWBAN is part of a telemedicine system, the medical system can alert medical personnel when life-threatening events occur, as presented in Figure 9.4. In addition, patients may benefit from continuous long-term monitoring as a part of a diagnostic procedure. We can achieve optimal maintenance of a chronic condition or can monitor the recovery period after an acute event or surgical procedure. The collected medical data may be a very good indicator of the

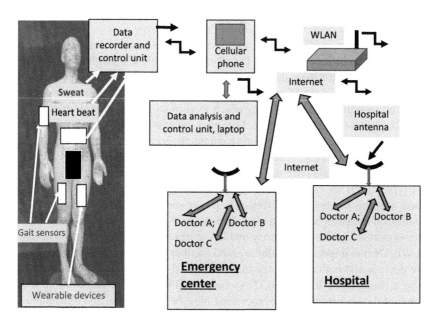

FIGURE 9.4   Wireless wearable body area network health monitoring system.

cardiac recovery of patients after heart surgery. Long-term monitoring can also confirm adherence to treatment guidelines or help monitor the effects of drug therapy. Health monitors can be used to monitor the physical rehabilitation of patients during stroke rehabilitation or brain trauma rehabilitation and after hip or knee surgeries. Many people use WBAN devices, such as wearable heart rate monitors, respiration rate monitor and pedometers, for medical reasons or as part of a fitness regime. WBANs may be attached to cotton shirts to measure respiratory activity, electrocardiograms, electromyograms and body posture.

9.7 WEARABLE RFID TECHNOLOGY AND ANTENNAS

9.7.1 INTRODUCTION

RFID is an electronic method of exchanging data via radio frequency (RF) waves. RFID uses electromagnetic fields to automatically identify and track tags attached to persons, animals and objects. There are four major components in a RFID system: the transponder, tag, antenna and a controller. The tags contain electronically stored information. Passive tags collect energy from a nearby RFID reader by employing radio waves. Active tags have a local power source (such as a battery) and may operate hundreds of meters from the RFID reader. Unlike a barcode, the tag should not be within the line of sight of the reader, so it may be embedded in the tracked person or object. The RFID tag, antenna and controller may be assembled on the same board. Printed small antennas are the best choice for RFID devices. Microstrip antennas were widely presented in books and papers in the last decade [47–53]. However, compact wearable printed antennas are not widely used at 13.5 MHz and in UHF RFID systems. RFID loop antennas are widely used. Several RFID loop antennas are presented in [47]. RFID loop antennas have low efficiency and narrow bandwidth. Wideband compact wearable printed antennas for RFID applications are presented in this chapter. RF transmission properties of human tissues have been investigated in papers [45, 46]. The effect of the human body on an antenna's performance for RFID applications is investigated in this chapter. The proposed antennas may be used as wearable antennas on persons or animals. The proposed antennas may be attached to cars, trucks, containers and other various objects. High frequency (HF) tags work best at close range but are more effective at penetrating non-metal objects especially objects with high water content.

9.7.2 RFID TECHNOLOGY

RFID technology has been available for more than 50 years. However, the progress in electronic, communication and materials technologies enhanced the development and production of low-cost RFID tags. A lack of standards in the RFID industry caused a delay in the development and production of multipurpose international RFID tags. RFID technology is an electronic method of exchanging data between persons, objects and products over RF waves. RFID tags track vehicles, airline passengers, patients with Alzheimer's disease, employees and animals. Data stored on RFID tags can be changed, updated and locked.

A RFID system consists of three parts.

- An antenna. Usually a printed antenna. The antenna receives and transmits the signal.
- A transceiver with a decoder to interpret the data. The transceiver is an integrated circuit for storing and processing information that modulates and demodulates RF signals.
- A transponder and a direct current (DC) power unit that collects power from the incident reader signal.

The antenna radiates RF fields in a relatively short range. When an RFID tag passes through the electromagnetic fields that the antenna radiates, the tag detects the activation signal. The activation signal "starts-up" the RFID chip, and it transmits the information on its microchip, which is picked up by the antenna. Tags may be read only, having a factory-assigned serial number that is used as a key into a database. There are also read tags and write tags. Object-specific data may be written into the tag by the system user. Field programmable tags may be write-once and read-multiple. The tag information is stored in a nonvolatile memory. The RFID tag includes either fixed or programmable logic for processing the transmission and sensor data, respectively. An RFID reader transmits an encoded radio signal to interrogate the tag. The RFID tag receives the message and then responds with its identification and other information. This can be a unique tag serial number or may be product information, such as a stock number, lot, batch number, production date and other specific information. Tags have individual serial numbers. The RFID system can scan several tags that might be within the range of the RFID reader and read them simultaneously. A passive RFID tag gets power from the reader through inductive coupling method. The reader consists of a coil connected to an alternating current (AC) supply such that a magnetic field is formed around it. The tag coil is placed in the vicinity of the reader coil and an electromotive force (EMF) is induced that is based on Faraday's law of induction. The EMF causes a flow of current in the coil, thus producing a magnetic field around it. By employing Lenz's law, the magnetic field of the tag coil opposes the reader's magnetic field and there will be a subsequent increase in the current through the reader coil. The reader intercepts this as the load information. This system may operate at very short distance communication. The tag AC voltage appearing across the tag coil is converted to DC by using a rectifier. In an active RFID system, the reader transmits a signal to the tag via the antenna. The tag receives this information and resends this information along with the information in its memory. The reader receives this signal and transmits to the processor for further processing. RFID frequency bands and applications are listed in Table 9.1. A RFID concept tag concept is shown in Figure 9.5.

Figure 9.6 presents a block diagram for one application of RFID tag reader. As shown in Figure 9.6, the block diagram consists of three main subsystems: a transceiver, a decoder, and a display module. The RFID tag will send its data as an amplitude modulated (AM) signal, so the reader will filter the signal to select and amplify

TABLE 9.1

Radio-frequency identification frequency bands and applications

Band	Regulations	Range	Data speed	ISO/IEC 18000 section	Applications
120–150 kHz (LF)	Unregulated	10 cm	Low	Part 2	Animal identification, factory data collection
13.56 MHz (HF)	ISM band worldwide	10 cm^{-1} m	Low to moderate	Part 3	Smart cards (ISO/IEC 15693, ISO/IEC 14443 A, B). Non-fully ISO compatible memory cards, microprocessor ISO compatible cards
433 MHz (UHF)	Short-range devices	1–100 m	Moderate	Part 7	Defense applications, with active tags
865–868 MHz (Europe) 902–928 MHz (North America) UHF	ISM band	1–12 m	Moderate to high	Part 6	EAN, various standards; used by railroads
2450–5800 MHz (microwave)	ISM band	1–2 m	High	Part 4	802.11 WLAN, Bluetooth standards
3.1–10 GHz (microwave)	Ultra-wide band	Max. 200 m	High	Not defined	Requires semi-active or active tags

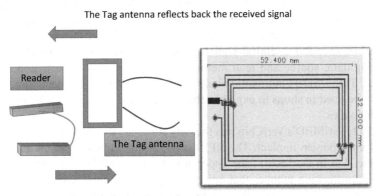

The Tag antenna reflects back the received signal

Reader

The Tag antenna

Reader antenna sends electric signal to the tag antenna

FIGURE 9.5 Radio-frequency identification concept. EAN European Article Number, ISM industrial, scientific and medical, WLAN wireless local area network.

the signal. After this filtering, the reader will demodulate the signal to obtain the signal transmitted by the tag. The decoder will decode the signal transmitted by the demodulator and perform the necessary transformations to get the identification data. Finally, this data is sent to a video display module to generate the necessary signals to output the data to a display.

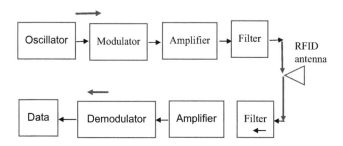

FIGURE 9.6   Block diagram of a radio-frequency identification tag reader.

9.7.3 RFID STANDARDS

The US Food and Drug Administration (FDA) gave final approval on 17 October 2004 to Applied Digital Solutions to sell their VeriChip RFID tags for implantation into patients in hospitals. The intent is to provide immediate positive identification of patients both in hospitals and in emergency services. Doctors, emergency-room personnel and ambulance crews could get immediate identification without a need to search wallets and purses for identification. If, for example, you had a pre-existing medical condition or allergy, this could be taken into account immediately.

Since RFID tags can be attached to cash, clothing and possessions, or implanted in people and animals, the possibility of reading personally linked information without consent has raised serious privacy concerns. These concerns resulted in standard specifications development addressing privacy and security issues. ISO/IEC 18000 and ISO/IEC 29167 use on-chip cryptography methods for un-traceability, tag and reader authentication, and over-the-air privacy. ISO/IEC 20248 specifies a digital signature data structure for RFID and barcodes providing data, source and read method authenticity. This work is done within ISO/IEC JTC 1/SC 31 automatic identification and data capture techniques. Tags can also be used in shops to expedite checkout, and to prevent theft by customers and employees.

In 2011, VeriMED's VeriChip was the only RFID tag that had been cleared by the US FDA for human implant. The RFID tags may be implanted in the fatty tissue under the skin in the back of the upper arm. Each veriMED microchip contains a unique identification number that emergency personnel may scan to immediately identify the patient and access the patient's personal health information. This process helps the medical team to treat the patient without delay. This is especially important for patients who cannot communicate with the medical team. Although the FDA approved the use of the device for anyone 12 years of age or older, it is very important for patients with diabetes, stroke, seizure disorders, dementia, Alzheimer's disease and organ transplants. The estimated life of the tags is 20 years. The subdermal RFID tag location is invisible to the naked eye. A unique verification number is transmitted to a suitable reader when the person is within range. In July 2017, a Wisconsin company was the first in the United States to offer implanted chips for opening doors and logging in to computers. A company's employees may agree to an

RFID chip being implanted in their hand. The tiny chip will use near-field communication (NFC) technology to allow employees to unlock doors, make vending machine purchases, login to computers and access office tools like photocopiers, all with a wave of the hand.

9.8 WEARABLE DUAL POLARIZED 13.5 MHZ COMPACT PRINTED ANTENNA

One of the most critical elements of any RFID system is the electrical performance of its antenna. The antenna is the main component for transferring energy from the transmitter to the passive RFID tags, receiving the transponder's replying signal and avoiding in-band interference from electrical noise and other nearby RFID components. Low-profile compact printed antennas are crucial in the development of RFID systems.

A compact microstrip-loaded dipole antenna has been designed at 13.5 MHz to provide horizontal polarization. The antenna consists of two layers. The first layer consists of flame retardant-4 (FR-4) 0.8 mm dielectric substrate. The second layer consists of Kapton 0.8 mm dielectric substrate. The substrate thickness determines the antenna bandwidth. A printed slot antenna provides a vertical polarization. The proposed antenna is dual polarized. The printed dipole and the slot antenna provide dual orthogonal polarizations. The dual polarized RFID antenna is shown in Figure 9.7. The antenna dimensions are 6.4 × 6.4 × 0.16 cm. The antenna may be attached to a customer's shirt on the customer's stomach or back zone. The antenna has been analyzed by using Keysight Advanced Design System (ADS) software [54].

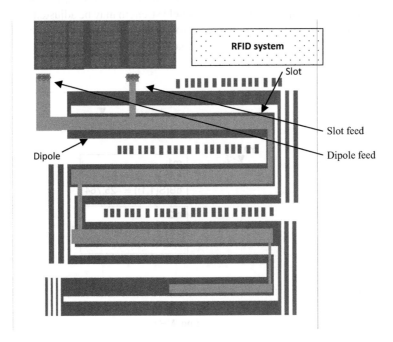

FIGURE 9.7   Printed compact dual polarized antenna, 64 × 64 × 1.6 mm.

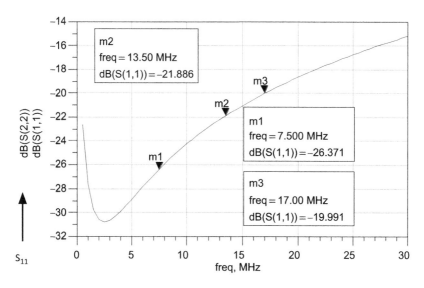

FIGURE 9.8 Computed S_{11} results.

The antenna S_{11} parameter is better than −21 dB at 13.5 MHz. The antenna gain is around −10 dBi. The antenna's beam width is around 160°. The computed S_{11} parameters are presented in Figure 9.8. There is a good agreement between measured and computed results. Figure 9.9 presents the antenna's measured S_{11} parameters on a human body. The antenna's cross-polarized field strength may be adjusted by varying the slot feed location. The computed radiation pattern is shown in Figure 9.10. The computed 3D radiation pattern of the antenna is shown in Figure 9.11.

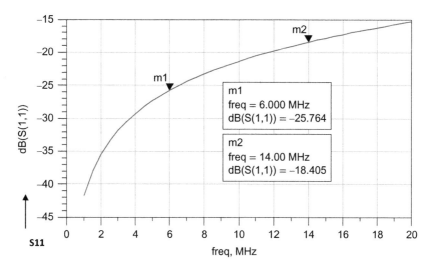

FIGURE 9.9 Measured S_{11} of the 13.5 MHz antenna on human body.

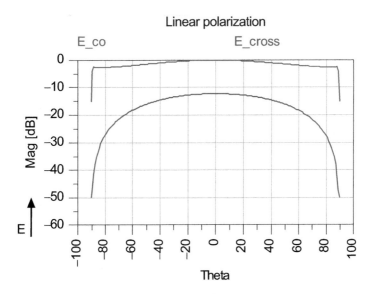

FIGURE 9.10 Radiation pattern of the 13.5 MHz antenna.

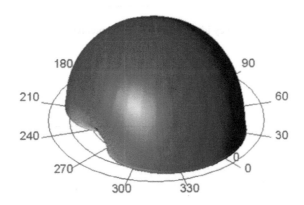

FIGURE 9.11 Three-dimensional antenna radiation pattern.

9.9 VARYING THE ANTENNA FEED NETWORK

Several designs with different feed networks have been developed. A compact antenna with a different feed network is shown in Figure 9.12. The antenna dimensions are 8.4 × 6.4 × 0.16 cm. Figure 9.13 presents the antenna's computed S_{11} on a human body. There is a good agreement between measured and computed results. The computed radiation pattern is shown in Figure 9.14. Table 9.2 compares the electrical performance of a loop antenna with the compact dual polarized antenna.

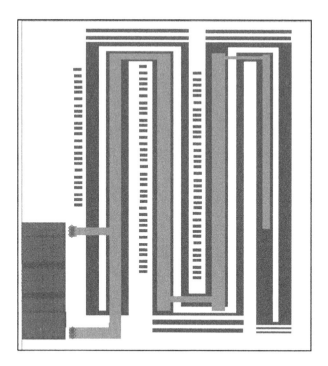

FIGURE 9.12   Radio-frequency identification printed antenna, $8.4 \times 6.4 \times 0.16$ cm.

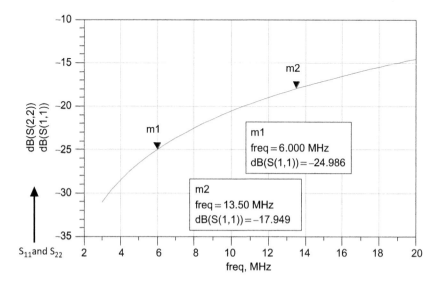

FIGURE 9.13   Radio-frequency identification antenna computed S_{11} and S_{22} results.

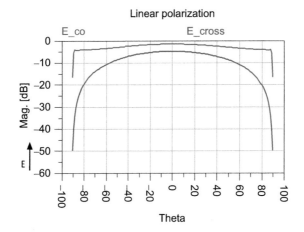

FIGURE 9.14   Compact antenna, 8.4 × 6.4 × 0.16 cm, radiation pattern.

TABLE 9.2
Comparison of loop antenna and microstrip antenna parameters

Antenna	3 dB Beam width (°)	Gain (dBi)	Voltage standing wave ratio
Loop antenna	140	−25	2:1
Microstrip antenna	160	−10	1.2:1

9.10   WEARABLE LOOP ANTENNAS FOR RFID APPLICATIONS

Several RFID loop antennas are presented in [47–53]. The disadvantages of loop antennas with a number of turns are low efficiency and narrow bandwidth. The real part of the loop antenna impedance approaches 0.5 Ω. The image part of the loop antenna impedance may be represented as high inductance. A matching network may be used to match the antenna to 50 Ω. The matching network consists of an RLC matching network. This matching network has narrow bandwidth. The loop antenna efficiency is lower than 1%.

A square four-turn loop antenna has been designed at 13.5 MHz by using Keysight ADS software. The antenna is printed on a FR-4 substrate. The antenna dimensions are 32 × 52.4 × 0.25 mm. The antenna layout is shown in Figure 9.15a. A photo of the antenna is shown in Figure 9.15b. S_{11} results of the printed loop antenna are shown in Figure 9.16. The antenna S_{11} parameter is better than −9.5 dB without an external matching network. The computed radiation pattern is shown in Figure 9.17. The computed radiation pattern takes into account an infinite ground plane.

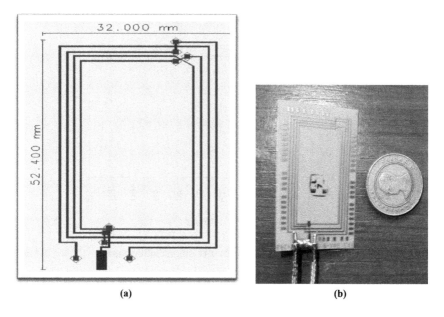

FIGURE 9.15 A square four-turn loop antenna. (a) Layout, (b) photo.

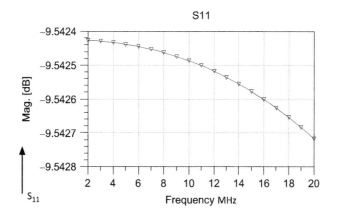

FIGURE 9.16 Loop antenna computed S_{11} results.

The microstrip antenna input impedance variation as a function of distance from the body has been computed by employing Momentum software. The analyzed structure is presented in Figure 9.18. Properties of human body tissues are listed in Table 9.3, see [45, 46]. These properties were used in the antenna design. S_{11} parameters for different human body thicknesses have been computed. We may note that

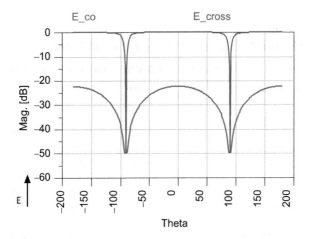

FIGURE 9.17   Loop antenna radiation patterns for an infinite ground plane.

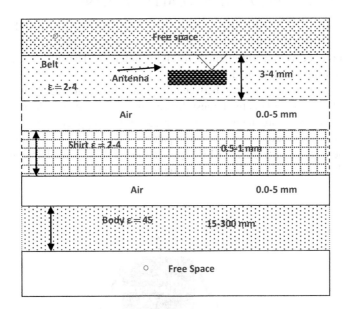

FIGURE 9.18   Antenna environment model.

the differences in the results for body thickness of 15–100 mm are negligible. S_{11} parameters for different positions relative to the human body have been computed. If the air spacing between the antenna and the human body is increased from 0 to 10 mm the antenna S_{11} parameters may change by less than 1%. The voltage standing wave ratio (VSWR) is better than 1.5:1.

TABLE 9.3
Electrical properties of body tissues [45, 46]

Tissue	Property	435 MHz	605 MHz	995 MHz
Human fat	σ	0.045	0.05	0.06
	ε	5.00	5	4.50
Human stomach	σ	0.67	0.75	0.79
	ε	42.9	41.40	39.05
Patient colon	σ	1.00	1.05	1.30
	ε	63.6	61.9	60.00
Skin	σ	0.57	0.6	0.63
	ε	41.6	40.45	40.25
Human lung	σ	0.27	0.27	0.27
	ε	38.4	38.4	38.4
Human kidney	σ	0.90	0.90	0.90
	ε	117.45	117.45	117.45

9.11 WEARABLE RFID ANTENNA APPLICATIONS

An application of the proposed antenna is shown in Figure 9.19. The RFID antennas may be assembled in a belt and attached to the customer's stomach. The antennas may be employed as transmitting or as receiving antennas. The antennas may receive or transmit information to medical systems.

In RFID systems the distance between the transmitting and receiving antennas is less than $2D^2/\lambda$, where D is the largest dimension of the antenna. The receiving and transmitting antennas are magnetically coupled. In these applications we refer to the near field and not to the far field radiation pattern. Figure 9.20 and Figure 9.21 present compact printed antenna for RFID applications. The presented antennas may be assembled in a belt and attached to the patient's stomach or back.

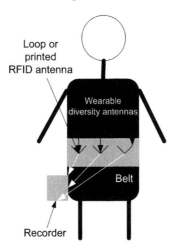

FIGURE 9.19 Wearable radio-frequency identification antenna.

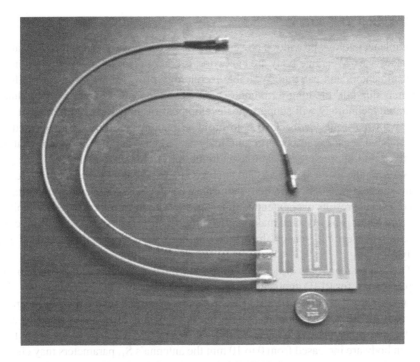

FIGURE 9.20   A microstrip antenna for radio-frequency identification applications.

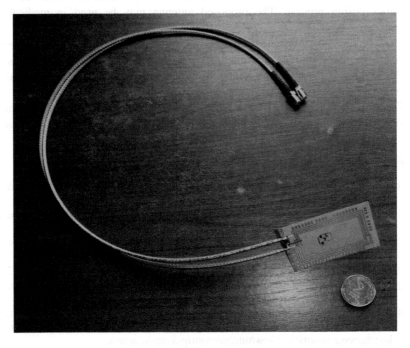

FIGURE 9.21   Loop antenna for radio-frequency identification applications.

9.12 CONCLUSIONS

Wearable technology provides a powerful new tool for medical and surgical rehabilitation services. Wearable body area network, is emerging as an important option for medical centers and patients. Wearable technology provides a convenient platform that may quantify the long-term context and physiological response of individuals. Wearable technology will support the development of individualized treatment systems with real-time feedback to help promote patients' health. Wearable medical systems and sensors can perform gait analysis and can measure body temperature, heartbeat, blood pressure, sweat rate and other physiological parameters of the person wearing the medical device. Gait analysis is a useful tool in clinical practice and biomechanical research. Gait analysis using wearable sensors provides quantitative and repeatable results over extended time periods with low cost and good portability, showing better prospects and making great progress in recent years. At present, commercialized wearable sensors have been adopted in various applications of gait analysis.

This chapter presents wideband compact printed antennas, and microstrip and loop antennas, for RFID applications. The antenna's beam width is around 160°. The antenna's gain is around −10 dBi. The proposed antennas may be used as wearable antennas on persons or animals. The proposed antennas may be attached to cars, trucks, and other various objects. If the air spacing between the antenna and the human body are increased from 0 to 10 mm the antenna's S_{11} parameters may change by less than 1%. The antenna's VSWR is better than 1.5:1 for all tested environments. The effect of the antenna's location on the human body should be considered in the antenna's design process. The proposed antenna may be used in medical RF systems.

A square four-turn loop antenna has been designed at 13.5 MHz by using Keysight ADS software. The antenna is printed on a FR-4 substrate. The antenna dimensions are 32 × 52.4 × 0.25 mm. The antenna S_{11} parameter is better than −9.5 dB without an external matching network.

The RFID antennas may be assembled in a belt and attached to the customer's stomach. The antennas may be employed as transmitting or as receiving antennas.

REFERENCES

[1] Sabban, A., *Low Visibility Antennas for Communication Systems*, Taylor & Francis Group, New York, 2015.
[2] Sabban, A., *Wideband RF Technologies and Antenna in Microwave Frequencies*, Taylor & Francis Group, New York, 2016.
[3] Sabban, A., New wideband printed antennas for medical applications, *IEEE Journal, Transactions on Antennas and Propagation*, 61(1), 84–91, January 2013.
[4] Mukhopadhyay, S.C. editor, *Wearable Electronics Sensors*, Springer, Switzerland, 2015.
[5] Sabban, A., Small wearable meta materials antennas for medical systems, *The Applied Computational Electromagnetics Society Journal*, 31(4), 434–443, April 2016.
[6] Sabban, A., Microstrip antenna arrays, *Microstrip Antennas*, Nasimuddin Nasimuddin (Ed.), InTech, Croatia, 361–384, 2011, ISBN: 978-953-307-247-0, http://www.intechopen.com/articles/show/title/microstrip-antenna-arrays.

[7] U.S Patent, Inventors: Albert Sabban, Dual polarized dipole wearable antenna, U.S Patent number: 8203497, June 19, 2012, USA.

[8] Bonfiglio, A. and De Rossi, D., *Wearable Monitoring Systems*, Springer, New York, 2011

[9] Gao, T., Greenspan, D., Welsh, M., Juang, R.R., and Alm, A., Vital signs monitoring and patient tracking over a wireless network, Proceedings of IEEE-EMBS 27th Annual International Conference of the Engineering in Medicine and Biology, Shanghai, China, 1–5 September 2005, 102–105.

[10] Otto, C.A., Jovanov, E., and Milenkovic, E.A., WBAN-based system for health monitoring at home, Proceedings of IEEE/EMBS International Summer School, Medical Devices and Biosensors, Boston, MA, USA, 4–6 September 2006, 20–23.

[11] Zhang, G.H., Poon, C.C.Y., Li, Y., and Zhang, Y.T., A Biometric Method to Secure Telemedicine Systems, Proceedings of the 31st Annual International Conference of the IEEE Engineering in Medicine and Biology Society, Minneapolis, MN, USA, September 2009, 701–704.

[12] Bao, S., Zhang, Y., and Shen, L., Physiological signal based entity authentication for body area sensor networks and mobile healthcare systems, Proceedings of the 27th Annual International Conference of the IEEE EMBS, Shanghai, China, 1–4 September 2005, 2455–2458.

[13] Ikonen, V. and Kaasinen, E., Ethical assessment in the design of ambient assisted living, Proceedings of Assisted Living Systems—Models, Architectures and Engineering Approaches, Schloss Dagstuhl, Germany, November 2008, 14–17.

[14] Srinivasan, V., Stankovic, J., and Whitehouse, K., Protecting your daily in home activity information from a wireless snooping attack, Proceedings of the 10th International Conference on Ubiquitous Computing, Seoul, Korea, 21–24 September 2008, 202–211.

[15] Casas, R., Blasco, M.R., Robinet, A., Delgado, A.R., Yarza, A.R., Mcginn, J., Picking, R., and Grout, V., User modelling in ambient intelligence for elderly and disabled people, Proceedings of the 11th International Conference on Computers Helping People with Special Needs, Linz, Austria, July 2008, 114–122.

[16] Jasemian, Y., Elderly comfort and compliance to modern telemedicine system at home, Proceedings of the Second International Conference on Pervasive Computing Technologies for Healthcare, Tampere, Finland, 30 January–1 February 2008, 60–63.

[17] Atallah, L., Lo, B., Yang, G.Z., and Siegemund, F., Wirelessly accessible sensor populations (WASP) for elderly care monitoring, Proceedings of the Second International Conference on Pervasive Computing Technologies for Healthcare, Tampere, Finland, 30 January–1 February 2008, 2–7.

[18] Hori, T., Nishida, Y., Suehiro, T., and Hirai, S., SELF-Network: Design and implementation of network for distributed embedded sensors, Proceedings of IEEE/RSJ International Conference on Intelligent Robots and Systems, Takamatsu, Japan, October–5 November 2000, 1373–1378.

[19] Mori, Y., Yamauchi, M., and Kaneko, K., Design and implementation of the vital sign box for home healthcare, Proceedings of IEEE EMBS International Conference on Information Technology Applications in Biomedicine, Arlington, VA, USA, November 2000, 104–109.

[20] Lauterbach, C., Strasser, M., Jung, S., and Weber, W., Smart clothes self-powered by body heat, Proceedings of Avantex Symposium, Frankfurt, Germany, May 2002, 5259–5263.

[21] Marinkovic, S. and Popovici, E., Network coding for efficient error recovery in wireless sensor networks for medical applications, Proceedings of International Conference on Emerging Network Intelligence, Sliema, Malta, 11–16 October 2009, 15–20.

[22] Schoellhammer, T., Osterweil, E., Greenstein, B., Wimbrow, M., and Estrin, D., Lightweight temporal compression of microclimate datasets, Proceedings of the 29th Annual IEEE International Conference on Local Computer Networks, Tampa, FL, USA, 16–18 November 2004, 516–524.

[23] Barth, A.T., Hanson, M.A., Powell, H.C., and Lach, J., Tempo 3.1: A body area sensor network platform for continuous movement assessment, Proceedings of the 6th International Workshop on Wearable and Implantable Body Sensor Networks, Berkeley, CA, USA, June 2009, 71–76.

[24] Gietzelt, M., Wolf, K.H., Marschollek, M., and Haux, R., Automatic self-calibration of body worn triaxial-accelerometers for application in healthcare, Proceedings of the Second International Conference on Pervasive Computing Technologies for Healthcare, Tampere, Finland, January 2008, 177–180.

[25] Gao, T., Greenspan, D., Welsh, M., Juang, R.R., and Alm, A., Vital signs monitoring and patient tracking over a wireless network, Proceedings of the 27th Annual International Conference of the IEEE EMBS, Shanghai, China, 1–4 September 2005, 102–105.

[26] Purwar, A., Jeong, D.U., and Chung, W.Y., Activity monitoring from realtime triaxial accelerometer data using sensor network, Proceedings of International Conference on Control, Automation and Systems, Hong Kong, 21–23 March 2007, 2402–2406.

[27] Baker, C., Armijo, K., Belka, S., Benhabib, M., Bhargava, V., Burkhart, N., Der Minassians, A., Dervisoglu, G., Gutnik, L., Haick, M., Ho, C., Koplow, M., Mangold, J., Robinson, S., Rosa, M., Schwartz, M., Sims, C., Stoffregen, H., Waterbury, A., Leland, E., Pering, T., and Wright, P., Wireless sensor networks for home health care, Proceedings of the 21st International Conference on Advanced Information Networking and Applications Workshops, Niagara Falls, Canada, 21–23 May 2007, 832–837.

[28] Schwiebert, L., Gupta, S.K.S., and Weinmann, J., Research challenges in wireless networks of biomedical sensors, Proceedings of the 7th Annual International Conference on Mobile Computing and Networking, Rome, Italy, 16–21 July 2001, 151–165.

[29] Aziz, O., Lo, B., King, R., Darzi, A., and Yang, G.Z., Pervasive body sensor network: An approach to monitoring the postoperative surgical patient, Proceedings of International Workshop on Wearable and Implantable Body Sensor Networks (BSN 2006), Cambridge, MA, USA, 2006, 13–18.

[30] Kahn, J.M., Katz, R.H., and Pister, K.S.J., Next century challenges: mobile networking for smart dust, Proceedings of the ACM MobiCom'99, Washington, DC, USA, August 1999, 271–278.

[31] Noury, N., Herve, T., Rialle, V., Virone, G., Mercier, E., Morey, G., Moro, A., and Porcheron, T., Monitoring behavior in home using a smart fall sensor, Proceedings of IEEE-EMBS Special Topic Conference on Micro-technologies in Medicine and Biology, Lyon, France, 12–14 October 2000, 607–610.

[32] Kwon, D.Y. and Gross, M., Combining body sensors and visual sensors for motion training, Proceedings of the 2005 ACM SIGCHI International Conference on Advances in Computer Entertainment Technology, Valencia, Spain, 15–17 June 2005, 94–101.

[33] Boulgouris, N.K.; Hatzinakos, D.; Plataniotis, K.N., Gait recognition: A challenging signal processing technology for biometric identification, *IEEE Signal Processing Magazine*, 22, 78–90, 2005.

[34] Kimmeskamp, S. and Hennig, E.M., Heel to Toe motion characteristics in parkinson patients during free walking, *Clinical Biomechanics*, 16, 806–812, 2001.

[35] Turcot, K., Aissaoui, R., Boivin, K., Pelletier, M., Hagemeister, N., and de Guise, J.A., New accelerometric method to discriminate between asymptomatic subjects and patients with medial knee osteoarthritis during 3-D gait, *IEEE Transactions on Biomedical Engineering*, 55, 1415–1422, 2008.

[36] Furnée, H., Real-time motion capture systems. In *Three-Dimensional Analysis of Human Locomotion*, Allard, P., Cappozzo, A., Lundberg, A., and Vaughan, C.L., Eds., John Wiley & Sons, Chichester, UK, 1997, 85–108.

[37] Bamberg, S.J.M., Benbasat, A.Y., Scarborough, D.M., Krebs, D.E., and Paradiso, J.A., Gait analysis using a shoe-integrated wireless sensor system, *IEEE Transactions on Information Technology in Biomedicine*, 12, 413–423, 2008.

[38] Choi, J.H., Cho, J., Park, J.H., Eun, J.M., and Kim, M.S., An efficient gait phase detection device based on magnetic sensor array, Proceedings of the 4th Kuala Lumpur International Conference on Biomedical Engineering, Kuala Lumpur, Malaysia, 25–28 June 2008, 21, 778–781.

[39] Hidler, J., Robotic-assessment of walking in individuals with gait disorders, Proceedings of the 26th Annual International Conference of the IEEE Engineering in Medicine and Biology Society, San Francisco, CA, USA, 1–5 September 2004, 7, 4829–4831.

[40] Wahab, Y. and Bakar, N.A, Gait analysis measurement for sport application based on ultrasonic system, Proceedings of the 2011 IEEE 15th International Symposium on Consumer Electronics, Singapore, 14–17 June 2011, 20–24.

[41] De Silva, B., Jan, A.N., Motani, M., and Chua, K.C., A real-time feedback utility with body sensor networks, Proceedings of the 5th International Workshop on Wearable and Implantable Body Sensor Networks (BSN 08), Hong Kong, 1–3 June 2008, 49–53.

[42] Salarian, A., Russmann, H., Vingerhoets, F.J.G., Dehollain, C., Blanc, Y., Burkhard, P.R., and Aminian, K., Gait assessment in Parkinson's disease: Toward an ambulatory system for long-term monitoring, *IEEE Transactions on Biomedical Engineering*, 51, 1434–1443, 2004.

[43] Atallah, L., Jones, G.G., Ali, R., Leong, J.J.H., Lo, B., and Yang, G.Z., Observing recovery from knee-replacement surgery by using wearable sensors, Proceedings of the 2011 International Conference on Body Sensor Networks, Dallas, TX, USA, 23–25 May 2011, 29–34.

[44] ElSayed, M., Alsebai, A., Salaheldin, A., El Gayar, N., and ElHelw, M., Ambient and wearable sensing for gait classification in pervasive healthcare environments, Proceedings of the 12th IEEE International Conference on e-Health Networking Applications and Services (Healthcom), Lyon, France, 1–3 July 2010, 240–245.

[45] Chirwa, L.C., Hammond, P.A., Roy, S., and Cumming, D.R.S., Electromagnetic Radiation from Ingested Sources in the Human Intestine between 150 MHz and 1.2 GHz, *IEEE Transaction on Biomedical Engineering*, 50(4), 484–492, April 2003.

[46] Werber, D., Schwentner, A., and Biebl, E.M., Investigation of RF transmission properties of human tissues, *Advanced Radio Science*, 4, 357–360, 2006.

[47] Fujimoto, K. and James, J.R., Eds., *Mobile Antenna Systems Handbook*, Artech House, Boston USA, 1994.

[48] Gupta, B., Sankaralingam, S., and Dhar, S., Development of wearable and implantable antennas in the last decade, Microwave Mediterranean Symposium (MMS), Guzelyurt, Turkey, August 2010, 251–267.

[49] Thalmann, T., Popovic, Z., Notaros, B.M, and Mosig, J.R., Investigation and design of a multi-band wearable antenna, 3rd European Conference on Antennas and Propagation, EuCAP 2009, Berlin, Germany, 462–465.

[50] Salonen, P., Rahmat-Samii, Y., and Kivikoski, M., Wearable antennas in the vicinity of human body, IEEE Antennas and Propagation Society Symposium, Monterey, USA, 2004, Vol. 1, 467–470.

[51] Kellomaki, T., Heikkinen, J., and Kivikoski, M., Wearable antennas for FM reception, First European Conference on Antennas and Propagation, EuCAP 2006, Hague, Netherlands, 2006, 1–6

[52] Sabban, A., Wideband printed antennas for medical applications, APMC 2009 Conference, Singapore, 12/2009.

[53] Lee, Y., Antenna Circuit Design for RFID Applications, Microchip Technology Inc., Microchip, Application Note 710c, AZ, USA.

[54] Keysight ADS software web, http://www.keysight.com/en/pc-1297113/advanced-design-system-ads?cc=IL&lc=eng.

10 Wearable Textile Systems and Antennas for IOT and Medical Applications

*Kashif Nisar Paracha, Ping Jack Soh,
Mohd Ridzuan Mohd Nor and
Mohammad Nazri Md Noor*

CONTENTS

10.1 INTRODUCTION

The extensive data demands from the multimedia communication-based society of today have been caused by the intensive use of smartphones and communication devices. This is expected to require an aggregate data rate 1000 times the current speed to cater for the increasing number of users, which will translate into tens of gigabyte per second (Gbps). Besides that, this also requires a highly dense low latency network with a round-trip time of less than 1 ms [1].

According to a Cisco annual report, the Cisco Virtual Networking Index (VNI, 2014–2019) [2], traffic from the Internet of Things (IOT) and wearable devices will increase to 277 petabytes per month by 2019, whereas the number of devices will increase to 578 million in 2019, which is five times the number of devices in 2014. A substantial portion of consumer electronics is expected to be a part of consumer outfits. Smart watches, virtual reality (VR) glasses, outfits and accessories used in rescue operations (e.g., fireproof clothing and helmet), activity trackers, stretchable sensors and flexible displays are several important examples of such devices [3–5].

Wearable systems are devices that can be worn by a person; they have the capability to connect with each other directly or through embedded wireless connectivity. To provide them with the wireless connectivity they require, conformal and flexible antennas need to be integrated with them. These antennas should be bendable, mechanically robust and environmentally resistant as part of future conformal/flexible wearable devices. As the wearable antennas are placed very close to the lossy medium of the users' body, a strong distortion in the radiation pattern and a shift in the resonant frequency are expected; hence, they need to be investigated and quantified.

On the other hand, the IOT is a technology through which all devices or "things," from household equipment to daily consumer electronics, will be interconnected wirelessly. According to a recent report by Machina Research [6], 71% of all IOT connections of today use a short-range technology (e.g. Wi-Fi and ZigBee). Cisco indicated that IOT will create US$14.4 trillion of value in the next decade, offering the opportunity to increase global corporate profits by 21% [7]. Consumer electronics, building security and automation are a major part of these IOT connections. Hence, wirelessly connected consumer electronics will play a vital role in the realization of IOT concepts supported by the future fifth generation (5G) networks. Wearable systems have diverse applications in our daily life. They are not limited to wristwatches, fitness bands, augmented reality glasses, but will also encompass many medical applications [8, 9].

This chapter will first explore the flexible materials, particularly textile materials, which are being used in the literature to build such systems and antennas; then, several important examples of textile systems and applications for communication and sensing for the IOT and medical applications are presented. Furthermore, wearable textile systems and antennas for purposes other than these two applications, such as location tracking and energy-harvesting, are also discussed.

10.2 TEXTILE MATERIALS

Wearable systems and antennas are built using different types of conductive and dielectric materials. These materials are carefully chosen to suit different applications, so that they may provide a reasonable amount of mechanical deformations (bending and wrapping) with minimal influence, based on different weather conditions (rain, snow, ice, etc.) and proper electromagnetic radiation protection. Recently, various types of fabric (textiles) and non-fabric materials have been used for wearable antennas. Textiles are mostly fabricated from polymer-based materials, which are not conductive in their original form. They are usually used without conducting elements, such as dielectric substrates, and need to be properly characterized prior to use [10]. On the other hand, the use of non-fabric and commercial flexible polymer-based materials, such as Kapton, polyethylene terephthalate (PET) and polyethylene naphthalate (PEN) substrates, is also increasing as they offer predefined stable dielectric properties.

Generally, for flexible antennas, the conductivity of the radiating part needs to be ideally high, whereas the substrate (textile fabric/polymer film) must be constant in thickness. Besides that, low permittivity (ε_r), a quantity to measure the ability of a material to store electrical energy in an electric field, and low loss tangent (tan δ) are

TABLE 10.1
The properties of several non-conductive (dielectric) materials

Non-conductive material (dielectric)	Relative dielectric constant (ε_r)	Dielectric loss (tan δ)
Nylon 6	3.5–5.0	0.15
Poly (ethylene terephthalate) (PET)	3.1–2.35	0.028
Poly-di-methyl-siloxane (PDMS) ceramic composite [14]	6.25	0.02
Poly-di-methyl-siloxane (PDMS) [15]	3.2	0.01
Ethylene-vinyl acetate (EVA) [16]	2.8	0.002
Mn-doped zinc ferrite [17]	7.5	0.025

preferred. It is well known that electro-textile antennas are limited in terms of gain, efficiency and bandwidth in comparison to their copper counterparts due to their relatively lower conductivity. Several types of popular conductive textiles include nickel-plated (corrosion resistant), silver-plated (strong and flexible), Flectron (copper-coated nylon fabric) and Nora conductive fabrics, and they feature different sheet resistances [11]. Providing high conductivity in electro-textile materials is the primary challenge in enabling continuous conduction current paths to minimize resistance losses and, hence, increasing antenna radiation efficiency [12]. A comprehensive review of these textile material is presented in [13] (Table 10.1).

The substrate used in a wearable antenna is of critical importance in terms of wearability, operation and fabrication. Most of the flexible substrates that are used are low in permittivity and loss tangent. This is to increase their efficiency in the presence of the human body, at the cost of larger antenna sizes. Felt, fleece, silk, and Cardura are a few examples of them. The measured dielectric values of different textile fabrics were provided in [18]. Other investigations, such as in [19], have also presented a complete inkjet-printed localization tracking system on different types of textile substrates for wearable applications.

The effects of different flexible substrate parameters on the performance of a patch antenna were investigated in [13, 17]. In [17], the researchers examined the magneto-dielectric flexible material properties under both flat and bending conditions. The flexible substrate exhibits high efficiency, adequate gain and stable radiation pattern without affecting bandwidth. Besides that, textile and flexible antennas are susceptible to deformation due to bending and stretching [20]. The main drawback of stretchable wearable antennas is their low radiation efficiency when being stretched [21]. To overcome this limitation, different stretchable conductive materials can be used with various conductive doping materials. Several examples of them include silver flakes embedded silicone [22, 23], silver loaded fluorine rubber [20], carbon nanotube (CNT)-based films [24], liquid metals in stretchable substrate [25] and, more importantly, the use of stretchable fabric itself (Figure 10.1.) [26, 27].

In [28], a textile patch antenna using Zelt as a conductive material and felt fabric as a substrate has been presented. Fabrics usually feature lower dielectric properties,

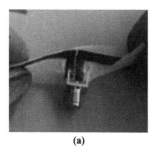
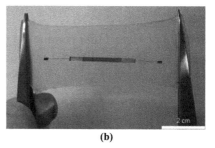

(a)   (b)

FIGURE 10.1   (a) Lycra fabric on porous film [24], (b) composite fiber mat [26].

which make them easily operational at lower frequencies. Besides that, low-cost foams have also been used as the substrate for a planar inverted-F antenna (PIFA) integrated in a life jacket in [29]. This antenna is aimed for use in a rescue system in one of the satellite bands. To produce the antenna, conductive inks such as the silver-based Harima NPS-J and Cabot CCI-300 have been used in this prototype, implemented using inkjet-printing technology.

10.3 TEXTILE SYSTEMS AND ANTENNAS FOR IOT AND MEDICAL APPLICATIONS

As aforementioned, IOT technology will enable millions of devices to be connected and to communicate with each other. It is forecasted that a significant number of these devices will be wearable systems and devices. Moreover, there would be a need for IOT devices and data to be aggregated and channeled to a user from different wireless networks, and this would ideally be done using wearable devices. In such a situation, wearable devices would provide users with a supervisory or monitoring role to access information sensed from their household equipment.

Wearable systems, with wearable antennas as one of their important components, offer ease of integration and offer comfort because they can be designed to be partial forms of clothing or accessories using textiles. The implementation of medical applications using wearable systems can be effective in real-time communication, sensing and location tracking of patients. Critical health parameters of the patients may also be monitored more efficiently. Examples include: a glucose monitoring system to monitor the sugar level of a patient; capsule endoscopy to examine the inner intestinal system; and wearable Doppler unit and thermometer to monitor heartbeat, blood pressure and temperature of the body [9].

Accessories on clothing, such as buttons and zippers, can commonly be integrated into clothes and items such as bags. They are effective radiators due to their metallic structure. This enables simple, low-cost and even tunable antennas to be realized via this method, due to their varying lengths. Instances of the use of this property in daily outfits for wireless local area network (WLAN) communication are demonstrated in [30–32]. The authors in [31] integrated a monopole antenna in the zip of a handbag and managed to obtain up to 5 dBi of realized gain at 2.45 GHz. The reported antenna

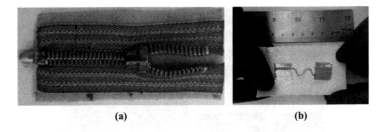

(a)   (b)

FIGURE 10.2   (a) Zip antenna [30], (b) antenna on stretchable fabric [34].

also showed reasonable performance even in the presence of a hand phantom. On the other hand, a zipper antenna sewed into a pure jeans substrate produced a measured realized gain of 0 dBi in [30], as shown in Figure 10.2.(a).

Next, a shoelace antenna fabricated using a thin flexible wire has been proposed for collision avoidance of the blind [33]. The shoelace, fed from the bottom of the laces, exhibited an acceptable performance even when tightened or loosened. Another potential application of wearable antennas for IOT applications is in footwear. In [35], a folded dipole is integrated in the soles of a pair of shoes to minimize the scuffing that would happen if a folded dipole was located on the outer surface. Besides that, the insole antenna design considered foot discomfort and the ability of the antenna to fit into different footwear and to be integrated with different types of biometric sensors. It was found that antennas with larger dimensions have better radiating efficiencies when operating in the 433 MHz industrial, scientific and medical (ISM) frequency band, with a maximum of 6 dBi realized gain.

For IOT to be effective, a communication mechanism needs to be implemented in wearable systems. They can be divided according to their frequency bands, which include the ISM band, Medical Implant Communication System (MICS) band, wireless communication to form a WLAN, wireless body area network (WBAN) and fourth generation Long-Term Evolution (4G LTE). Communication in medical applications is critical in enabling the transfer of continuous and reliable health data from patients to healthcare providers. Such data can be channeled into an online monitoring system as a preventive step in early detection or monitoring of diseases. In such systems, the sensors will first collect vital health signs, such as the heart rate, blood glucose, blood pressure and electrocardiogram (ECG), and transmit this information to an aggregator (collector) through one or several on-body communication link(s). After the information from the body-worn sensors regarding various physiological parameters is received, it is transmitted to the nearest receiving node through an off/ body communication link. To effectively communicate within these bands, these wearable systems need to be integrated with wearable antennas made of flexible materials.

Several examples of wearable antennas developed for these systems include a stretchable antenna using metal/polymer textile material in [34] for long distance communication. The antenna showed robustness against stretching and bending position for different body parts. It is capable of communication in the 2.45 GHz band under any strain condition (up to 30%), as shown in Figure 10.3.(a).

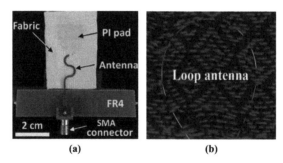

(a)   (b)

FIGURE 10.3   (a) Stretchable antenna on fabric [34], (b) multimaterial fabric loop antenna [36].

Next, a multimaterial textile fiber, as shown in Figure 10.3(b), has been used for low-power and short-range wireless network applications in the same frequency band [36]. A loop and dipole antenna have been weaved into the fabric using multi-material fiber threads made of hollow-core silica fiber coated with a polyimide layer. The polyimide layer along with silica glass are chosen due to their chemical stability, mechanical robustness and thermal robustness in high moisture conditions. Results of their evaluation indicated satisfactory radio frequency (RF) performance in the desired frequency, thus allowing easy integration with clothing in a cost-effective and minimally invasive manner.

It is also expected that the planar orientation of the antenna when placed on the body, although mechanically acceptable, may actually limit its capability in radiation toward specific directions. This is especially evident considering that a wearable node at a specific on-body location may need to communicate with other on-body nodes (at the end-fire direction), and, alternately, need to channel sensed/aggregated data to a base station located off-body (at the broadside direction) or even collect data from an in-body implant. This requires the wearable antennas, ideally, to be switch-able. Thus, a reconfigurable (switchable) feature in wearable antennas with operation in multiband frequencies is attractive for medical applications. These wearable antennas are also capable of merging numerous devices for different applications (from healthcare to entertainment) operating in various wireless standards into a single radiating platform.

Such reconfigurable features are commonly implemented using different types of surface-mounted switches. Examples of them include positive-intrinsic-negative (PIN) diodes, field-effect transistors (FETs), inductors and varactor diodes [37–39]. For example, a frequency reconfigurable textile antenna on felt substrate using Shieldex fabric has been proposed in [40]. The antenna operates from 1.54 to 2.82 GHz. Tuning of this corner-truncated circularly polarized (CP) patch antenna is facil-itated by changing the slot length on the ground plane using three embedded RF switches. Next, the work in [41] proposed a textile antenna for both on-body/off-body communication on felt substrate. The patch type radiator using a RF switching diode exhibits a monopole-like radiation pattern when operating in the zero-order resonance (ZOR) mode, and shows a patch-type radiation pattern when it resonates in +1 mode. Measurements indicated a radiation efficiency of 38% and 45% for the two modes due to the lossy textile conductive material used.

10.4 TEXTILE SYSTEMS AND ANTENNAS FOR SENSING

Besides communicating the readings obtained from patients, another important aspect of wearable systems for medical applications is the sensing mechanism. In [42], new applications for body-centric systems were proposed where flexible, miniaturized and conformal antennas are required for placement over the human body and implantation inside limbs and internal organs. The authors proposed design and experimentation of tags for orthopedic limb prosthesis and for vascular implants: (i) dipole and loop radio-frequency identification (RFID) tags for limb implant, and (ii) conformal antenna for vascular monitoring. RFID tag antennas based on dipole and loops can be implanted into a limb with achievable read-range performance. Such antennas are selected because they use a magnetic or an electric radiator. An example is the square loop tag integrated over a 3 mm-thick polyvinyl chloride (PVC) substrate illustrated in Figure 10.4. It has been optimized to operate within a muscle-like phantom resembling a human limb, which was modeled based on the human dielectric parameters at 870 MHz UHF band. On the other hand, a dipole tag has been designed with a length of 22 mm, which is equal to the diagonal length of the loop. This is to ensure that the two antennas have the same footprint, besides being

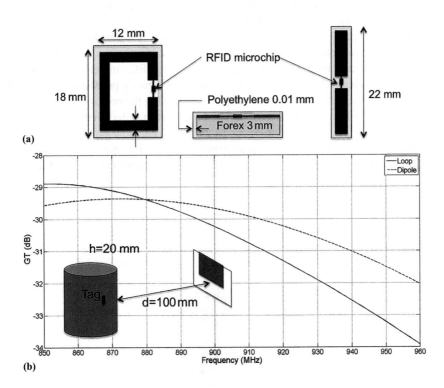

FIGURE 10.4 (a) Reference antennas (square loop and dipole antennas) for the feasibility study of implantable radio-frequency identification system. (b) Comparison between transducer gains numerically evaluated for depth of implant $h = 20$ mm and reader–body distance $d = 100$ mm [42].

fabricated using the same materials. These antennas are encapsulated by 0.01-mm polyethylene coating to be electrically insulated from the biological tissues. Results of the read range are illustrated in Figure 10.4(b), where gains were evaluated with both antennas placed at 20 mm of depth within the phantom, whereas the reader to phantom distance is set at 100 mm. It is noticed that the two antennas performed similarly in terms of gains in the intended UHF band, with less than 1 dB of difference.

Another interesting antenna used for medical applications is a conformal antenna for vascular monitoring, which was proposed in [43]. In this work, a conformal antenna takes the form of a loop or helical structure integrated into a blood vessel. Alternatively, it can also be implanted into devices, such as a prosthesis, sutures and surgical implants, orthopedic fixing and vascular stents without an additional radiating element. Bio-compatible metallic alloys, such as nickel–titanium (Nitinol), can be used with satisfactory conductivity to form the vascular stent as a radiating element. Radiators in the form of a helical coil, continuous/slotted cylinder, stacked loops or folded meander line can then be used to set up a transcutaneous wireless telemetry system to be used alongside the stent.

Another research on a textile-based system is an integrated array of electrodes proposed in [44] for rehabilitating stroke survivors. The electronic sleeve (e-sleeve), as shown in Figure 10.5, is aimed at training an arm's muscle by stimulating the underlying nerves to practice specific movements. The proposed e-sleeve is integrated with textile electrodes, which are commonly used in measuring bio-potentials, similar to those used in ECG, electroencephalography (EEG) and electromyography (EMG). These electrodes, in the form of an array, are directly printed onto everyday clothing using the screen printing technique. In assessing the prototype, interactions with end users were performed to optimize their comfort, effectiveness and ease of use. These sleeves are designed to be robust enough to enable a sufficient number of bending cycles to guarantee an acceptable lifetime usage. Besides that, an evaluation of different cleaning methods (washing and wiping) was performed to enable reuse

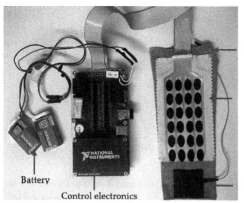

FIGURE 10.5   Electronic sleeve and electronics connection [44].

of the e-sleeve after contamination. Its application has also been demonstrated via muscle stimulation on the upper limb to achieve functional tasks (e.g., hand opening, pointing) for eight stroke survivors.

Next, a robust broadband textile-based UHF RFID tag antenna was proposed in [45]. The main feature of the proposed antenna is that it is capable of operating with minimal susceptibility to objects and materials in its vicinity. This alleviates the problem of conventional RFID tag antennas, which are typically designed to operate in a narrow frequency bandwidth, and must be re-optimized when they need to be used with different materials. In contrast, the proposed antenna featured good elasticity, flexibility and improved mechanical strength with the use of a highly conductive textile material (E-fiber). Most importantly, the designed tag antenna is capable of operating on a wide range of dielectric materials and various environments. Several implementations of the elastic materials are shown, such as on automotive tires, which demonstrated much better performance compared with commercially available tags.

Sensing of different parameters can also be performed using textile-based systems, besides location tracking and identification. This was demonstrated in [46], where an antenna-impregnated fabric for unobtrusive height measurement was proposed. This system aims to provide an alternative to conventional technologies (infantometer and stadiometer) in the detection and monitoring of diseases such as Crohn's disease, Turner syndrome, growth hormone deficiency, short stature, celiac disease and obesity. These conventional systems are limited in their capability to measure height due to sporadic intervals. On the other hand, multiple dipole antennas are located at known distances from each other to form the proposed system. In practice, these antennas will be detuned when an individual lies down on top of the fabric. This then introduces losses between the predefined wireless transmission paths. The individual's height can then be calculated by the degraded transmission coefficients between the two antennas located at the head (starting point) and feet (end point). To validate this system, a piece of fabric sized at 52 cm × 43 cm has been fabricated to be operated at 915 MHz. A phantom with a height of 16 cm is then placed onto the fabric at random locations. Results obtained from the measurements indicated the feasibility of estimating the height of a phantom regardless of its exact location. Besides that, the proposed fabric conforms to the Federal Communication Commission guidelines for the Specific Absorption Rate (SAR); the fabric can potentially be implemented onto baby cribs, bed sheets or rollable mats. Most importantly, the proposed method and fabric enables regular height monitoring with minimum impact on the individual's activity.

10.5 TEXTILE SYSTEMS AND ANTENNAS FOR LOCATION TRACKING

Wearable devices are potentially most effective in indoor and outdoor location tracking systems. The design of CP wearable antennas that can function with the least disruption to their operation, given the dynamics of the human body, enables their effective implementation. A technology which can effectively provide indoor location tracking is the passive RFID. RFID technology can be categorized by its operating frequencies, as follows: the low frequency (LF), which ranges from 120 to 145 KHz; the high frequency (HF) centered at 13.56 MHz; or the UHF band between 866 and 956 MHz.

For instance, low-cost RFID tag antennas were fabricated directly onto a cotton substrate in [47], by depositing graphene-based ink. The doctor-blade technique was used in the deposition on the cotton substrate to ensure the final prototype is suitable for wearable applications. Upon the complete attachment of the chip, the overall performance of the tag is evaluated: in conditions of before and after bending and stretching; and when operated in a high-humidity environment. The read range of the final prototype is experimentally evaluated to be below 1 m. In the planar condition, the read range of the tag decreases as the number of bending cycles increases. After 100 bending cycles, the final peak read range is about 1.1 m, which indicated that the tag is capable of enduring repeated bending. By contrast, the tag is not robust under stretching, as it failed to work upon the first stretching test.

In terms of outdoor location tracking, an alternative to RFID technology is the use of the Global Navigation Satellite System (GNSS)-based system. The more accurate GNSS systems require that the antenna is CP and broadband to support the multiple satellite constellations. A CP antenna improves polarization mismatch during operation and is more reliable in overcoming the problem of multipath fading. Due to the transmission distance from a satellite and low power received, up to 50% additional power (3 dB) can be achieved if a CP antenna is employed at the receiving end. Moreover, CP antennas also provide inherent immunity toward time-varying orientations between transmitter and receiver in an off-body wearable application. Most importantly, the ability of an antenna to be operated in the GNSS L1/E1 band with circular polarization is a key feature for outdoor wearable applications.

Now, several wearable antennas for Global Positioning Systems (GPSs) will be presented and discussed, including their designs, sizes and the types of material used. Besides that, several other considerations that should be taken into account in the design of such wearable antennas include the integration of null-steering algorithms to maximize anti-jamming feature, system power consumption and data rate limitations of the GPS receiver. First, a textile-integrated pin-fed GPS patch antenna is proposed in [48]. Simulations and measurements indicated that it performed reasonably well on the human body. The antenna was fabricated on three layers of Cordura fabric (ε_r =1.9, tanδ =0.0098) substrate with total thickness of 48 mil in order to increase the bandwidth of the antenna with conductive part (patch and ground) fabricated using Shieldex fabric.

Upon receipt of signals by the antenna array, out-of-band signals are filtered out by the RF front end prior to input into the GPS receiver. To adjust the reception pattern of the array and to spatially filter out undesired in-band signals, null-steering algorithms are implemented. This is done by calculating the amplitude and phase adjustment weight values for each array element using a basic optimization method, such as the minimum mean square error (MMSE). This is to simultaneously minimize the jammer signal power at the combined received output signal and maintain one of the array elements at a constant to ensure the reception and processing of valid GPS signals. Moreover, lumped components have also been integrated into textile materials for a rescue and tracking system in [49]. RF switches were implemented to

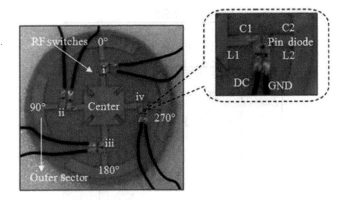

FIGURE 10.6 Textile antenna with lumped elements [49].

enable dual-polarized characteristics on the probe-fed GPS antenna, which is designed to be part of rain jackets, as illustrated in Figure 10.6.

Next, a simple diagonally-fed patch antenna fabricated using commercial textiles was designed and integrated onto a jacket sleeve badge for security personnel as presented in [50]. The antenna exhibited right-handed circular polarization (RHCP) with a wide fractional bandwidth of 5.6% and a gain of 4.7 dBi. An all-textile wearable antenna using conductive fiber implemented on a felt substrate, integrated with shorted pins and proposed for indoor/outdoor positioning systems was presented in [51]. A corner-truncated patch slot was used to generate the required phase shift in the current distribution to achieve CP. Two separate ports were used to excite the antenna for CP and linear polarization (LP) radiations at the same time.

Next, a wearable system based on GPS and WBAN/WLAN was proposed and designed in [52], aimed at indoor and outdoor location tracking of patients and elderly individuals (Figure 10.7). An antenna backing based on an artificial magnetic conductor plane operating at 2.45 GHz band is introduced to reduce the backward radiation and to improve antenna gain. This artificial magnetic conductor plane consists of a 3×3 array of unit cells. Each unit cell is formed using a square-shaped patch with a square-shaped ring. Four square slits are also integrated with each of the square-shaped patches. The dual-band and dual-polarized operation of the patch located on the top of this plane is enabled by truncating two corners of this patch, and four rectangular slits are incorporated at each corner of the same patch. The final measurements of the antenna validated that the antenna operated with linear polarization in the 2.45 GHz band and with circular polarization in the 1.575 GHz band, with a realized gain of 1.94 dBi and 1.98 dBic, respectively, as shown in Figure 10.8. The overall antenna is a fully designed textile, with felt as the substrate and ShieldIt as its conductive textile, and it is fed using a coaxial line. Prior to the incorporation of the antenna into the complete system and further testing, it was first assessed in terms of bending and SAR in the vicinity of the human body to ensure safety.

The designed wearable antenna operates at 1.575 GHz with acceptable impedance and axial ratio bandwidths. The antenna is validated to be safe for on-body use with

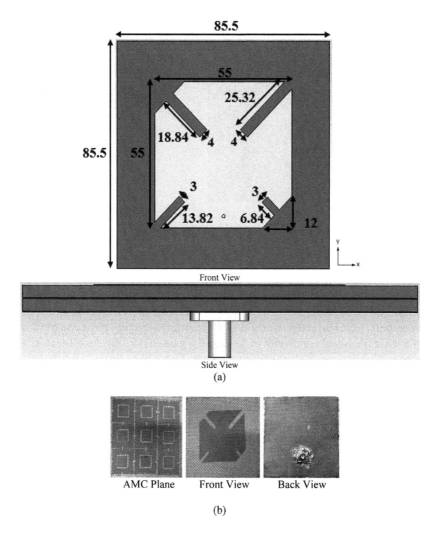

FIGURE 10.7 Structure of the proposed wearable circularly polarized antenna. (a) Topology of the dual-polarized planar textile antenna (all dimensions in mm) and (b) fabricated prototype [52].

SAR values of 1.478 W/kg (averaged over 1g of tissue) and 0.356 W/kg (averaged over 10 g of tissue).

10.6 TEXTILE RECTENNA SYSTEMS FOR ENERGY HARVESTING

Another main challenge to the effective implementation of wearable systems and antennas is in ensuring a sustainable energy supply to these nodes, besides the reduction of battery consumption. Moreover, as with any portable device, wearable nodes are ideally required to be autonomous – they are expected to have enough resources

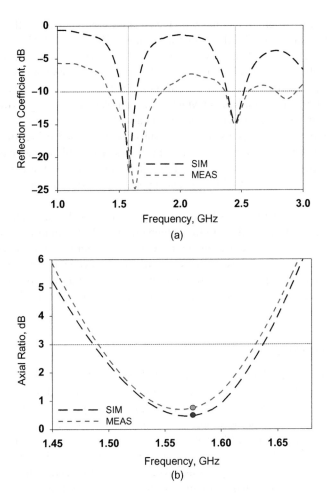

FIGURE 10.8 Simulated and measured results of the antenna performance: (a) reflection coefficient; and (b) axial ratio [52].

(in terms of processing power and energy) to perform their tasks for the longest pos-sible duration, independently, without charging. One of the most effective methods to ensure energy sustainability for wearable devices is to integrate energy-harvesting components into them. Wearable antennas in these devices can be deemed the most suitable due to the relatively large unobstructed space available for the integration of RF or even solar energy-harvesting components.

The primary motivation behind battery-free and maintenance-free wearables is the availability of energy harvesting from the wireless spectrum, the recent advance-ments in low-power electronics, the need for sustainable power sources and the quest for an autonomous lifestyle [53]. RF energy at different frequency bands is widely available to be harvested to achieve this objective of battery-independent wearable devices. An instance of such a system is a multilayered structure with a slotted annular-ring patch on a low-loss pile textile substrate backed by a full ground plane [54]. This structure operates with the capability of harvesting RF energy from the Global System for Mobile Communications (GSM) (900 MHz and 1.8 GHz) and WLAN (2.45 GHz) bands. A single-stage power converter has been integrated with a broadband phase shifter to enable the designed CP rectenna to harvest RF energy independently from all directions, polarization and orientations.

Next, a similar rectenna system which is fully implemented using textile materials is proposed in [55]. The patch of the rectenna is designed on a frayless nonwoven conductive textile, whereas its bi-layered substrate is formed using pile and jeans. The rectenna consists of a full-bridge rectifier on its ground plane that operates in the UHF band (between the frequencies of 860 and 918 MHz) with a conversion efficiency of more than 20%. A 1 kΩ load and a power of 14 μW/cm^2 was incident on the rectenna.

A flexible antenna for a wristband was proposed in [56]. A coupled microstrip line is used to connect the rectifier to the antenna designed on a multilayered polyester and woven fabric substrate. Results indicate that such fabrication techniques for the lines using flexible conductive fabrics resulted in a conversion efficiency of 33.6% with a −20 dBm of input power. Owing to the existence of a full ground plane on its rear side, there is minimal coupling to the human body when practically worn on the wrist.

10.7 SUMMARY

The evolution of wireless technologies and electronics have continuously impacted the development of wearable systems and antennas. This includes the medical field and IOT, two of its main application domains. More specifically, these fields are expected to thrive via the development of communication and sensing capabilities using textile-based devices. Alternative functionalities for these two applications will include location tracking and energy-harvesting capabilities, which have also been discussed in this chapter. Despite this, more advances are needed; this progress has already opened up the potential to improve the capability of future medical devices, simultaneously enhancing the quality of healthcare and the quality of life for patients and elderly individuals.

REFERENCES

[1] Agiwal, M., Roy, A., and Saxena, N.: 'Next Generation 5G Wireless Networks: A Comprehensive Survey', *IEEE Communications Surveys & Tutorials*, 2016, pp. 1-1.

[2] CISCO: '*Cisco Visual Networking Index: Global Mobile Data Traffic Forecast Update, 2014–2019*', pp. 1–48.

[3] Hertleer, C., Rogier, H., Vallozzi, L., and Van Langenhove, L.: 'A Textile Antenna for Off-Body Communication Integrated Into Protective Clothing for Firefighters', *IEEE Transactions on Antennas and Propagation*, 2009, 57 (4), pp. 919–925.

[4] Cihangir, A., Whittow, W.G., Panagamuwa, C.J., Ferrero, F., Jacquemod, G., Gianesello, F., and Luxey, C.: 'Feasibility Study of 4G Cellular Antennas for Eyewear Communicating Devices', *IEEE Antennas and Wireless Propagation Letters*, 2013, 12, pp. 1704–1707.

[5] Khaleel, H.R., Al-Rizzo, H.M., and Rucker, D.G.: 'Compact Polyimide-Based Antennas for Flexible Displays', *Journal of Display Technology*, 2012, 8 (2), pp. 91–97.

[6] Buckland, E., Ranken, M., Arnott, M., and Owen, P.: '*IOT global forecast & analysis 2015–2025*', Machina Research.

[7] Amyx, S.: "Haute IoT: Smart Garments and e-textiles", 'https://internetofthingsagenda. techtarget.com/blog/IoT-Agenda/Haute-IoT-Smart-garments-and-e-textiles', 2019.

[8] Bhattacharyya, S., Das, N., Bhattacharjee, D., and Mukherjee, A.: *Handbook of Research on Recent Developments in Intelligent Communication Application* (IGI Global, Hershey, Pennsylvania, USA, 2016).

[9] Chan, M., Estève, D., Fourniols, J.-Y., Escriba, C., and Campo, E.: 'Smart wearable systems: Current status and future challenges', *Artificial Intelligence in Medicine*, 2012, 56 (3), pp. 137–156.

[10] Locher, I., Klemm, M., Kirstein, T., and Trster, G.: 'Design and Characterization of Purely Textile Patch Antennas', *IEEE Transactions on Advanced Packaging*, 2006, 29 (4), pp. 777–788

[11] Nepa, P., and Rogier, H.: 'Wearable antennas for off-body radio links at VHF and UHF bands: Challenges, the state of the art, and future trends below 1 GHz', *IEEE Antennas and Propagation Magazine*, 2015, 57 (5), pp. 30–52.

[12] Balint, R., Cassidy, N.J., and Cartmell, S.H.: 'Conductive polymers: Towards a smart biomaterial for tissue engineering', *Acta Biomaterialia*, 2014, 10 (6), pp. 2341–2353.

[13] Salvado, R., Loss, C., Gonçalves, R., and Pinho, P.: 'Textile materials for the design of wearable antennas: a survey', *Sensors*, 2012, 12 (11), pp. 15841–15857.

[14] Simorangkir, R.B.V.B., Yang, Y., Hashmi, R.M., Bjorninen, T., Esselle, K.P., and Ukkonen, L.: 'Polydimethylsiloxane-embedded conductive fabric: Characterization and application for realization of robust passive and active flexible wearable antennas', *IEEE Access*, 2018, 6, pp. 48102–48112.

[15] Liyakath, R.A., Takshi, A., and Mumcu, G.: 'Multilayer stretchable conductors on polymer substrates for conformal and reconfigurable antennas', *IEEE Antennas and Wireless Propagation Letters*, 2013, 12, pp. 603–606.

[16] Nivethika, S.D., Sreeja, B.S., Manikandan, E., and Radha, S.: 'A stretchable smart and highly efficient radio frequency antenna on low cost substrate', *Microwave and Optical Technology Letters*, 2018, 60 (7), pp. 1798–1803.

[17] Rahman, A., Islam, M.T., Singh, M.J., and Misran, N.: 'Sol-gel synthesis of transition-metal doped ferrite compounds with potential flexible, dielectric and electromagnetic properties', *RSC Adv*, 2016, 6 (88), pp. 84562–84572.

[18] Roshni, S.B., Jayakrishnan, M.P., Mohanan, P., and Surendran, K.P.: 'Design and fabrication of an E-shaped wearable textile antenna on PVB-coated hydrophobic polyester fabric', *Smart Materials and Structures*, 2017, 26 (10), p. 105011.

[19] Krykpayev, B., Farooqui, M.F., Bilal, R.M., Vaseem, M., and Shamim, A.: 'A wearable tracking device inkjet-printed on textile', *Microelectronics Journal*, 2017, 65 (Supplement C), pp. 40–48.

[20] Kumar, A., Saghlatoon, H., La, T.-G., Mahdi Honari, M., Charaya, H., Abu Damis, H., Mirzavand, R., Mousavi, P., and Chung, H.-J.: 'A highly deformable conducting traces for printed antennas and interconnects: silver/fluoropolymer composite amalgamated by triethanolamine', *Flexible and Printed Electronics*, 2017, 2 (4), p. 045001.

[21] Salonen, P., Rahmat-Samii, Y., Schaffrath, M., and Kivikoski, M.: 'Effect of textile materials on wearable antenna performance: a case study of GPS antennas', *IEEE Antennas and Propagation Society Symposium, 2004*, Monterey, CA, 2004, Vol. 1, pp. 459–462, doi: 10.1109/APS.2004.1329673.

[22] Rai, T., Dantes, P., Bahreyni, B., and Kim, W.S.: 'A stretchable RF antenna with silver nanowires', *IEEE Electron Device Letters*, 2013, 34 (4), pp. 544–546.

[23] Song, L., Myers, A.C., Adams, J.J., and Zhu, Y.: 'Stretchable and reversibly deformable radio frequency antennas based on silver nanowires', *ACS Applied Materials and Interfaces*, 2014, 6 (6), pp. 4248–4253.

[24] Xiaohui, G., Ying, H., Can, W., Leidong, M., Yue, W., Zhicheng, X., Caixia, L., and Yugang, Z.: 'Flexible and reversibly deformable radio-frequency antenna based on stretchable SWCNTs/PANI/Lycra conductive fabric', *Smart Materials and Structures*, 2017, 26 (10), p. 105036.

[25] So, J.H., Thelen, J., Qusba, A., Hayes, G.J., Lazzi, G., and Dickey, M.D.: 'Reversibly deformable and mechanically tunable fluidic antennas', *Advanced Functional Materials*, 2009, 19 (22), pp. 3632–3637.

[26] Park, M., Im, J., Shin, M., Min, Y., Park, J., Cho, H., Park, S., Shim, M.-B., Jeon, S., Chung, D.-Y., Bae, J., Park, J., Jeong, U., and Kim, K.: 'Highly stretchable electric circuits from a composite material of silver nanoparticles and elastomeric fibres', *Nature Nanotechnology*, 2012, 7 (12), pp. 803–809.

[27] Ashayer-Soltani, R., Hunt, C., and Thomas, O.: 'Fabrication of highly conductive stretchable textile with silver nanoparticles', *Textile Research Journal*, 2015, 86 (10), pp. 1041–1049.

[28] Carter, J., Saberin, J., Shah, T., and Furse, C.: 'Inexpensive fabric antenna for off-body wireless sensor communication', in IEEE Antennas and Propagation Society International Symposium, 2010, pp. 1–4.

[29] Lilja, J., Pynttäri, V., Kaija, T., Mäkinen, R., Halonen, E., Sillanpää, H., Heikkinen, J., Mäntysalo, M., Salonen, P., and de Maagt, P.: 'Body-worn antennas making a splash: lifejacket-integrated antennas for global search and rescue satellite system', *IEEE Antennas and Propagation Magazine*, 2013, 55, (2), pp. 324-341. https://ieeexplore. ieee.org/document/6529385

[30] Mantash, M., Tarot, A.-C., Collardey, S., and Mahdjoubi, K.: 'Wearable monopole zip antenna', *Electronics Letters*, 2011, 47 (23), pp. 1266–1267.

[31] Li, G., Huang, Y., Gao, G., Wei, X., Tian, Z., and Bian, L.-A.: 'A handbag zipper antenna for the applications of body-centric wireless communications and internet of things', *IEEE Transactions on Antennas and Propagation*, 2017, 65 (10), pp. 5137–5146.

[32] Haydar, W., AlSayah, S., and Sarkis, R.: 'Design and analysis of conformal antenna for smart shoes', *IET Proceedings*, 2018 (1), pp. 53–55.

[33] Li, G., Tian, Z., Gao, G., Zhang, L., Fu, M., and Chen, Y.: 'A shoelace antenna for the application of collision avoidance for the blind person', *IEEE Transactions on Antennas and Propagation*, 2017, 65 (9), pp. 4941–4946.

[34] Hussain, A.M., Ghaffar, F.A., Park, S.I., Rogers, J.A., Shamim, A., and Hussain, M.M.: 'Metal/polymer based stretchable antenna for constant frequency far-field communication in wearable electronics', *Advanced Functional Materials*, 2015, 25 (42), pp. 6565–6575.

[35] Gaetano, D., McEvoy, P., Ammann, M.J., John, M., Brannigan, C., Keating, L., and Horgan, F.: 'Insole antenna for on-body telemetry', *IEEE Transactions on Antennas and Propagation*, 2015, 63 (8), pp. 3354–3361.

[36] Gorgutsa, S., Blais-Roberge, M., Viens, J., LaRochelle, S., and Messaddeq, Y.: 'User-interactive and wireless-communicating RF textiles', *Advanced Materials Technologies*, 2016, 1 (4), p. 1600032.

[37] Yang, X., Lin, J., Chen, G., and Kong, F.: 'Frequency reconfigurable antenna for wireless communications using GaAs FET switch', *IEEE Antennas and Wireless Propagation Letters*, 2015, 14, pp. 807–810.

[38] Zhai, H., Liu, L., Zhan, C., and Liang, C.: 'A frequency-reconfigurable triple-band antenna with lumped components for wireless applications', *Microwave and Optical Technology Letters*, 2015, 57 (6), pp. 1374–1379.

[39] Young, M.W., Yong, S., and Bernhard, J.T.: 'A miniaturized frequency reconfigurable antenna with single bias, dual varactor tuning', *IEEE Transactions on Antennas and Propagation*, 2015, 63 (3), pp. 946–951.

[40] Salleh, S.M., Jusoh, M., Ismail, A.H., Kamarudin, M.R., Nobles, P., Rahim, M.K.A., Sabapathy, T., Osman, M.N., Jais, M.I., and Soh, P.J.: 'Textile Antenna With Simultaneous Frequency and Polarization Reconfiguration for WBAN', *IEEE Access*, 2018, 6, pp. 7350–7358.

[41] Yan, S., and Vandenbosch, G.A.E.: 'Radiation pattern-reconfigurable wearable antenna based on metamaterial structure', *IEEE Antennas and Wireless Propagation Letters*, 2016, 15, pp. 1715–1718.

[42] Lodato, R., Manzari, S., Occhiuzzi, C., and Marrocco, G.: 'Flexible and conformable antennas for body-centric radiofrequency identification', *WIT Transactions on State-of-the-Art in Science and Engineering*, 2014, 82, p. 123.

[43] Occhiuzzi, C., Contri, G., and Marrocco, G.: 'Design of implanted RFID tags for passive sensing of human body: The STENTag', *IEEE Transactions on Antennas and Propagation*, 2012, 60 (7), pp. 3146–3154.

[44] Yang, K., Meadmore, K., Freeman, C., Grabham, N., Hughes, A.M., Wei, Y., Torah, R., Glanc-Gostkiewicz, M., Beeby, S., and Tudor, J.: 'Development of user-friendly wearable electronic textiles for healthcare applications', *Sensors*, 2018, 18, Article no. 8. https://www.mdpi.com/about/announcements/784.

[45] Shao, S., Kiourti, A., Burkholder, R.J., and Volakis, J.L.: 'Broadband textile-based passive UHF RFID tag antenna for elastic material', *IEEE Antennas and Wireless Propagation Letters*, 2015, 14, pp. 1385–1388.

[46] Zhu, K., Militello, L., and Kiourti, A.: 'Antenna-impregnated fabrics for recumbent height measurement on the go', *IEEE Journal of Electromagnetics, RF and Microwaves in Medicine and Biology*, 2018, 2 (1), pp. 33–39.

[47] Akbari, M., Virkki, J., Sydanheimo, L., and Ukkonen, L.: 'Toward graphene-based passive UHF RFID textile tags: A reliability study', *IEEE Transactions on Device and Materials Reliability*, 2016, 16 (3), pp. 429-431

[48] Keller, S.D., Weiss, S.J., Maloney, J.A., Kwon, D.-H., Janaswamy, R., and Morley, J.: 'Design considerations for a wearable anti-jam gps antenna', 2nd URSI Atlantic Radio Science Meeting (AT-RASC), Gran Canaria, Spain, 2018, pp. 1–4

[49] Jais, M.I., Jamlos, M.F., Jusoh, M., Sabapathy, T., and Kamarudin, M.R.: 'A Novel 1.575-GHz Dual-Polarization Textile Antenna for GPS Application', *Microwave and Optical Technology Letters*, 2013, 55 (10), pp. 2414–2420.

[50] Joler, M., and Boljkovac, M.: 'A Sleeve-Badge Circularly Polarized Textile Antenna', *IEEE Transactions on Antennas and Propagation*, 2018, 66 (3), pp. 1576–1579.

[51] Lee, H., Tak, J., and Choi, J.: 'Wearable Antenna Integrated into Military Berets for Indoor/Outdoor Positioning System', *IEEE Antennas and Wireless Propagation Letters*, 2017, 16, pp. 1919–1922.

[52] R. Joshi, E. F. N. M. Hussin, P. J. Soh, M. F. Jamlos, H. Lago, A. A. Al-Hadi, S. K. Podilchak, "Dual-Band, Dual-Sense Textile Antenna with AMC Backing for Localization using GPS and WBAN/WLAN", *IEEE Access*, May 2020, 8, pp. 89468–89478. doi: 10.1109/ACCESS.2020.2993371.

[53] Lemey, S., Agneessens, S., and Rogier, H.: 'Wearable smart objects: Microwaves propelling smart textiles: A review of holistic designs for wireless textile nodes', *IEEE Microwave Magazine*, 2018, 19 (6), pp. 83–100.

[54] Masotti, D., Costanzo, A., and Adami, S.: 'Design and realization of a wearable multifrequency RF energy harvesting system', Proceedings of the 5th European Conference on Antennas and Propagation (EUCAP), Rome, Italy, 2011, pp. 517–520.

[55] Monti, G., Corchia, L., and Tarricone, L.: 'UHF wearable rectenna on textile materials', *IEEE Transactions on Antennas and Propagation*, 2013, 61 (7), pp. 3869–3873.

[56] Adami, S.-E., Proynov, P., Hilton, G.S., Yang, G., Zhang, C., Zhu, D., Li, Y., Beeby, S.P., Craddock, I.J., and Stark, B.H.: 'A flexible 2.45-GHz power harvesting wristband with net system output from −24.3 dBm of RF power', *IEEE Transactions on Microwave Theory and Techniques*, 2018, 66 (1), pp. 380–395.

11 Development of Wearable Body Area Networks for 5G and Medical Communication Systems

Albert Sabban

CONTENTS

11.1 INTRODUCTION

Wearable technology provides a powerful new tool to medical and surgical rehabilitation services. Wireless body area networks (WBANs) can record electrocardiograms and can measure body temperature, blood pressure, heartbeat rate, arterial blood pressure, electro-dermal activity and other healthcare parameters. The recorded and collected data may be stored and analyzed by employing cloud storage and cloud computing services. Wearable systems have several applications in personal communication devices, Internet of Things (IOT) and medical devices as presented in [1–8]. The biomedical industry has been in continuous growth in the past decade. Several medical devices and systems were developed to monitor patients' health as presented in several books and papers [1–43]. Applications of body area network (BAN) and

WBAN systems are presented in Chapter 9, see Figures 9.1–9.3. WBANs can record electrocardiograms and can measure body temperature, blood pressure, heartbeat rate, arterial blood pressure, electro-dermal activity and other healthcare parameters in an efficient way. The medical data can be stored in computer centers or in cloud data centers. These medical data can be used to analyze a patient's health at any time.

11.2 CLOUD STORAGE AND COMPUTING SERVICES FOR WBANs

Cloud storage is a service package in which data is stored, managed, backed up remotely and made available to users over a network and internet services. There are three main cloud-based storage architecture models: public, private and hybrid.

Public cloud storage services provide a multi-customer storage environment that is most suited for data storage. Data is stored in global data centers with storage data spread across multiple regions or continents.

Private cloud storage provides local storage services to a dedicated environment protected behind an organization's firewall. Private clouds are appropriate for users who need customization and more control over their data.

Hybrid cloud storage is a mix of private cloud and third-party public cloud services with synchronization between the platforms. The model offers businesses flexibility and more data deployment options. An organization might, for example, store actively used and structured data in a local cloud, and unstructured and archival data in a public cloud. In recent years, a greater number of customers have adopted the hybrid cloud model. Despite its benefits, a hybrid cloud presents technical, business and management challenges. For example, private workloads must access and interact with public cloud storage providers, so compatibility and solid network connectivity are very important factors.

This service is a useful service for wearable communication systems. Medical data can be stored in cloud data centers. Cloud storage is based on a virtualized infrastructure with accessible interfaces. Cloud-based data is stored in servers located in data centers managed by a cloud provider. A file and its associated metadata are stored in the server by using an object storage protocol. The server assigns an identification number (ID) to each stored file. When a file needs to be retrieved, the user presents the ID to the system and the content is assembled with all its metadata, authentication and security. The most common use of cloud services is cloud backup, disaster recovery and archiving infrequently accessed data. Cloud storage providers are responsible for keeping the data available and accessible, and the physical environment protected and running. Customers buy or lease storage capacity from the providers to store and archive data files. Cloud storage services may be accessed via cloud computers and web services that use an application programming interface (API), such as cloud desktop storage and cloud storage gateways.

11.2.1 ADVANTAGES OF CLOUD STORAGE

- Cloud storage can provide the benefits of greater accessibility and reliability, rapid deployment and strong protection for data backup, archival and disaster recovery purposes.

- Cloud storage is used as a natural disaster-proof backup. Usually there are at least two backup servers located in different places around the world.
- Cloud storage can be used for copying virtual machine images from the cloud to a desired location or to import a virtual machine image from any designated location to the cloud image library. Cloud storage can also be used to move virtual machine images between user accounts or between data centers.
- Cloud storage provides users with immediate access to a broad range of resources hosted in the infrastructure of another organization via a web service interface.
- By using cloud storage, companies can cut computing expenses, such us storage maintenance tasks and purchasing additional storage capacity, and cut their energy consumption.
- Storage availability and data protection are provided by cloud storage services. So, depending on the application, the additional technology efforts and cost to ensure availability and protection of data storage can be eliminated.

11.2.2 Disadvantages of Cloud Storage

- Increasing the risk of unauthorized physical access to the data.
- In cloud-based architecture, data is replicated and moved frequently so the risk of unauthorized data recovery increases dramatically.
- Decrease in the security level of the stored data.
- It increases the number of networks over which the data travels (instead of just a local area network (LAN) or storage area network). Data stored on a cloud requires a wide area network to connect storage area networks and LAN devices.
- A cloud storage company has many customers and thousands of servers. Therefore, it has a large team of technical staff with physical and electronic access to almost all the data at the entire facility. Encryption keys that are kept by the service user, as opposed to the service provider, limit the access to data by service provider employees. Many keys must be distributed to users via secure channels for decryption. The keys must be securely stored and managed by the users in their devices. Storing these keys requires expensive secure storage.
- Cloud storage companies are not permanent and the services and products they provide can change.
- Cloud storage companies can be purchased by other foreign larger companies, can go bankrupt and suffer from an irrecoverable disaster.
- Cloud storage is a rich resource for both hackers and national security agencies. The cloud stores data from many different users. Hackers see it as a very valuable target.
- Cloud storage sites have faced lawsuits from the owners of intellectual property that has been uploaded and shared on the site. Piracy and copyright problems may be enabled by sites that permit file sharing.

11.2.3 CLOUD COMPUTING

Cloud computing is a type of Internet computing service that provides shared computer processing resources and data to computers and other devices on demand.

Cloud computing enables on-demand access to a shared pool of configurable computing resources, such as computer networks, servers, storage, applications and services. Cloud computing relies on the sharing of computing resources.

Cloud computing services can be rapidly provisioned and released with minimal management effort. Cloud computing and storage solutions provide users and enterprises with various capabilities to store and process their data in privately owned data centers.

Cloud computing allows companies to avoid high infrastructure costs, such as servers and expensive software. Cloud computing allows organizations to focus on their core businesses instead of spending time and money on computer networks.

Cloud computing allows companies to get their applications up and running faster, with improved manageability and less maintenance costs. Information technology teams can rapidly adjust resources to meet unpredictable business demands. Cloud computing applies high-performance computing power to perform tens of trillions of computations per second.

11.3 RECEIVING CHANNEL FOR COMMUNICATION AND MEDICAL APPLICATIONS

A medical system may be implanted or inserted in the human body as a swallowed capsule. The medical device will transmit medical data to a recorder. The medical data may be analyzed by medical staff online or stored as medical data about the patient. The receiving channel is part of the recorder and consists of receiving wearable antennas, radio frequency heads and a signal processing unit. A block diagram of the receiver is shown in Figure 11.1. Receiving channel main specifications are listed in Table 11.1. A block diagram of the receiving channel is shown in Figure 11.2. The receiving channel consists of an uplink channel at 434 MHz and a downlink at 13.56 MHz. The uplink channel consists of a switching matrix, low noise amplifier (LNA) and filter. The switching matrix losses are around 2 dB. The LNA noise figure is around 1 dB with 21 dB gain. The downlink channel consists of a transmitting

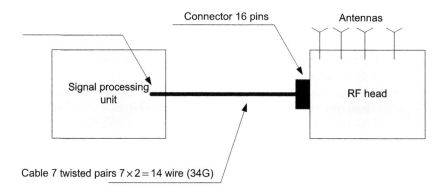

FIGURE 11.1 Recorder block diagram.

TABLE 11.1
Receiving channel main specifications

Requirement	Specification
Frequency range Up link	430–440 MHz
Return loss (dB)	–9
Group delay	Maximum 50 ns for 12 MHz BW
Input power	–30 dBm to –60 dBm
Signal-to-noise ratio	>20 dB
Current consumption (mA)	50
Dimensions (cm)	$12 \times 12 \times 5$
Frequency range Down link	13.56 MHz
Received power Down link	–8 dBm ± 1 dB
Current consumption (mA)	100–110

antenna, antenna matching network and differential amplifier. The downlink channel transmits commands to the medical system. Receiving channel gain and noise figure budget are shown in Figure 11.3. The receiving channel noise figure is around 3.5 dB. A receiving channel with lower noise figure values is shown in Figure 11.4. The LNA is connected to the receiving antenna.

Receiving channel gain and noise figure budget is shown in Figure 11.5. The receiving channel noise figure is around 3.5 dB.

Four printed folded dipole or loop antennas may be assembled in a belt and attached to a patient's stomach as shown in Figure 11.6a and 11.6b. Printed loop antennas were presented in [1]. The cable from each antenna is connected to a recorder. The received signal is routed to a switching matrix. The signal with the highest level is selected during the medical test. The antennas receive a signal that is transmitted from various positions in the human body. Wearable antennas may also be attached to the patient's back in order to improve the level of the received signal from different locations in the human body.

11.4 DEVELOPMENT PROCESS OF WEARABLE MEDICAL AND IOT SYSTEMS

The development process of wearable medical and IOT systems is a process in which basic sciences, electrical engineering, mechanical engineering, biomedical engineering and system engineering, are used to convert resources optimally to meet the system specifications. System engineering design is the process of developing a system, component or process to meet a customer's desired requirements, see [44–49]. It is an iterative decision-making process, in which basic sciences, mathematics, physics, biology, biomedical engineering and other engineering sciences are applied to convert resources optimally to meet the system specifications. Among the fundamental steps of the design process are the establishment of requirements and criteria, synthesis, analysis, construction, testing and evaluation of the system characteristics.

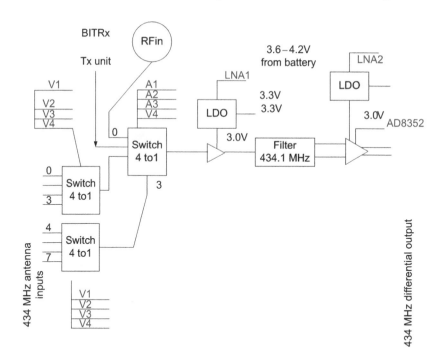

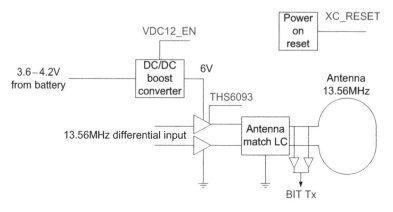

FIGURE 11.2 Receiving channel block diagram.

The system engineering process transforms needs and requirements into a set of system products and process descriptions, and generates information for investors and customers. The process is applied sequentially, one level at a time, top-down, to develop a system, by integrated teams. For example, for wearable medical systems the customers are patients, physicians, medical staff, medical centers system developers and distributors. Patients' and physicians' requirements should be translated by employing functional analysis tools to system technical specifications.

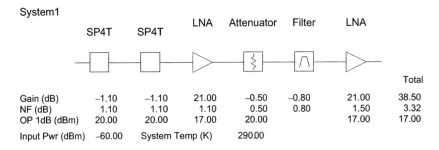

	SP4T	SP4T	LNA	Attenuator	Filter	LNA	Total
Gain (dB)	−1.10	−1.10	21.00	−0.50	−0.80	21.00	38.50
NF (dB)	1.10	1.10	1.10	0.50	0.80		1.50	3.32
OP 1dB (dBm)	20.00	20.00	17.00	20.00		17.00	17.00
Input Pwr (dBm)	−60.00	System Temp (K)		290.00			

FIGURE 11.3   Receiving channel gain and noise figure budget.

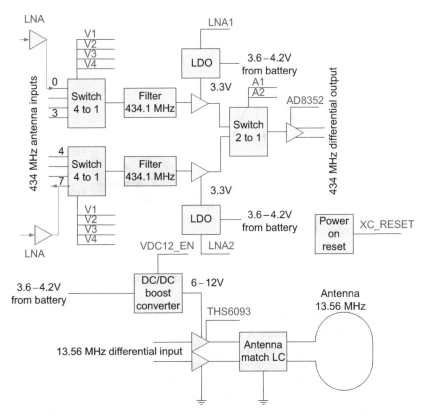

FIGURE 11.4   Receiving channel block diagram with low noise amplifier connected to the antennas.

11.4.1 STEPS IN SYSTEM ENGINEERING PROCESS

The major steps in the system engineering process are: **requirements analysis, system analysis control, functional analysis, design synthesis** and **project management**. System engineering development process is presented in Figure 11.7.

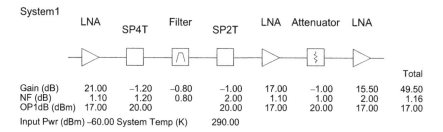

	LNA	SP4T	Filter	SP2T	LNA	Attenuator	LNA	Total
Gain (dB)	21.00	−1.20	−0.80	−1.00	17.00	−1.00	15.50	49.50
NF (dB)	1.10	1.20	0.80	2.00	1.10	1.00	2.00	1.16
OP1dB (dBm)	17.00	20.00		20.00	17.00	20.00	17.00	17.00

Input Pwr (dBm) −60.00 System Temp (K)   290.00

FIGURE 11.5   Receiving channel with low noise amplifier connected to the antennas: gain and noise figure budget.

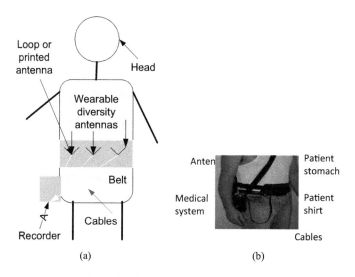

FIGURE 11.6   (a) Wearable medical system, (b) medical system on patient.

11.4.1.1 Requirements Analysis

Requirements analysis examines, evaluates and translates customer requirements into a set of functional and performance requirements that are the basis for the **functional analysis.** The goal of requirements analysis is to determine the system specifications that assist the system engineers to develop a system that satisfies the overall customer requirements. **Requirements analysis** is used to establish system functions, system specifications and design constraints. The output of this process generates functional requirements, system specifications and system architecture. This process is an iterative decision-making process that examines original system requirements. The result of this process leads to some modification of the original system requirements and reflects the customer's needs that can be impractical or excessively costly. The quality function deployment (QFD) method is employed to optimize this process. The output of the requirements analysis is a set of design requirements and functional definitions that are the starting milestones in the system development process. This iterative process determines how firm the requirements

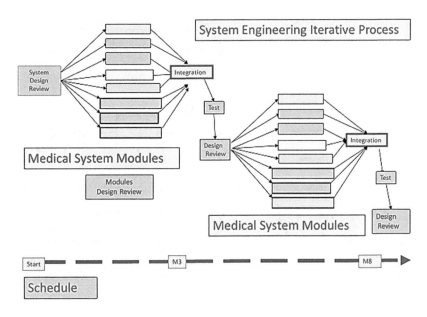

FIGURE 11.7   Medical wearable system engineering development process.

are for items that significantly affect the system cost, schedule, performance and risk. At the final stage of this process, detailed system characteristics are compared against the customer requirements to verify that the customer's major needs are being met. System engineering iterative development process for a wearable system, first cycle, is presented in Figure 11.8. System engineering development process for a wearable system, second cycle, is presented in Figure 11.9.

11.4.1.2 System Analysis Control

System analysis and control manages and controls the system development process. This activity determines the work required to develop the system and the development schedule. This process estimates the cost of the system development and manufacturing. It helps the project manager to coordinate all activities and assures that all the project's activities meet the same set of requirements, agreements and design constraints. This process results in a final system engineering plan of all the system activities. It determines when the result of one activity requires the action of another activity. The system engineering plan can be updated as needed during the development process. As the system engineering process progresses, trade-off studies and system cost effectiveness analyses are performed to support the evaluation and selection processes of all the development activities. Risk identification studies are performed as part of a risk management plan. System analysis and control perform configuration management and identify critical milestones to be used in the project progress control. In this process, interface management and data management are performed. The system engineer specifies the performance parameters to be tracked in the progress control plan. The system control plan consists of design reviews,

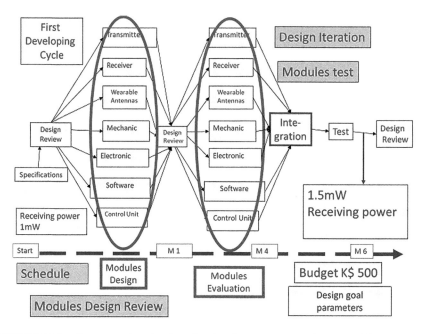

FIGURE 11.8   System engineering iterative development process for a wearable system.

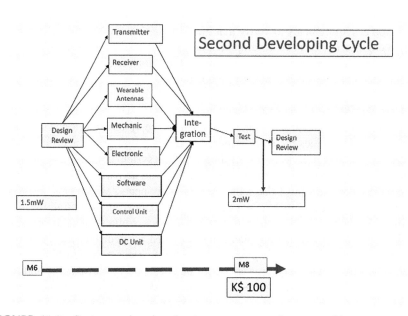

FIGURE 11.9   System engineering development process for a wearable system, second cycle.

TABLE 11.2
Medical wearable system project design goals list

#	Parameter	Value (units)	Testing conditions
1	Transmitted output power	25 dBm	25°C
2	Received input power	−12 dBm	25°C
3	DC voltage	5 V	25°C
4	Medical tests	$12000	Test on 25 patients
5	Food and Drug Administration approval process	$12000	24 months
6	Fabrication cost	$180	Fabrication of 1000 units

TABLE 11.3
Wearable medical system development plan

#	Task	Time (Hours)	Materials costs $	Schedule
1	Requirements analysis and **customer survey**	110		1.1.2020
2	**System design review**	50	4000	1.2.2020
3	**System analysis control**	110	1000	1.3.2020
4	**Functional analysis**	110	1000	1.4.2020
5	**System design**	310		1.6.2020
6	**Critical design review**	100		15.6.2020
7	**Material purchase**	45	9000	1.8.2020
8	**Modules fabrication**	110	5000	15.9.2020
9	**Modules test**	210	1000	1.10.2020
10	**Modules improvements**	110	3000	1.11.2020
11	**Modules design review**	45		15.11.2020
12	**Modules integration**	110	2000	1.12.2020
13	**System tests**	210		5.12.2020
14	**Final design review**	50	2000	15.12.2020
15	**Production of 100 units**	1000	10000	1.4.2021
	Total	2680	38000	

progress reports, risk management, budget control and schedule management. A project design goals list is given in Table 11.2. A system development plan is given in Table 11.3.

11.4.1.3 Functional Analysis

Functional analysis is a process of translating system requirements (customer needs) into detailed functional and performance design specifications. The result of the process is a defined architectural model that identifies system functions and their interactions. A functional analysis process is presented in Figure 11.10. It defines how the functions will operate together to perform the system functions. More than one

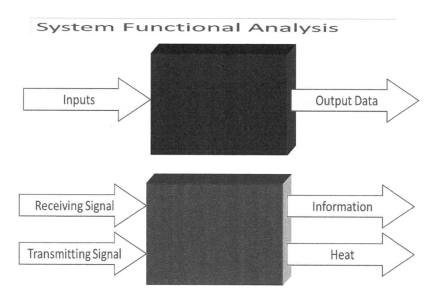

FIGURE 11.10   System functional analysis.

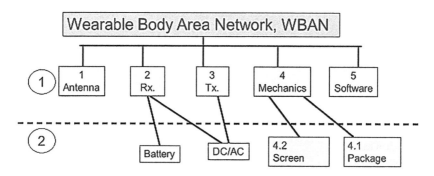

FIGURE 11.11   System engineering modular architecture for a wearable body area network system.

architecture can satisfy the customer's needs. However, each architecture has its set of associated requirements, different cost values, schedule, performance and risk implications. System engineering modular architecture for a wearable BAN system is presented in Figure 11.11. The functional architecture is used to define functional and evaluation development tests. The initial step in the process of functional analysis is to identify the basic functions required to perform the various system missions. As this is accomplished, the system specifications are defined, and functional architectures are developed for each module and for the system. These activities are continually validated and optimized to get the best system architecture. The functional

Role of Architecture in Product Development

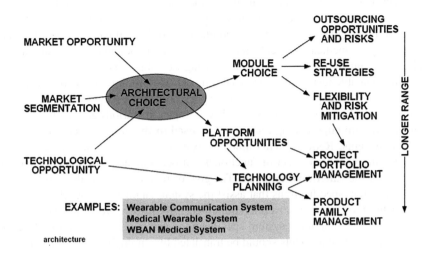

FIGURE 11.12   Role of architecture in system and product development.

system architecture and the functional system requirements are the input to the synthesis and development process.

11.4.1.4 Design Synthesis

The **synthesis process translates** functional architectures and requirements into physical architectures and to one or more physical sets of hardware, software and personnel solutions. The design results are compared to the original requirements, developed during the **requirements analysis** process, to verify that the developed system meets all the system specifications. The output of this activity is a checklist that verifies that the system specifications are met and a database which documents the design process. The role of system architecture in system and product development is presented in Figure 11.12. System architecture affects the technology that we will use in the development process. System architecture affects almost every stage of the system life cycle such as design, marketing, technology and system production.

The continuous growth in wearable communication and medical system complexity in the last 50 years created the demand to develop system engineering tools that help system engineers to optimize the development process of products and systems, see [49]. A system engineer transfers customer needs and requirements into technical specifications for new product and service development.

11.5 CONCLUSIONS

Wearable technology provides a powerful new tool to medical and surgical rehabilitation services. A wearable BAN is emerging as an important option for medical centers and patients. Wearable technology provides a convenient platform that may quantify the long-term context and physiological response of individuals. Wearable technology will support the development of individualized treatment systems with real-time feedback to help promote patients' health. Wearable medical systems and sensors can perform gait analysis and can measure body temperature, heartbeat, blood pressure, sweat rate and other physiological parameters of the person wearing the medical device. Gait analysis is a useful tool both in clinical practice and biomechanical research. Gait analysis using wearable sensors provides quantitative and repeatable results over extended time periods with low cost and good portability, showing better prospects and making great progress in recent years. At present, commercialized wearable sensors have been adopted in various applications of gait analysis. **System engineering tools must be used in the development of** wearable **medical systems. Requirements analysis** examines, evaluates and translates customer requirements into a set of functional and performance requirements that are the basis for the **functional analysis.** The goal of requirements analysis is to determine the system specifications that assist the system engineers to develop a system that satisfies the customer's overall requirements. **Requirements analysis** is used to establish system functions, system specifications and design constraints. Patients' and physicians' requirements should be translated by employing functional analysis tools to system technical specifications.

REFERENCES

[1] Sabban, A., *Wideband RF Technologies and Antenna in Microwave Frequencies*, Wiley Sons, New York, July 2016.

[2] Sabban, A., *Low Visibility Antennas for Communication Systems*, Taylor & Francis Group, New York, 2015.

[3] A. Sabban, Small Wearable Meta Materials Antennas for Medical Systems, *The Applied Computational Electromagnetics Society Journal*, 31(4), 434–443. April 2016.

[4] Sabban, A., Microstrip antenna arrays, *Microstrip Antennas*, Nasimuddin Nasimuddin (Ed.), InTech, Croatia, 361–384, 2011, ISBN: 978-953-307-247-0, http://www.intecho-pen.com/articles/show/title/microstrip-antenna-arrays.

[5] Sabban, A., New Wideband Printed Antennas for Medical Applications, *IEEE Journal, Transactions on Antennas and Propagation*, 61(1), 84–91, January 2013.

[6] U.S Patent, Inventors: Albert Sabban, Dual polarized dipole wearable antenna, U.S Patent number: 8203497, June 19, 2012, USA.

[7] Mukhopadhyay, S.C. editor, *Wearable Electronics Sensors*, Springer, Switzerland, 2015.

[8] Bonfiglio, A. and De Rossi, D., *Wearable Monitoring Systems*, Springer, New York, 2011

[9] Gao, T., Greenspan, D., Welsh, M., Juang, R.R., and Alm, A., Vital signs monitoring and patient tracking over a wireless network, Proceedings of IEEE-EMBS 27th Annual International Conference of the Engineering in Medicine and Biology, Shanghai, China, 1–5 September 2005, 102–105.

[10] Otto, C.A., Jovanov, E., and Milenkovic, E.A., WBAN-based system for health monitoring at home, Proceedings of IEEE/EMBS International Summer School, Medical Devices and Biosensors, Boston, MA, USA, 4–6 September 2006, 20–23.

[11] Zhang, G.H., Poon, C.C.Y., Li, Y., and Zhang, Y.T., A biometric method to secure telemedicine systems, Proceedings of the 31st Annual International Conference of the IEEE Engineering in Medicine and Biology Society, Minneapolis, MN, USA, September 2009, 701–704.

[12] Bao, S., Zhang, Y., and Shen, L., Physiological signal based entity authentication for body area sensor networks and mobile healthcare systems, Proceedings of the 27th Annual International Conference of the IEEE EMBS, Shanghai, China, 1–4 September 2005, 2455–2458.

[13] Ikonen, V. and Kaasinen, E., Ethical assessment in the design of ambient assisted living, Proceedings of Assisted Living Systems—Models, Architectures and Engineering Approaches, Schloss Dagstuhl, Germany, November 2008, 14–17.

[14] Srinivasan, V., Stankovic, J., and Whitehouse, K., Protecting your daily in home activity information from a wireless snooping attack, Proceedings of the 10th International Conference on Ubiquitous Computing, Seoul, Korea, 21–24 September 2008, 202–211.

[15] Casas, R., Blasco, M.R., Robinet, A., Delgado, A.R., Yarza, A.R., Mcginn, J., Picking, R., and Grout, V., User modelling in ambient intelligence for elderly and disabled people, Proceedings of the 11th International Conference on Computers Helping People with Special Needs, Linz, Austria, July 2008, 114–122.

[16] Jasemian, Y., Elderly comfort and compliance to modern telemedicine system at home, Proceedings of the Second International Conference on Pervasive Computing Technologies for Healthcare, Tampere, Finland, 30 January–1 February 2008, 60–63.

[17] Atallah, L., Lo, B., Yang, G.Z., and Siegemund, F., Wirelessly accessible sensor populations (WASP) for elderly care monitoring, Proceedings of the Second International Conference on Pervasive Computing Technologies for Healthcare, Tampere, Finland, 30 January–1 February 2008, 2–7.

[18] Hori, T., Nishida, Y., Suehiro, T., and Hirai, S., SELF-Network: Design and implementation of network for distributed embedded sensors, Proceedings of IEEE/RSJ International Conference on Intelligent Robots and Systems, Takamatsu, Japan, October–5 November 2000, 1373–1378.

[19] Mori, Y., Yamauchi, M., and Kaneko, K., Design and implementation of the vital sign box for home healthcare, Proceedings of IEEE EMBS International Conference on Information Technology Applications in Biomedicine, Arlington, VA, USA, November 2000, 104–109.

[20] Lauterbach, C., Strasser, M., Jung, S., and Weber, W., Smart clothes self-powered by body heat," Proceedings of Avantex Symposium, Frankfurt, Germany, May 2002, 5259–5263.

[21] Marinkovic, S. and Popovici, E., Network coding for efficient error recovery in wireless sensor networks for medical applications, Proceedings of International Conference on Emerging Network Intelligence, Sliema, Malta, 11–16 October 2009, 15–20.

[22] Schoellhammer, T., Osterweil, E., Greenstein, B., Wimbrow, M., and Estrin, D., Lightweight temporal compression of microclimate datasets, Proceedings of the 29th Annual IEEE International Conference on Local Computer Networks, Tampa, FL, USA, 16–18 November 2004, 516–524.

[23] Barth, A.T., Hanson, M.A., Powell, H.C., and Lach, J., Tempo 3.1: A body area sensor network platform for continuous movement assessment, Proceedings of the 6th International Workshop on Wearable and Implantable Body Sensor Networks, Berkeley, CA, USA, June 2009, 71–76.

[24] Gietzelt, M., Wolf, K.H., Marschollek, M., and Haux, R., Automatic self-calibration of body worn triaxial-accelerometers for application in healthcare, Proceedings of the Second International Conference on Pervasive Computing Technologies for Healthcare, Tampere, Finland, January 2008, 177–180.

[25] Gao, T., Greenspan, D., Welsh, M., Juang, R.R., and Alm, A., Vital signs monitoring and patient tracking over a wireless network, Proceedings of the 27th Annual International Conference of the IEEE EMBS, Shanghai, China, 1–4 September 2005, 102–105; Purwar, A, Jeong, D.U., and Chung, W.Y., Activity monitoring from realtime triaxial

accelerometer data using sensor network, Proceedings of International Conference on Control, Automation and Systems, Automation and Systems, Hong Kong, 21–23 March 2007, 2402–2406.

[26] Baker, C., Armijo, K., Belka, S., Benhabib, M., Bhargava, V., Burkhart, N., Der Minassians, A., Dervisoglu, G., Gutnik, L., Haick, M., Ho, C., Koplow, M., Mangold, J., Robinson, S., Rosa, M., Schwartz, M., Sims, C., Stoffregen, H., Waterbury, A., Leland, E., Pering, T., and Wright, P., Wireless sensor networks for home health care, Proceedings of the 21st International Conference on Advanced Information Networking and Applications Workshops, Niagara Falls, Canada, 21–23 May 2007, 832–837.

[27] Schwiebert, L., Gupta, S.K.S., and Weinmann, J., Research challenges in wireless networks of biomedical sensors, Proceedings of the 7th Annual International Conference on Mobile Computing and Networking, Rome, Italy, 16–21 July 2001, 151–165.

[28] Aziz, O., Lo, B., King, R., Darzi, A., and Yang, G.Z., Pervasive body sensor network: An approach to monitoring the postoperative surgical patient, Proceedings of International Workshop on Wearable and Implantable Body Sensor Networks (BSN 2006), Cambridge, MA, USA, 2006, 13–18.

[29] Kahn, J.M., Katz, R.H., and Pister, K.S.J., Next century challenges: mobile networking for smart dust, Proceedings of the ACM MobiCom'99, Washington, DC, USA, August 1999, 271–278.

[30] Noury, N., Herve, T., Rialle, V., Virone, G., Mercier, E., Morey, G., Moro, A., and Porcheron, T., Monitoring behavior in home using a smart fall sensor, Proceedings of IEEE-EMBS Special Topic Conference on Micro-technologies in Medicine and Biology, Lyon, France, 12–14 October 2000, 607–610.

[31] Kwon, D.Y. and Gross, M., Combining body sensors and visual sensors for motion training, Proceedings of the 2005 ACM SIGCHI International Conference on Advances in Computer Entertainment Technology, Valencia, Spain, 15–17 June 2005, 94–101.

[32] Boulgouris, N.K., Hatzinakos, D., and Plataniotis, K.N., Gait recognition: A challenging signal processing technology for biometric identification, *IEEE Signal Processing Magazine*, 22, 78–90, 2005.

[33] Kimmeskamp, S. and Hennig, E.M., Heel to Toe motion characteristics in parkinson patients during free walking, *Clinical Biomechanics*, 16, 806–812, 2001.

[34] Turcot, K., Aissaoui, R., Boivin, K., Pelletier, M., Hagemeister, N., and de Guise, J.A., New accelerometric method to discriminate between asymptomatic subjects and patients with medial knee osteoarthritis during 3-D gait, *IEEE Transactions on Biomedical Engineering*, 55, 1415–1422, 2008.

[35] Furnée, H., Real-time motion capture systems. In *Three-Dimensional Analysis of Human Locomotion*, Allard, P., Cappozzo, A., Lundberg, A., and Vaughan, C.L., Eds., John Wiley & Sons, Chichester, UK, 1997, 85–108.

[36] Bamberg, S.J.M., Benbasat, A.Y., Scarborough, D.M., Krebs, D.E., and Paradiso, J.A., Gait analysis using a shoe-integrated wireless sensor system, *IEEE Transactions on Information Technology in Biomedicine* 12, 413–423, 2008.

[37] Choi, J.H., Cho, J., Park, J.H., Eun, J.M., and Kim, M.S., An efficient gait phase detection device based on magnetic sensor array, Proceedings of the 4th Kuala Lumpur International Conference on Biomedical Engineering, Kuala Lumpur, Malaysia, 25–28 June 2008, 21, 778–781.

[38] Hidler, J., Robotic-assessment of walking in individuals with gait disorders, Proceedings of the 26th Annual International Conference of the IEEE Engineering in Medicine and Biology Society, San Francisco, CA, USA, 1–5 September 2004, 7, 4829–4831.

[39] Wahab, Y. and Bakar, N.A, Gait analysis measurement for sport application based on ultrasonic system, Proceedings of the 2011 IEEE 15th International Symposium on Consumer Electronics, Singapore, 14–17 June 2011, 20–24.

[40] De Silva, B., Jan, A.N., Motani, M., and Chua, K.C., A real-time feedback utility with body sensor networks, Proceedings of the 5th International Workshop on Wearable and Implantable Body Sensor Networks (BSN 08), Hong Kong, 1–3 June 2008, 49–53.

[41] Salarian, A., Russmann, H., Vingerhoets, F.J.G., Dehollain, C., Blanc, Y., Burkhard, P.R., and Aminian, K., Gait assessment in Parkinson's disease: Toward an ambulatory system for long-term monitoring. *IEEE Transactions on Biomedical Engineering*, 51, 1434–1443, 2004.

[42] Atallah, L., Jones, G.G., Ali, R., Leong, J.J.H., Lo, B., and Yang, G.Z., Observing recovery from knee-replacement surgery by using wearable sensors, Proceedings of the 2011 International Conference on Body Sensor Networks, Dallas, TX, USA, 23–25 May 2011, 29–34.

[43] ElSayed, M., Alsebai, A., Salaheldin, A., El Gayar, N., and ElHelw, M., Ambient and wearable sensing for gait classification in pervasive healthcare environments, Proceedings of the 12th IEEE International Conference on e-Health Networking Applications and Services (Healthcom), Lyon, France, 1–3 July 2010, 240–245.

[44] Cross, N., *Engineering Design Methods: Strtegies for Product Design*, 3rd ed., John Wiley & Sons, NJ, USA, 2000.

[45] Buede, D.M., *The Engineering Design of Systems Models & Methods*, John Wiley & Sons, 1999.

[46] Delbecq, A.L. and VandeVen, A.H, A group process model for problem identification and program planning, *Journal of Applied Behavioral Science*, VII, 466–491, July/August, 1971.

[47] Rechtin, E., *The Art of Systems Architecting*, 2nd ed., with Mark W. Maier, CRC Press LLC, 2000.

[48] Blanchard, B.S. and Fabrycky, W.J., *Systems Engineering and Analysis*, 4th ed., Prentice-Hall, 2005.

[49] Sabban, A., *Wearable Communication Systems and Antennas for Commercial, Sport, and Medical Applications*, IET Publication, December 2018.

12 Efficient Wearable Metamaterial Antennas for Wireless Communication, IOT, 5G and Medical Systems

CONTENTS

INTRODUCTION

Small printed antennas suffer from low efficiency. Low profile efficient antennas are crucial in the development of commercial caompact 5G communication and Internet of things (IOT) systems. The communication, IOT and biomedical industries have been in rapid growth in recent years. It is important to develop efficient high gain compact antennas for 5G communication and IOT systems. Metamaterials and fractal structures may be used to improve the efficiency of compact printed antennas. In this chapter metamaterial antennas will be presented. Metamaterial technology is used to design small wideband wearable antennas with high efficiency. Design considerations and the computed and measured results of printed metamaterial antennas with high efficiency are presented in this chapter. The proposed antenna may be used in communication and medical systems. The antennas' S11 results for different positions on the human body are ahigher by 2.5 dB than the patch antenna without SRR. The resonant frequency of the antenna with SRR on human body is shifted by 3%.

12.1 WEARABLE SMALL METAMATERIAL ANTENNAS FOR WIRELESS COMMUNICATION AND MEDICAL APPLICATIONS

12.1.1 INTRODUCTION

Metamaterial technology is used to design small wideband wearable antennas with high efficiency. Design considerations and the computed and measured results of printed metamaterial antennas with high efficiency are presented in this chapter. The proposed antenna may be used in communication and medical systems. The antennas' S11 results for different positions on the human body are presented in this chapter. The gain and directivity of the patch antenna with split ring resonator (SRR) is higher by 2.5 dB than the patch antenna without SRR. The resonant frequency of the antenna with SRR on human body is shifted by 3%.

Microstrip antennas are widely used in communication systems. Microstrip antennas have several advantages such as low profile, flexibility, light weight, small volume and low production cost. Compact printed antennas are presented in journals and books, as referred to in [1–4]. However, small printed antennas suffer from low efficiency. Metamaterial technology is used to design small printed antennas with high efficiency. Printed wearable antennas were presented in [5]. Artificial media with negative dielectric permittivity were presented in [6]. Periodic SRR and metallic posts structures may be used to design materials with dielectric constant and permeability less than 1 as presented in [6–14]. In this chapter, metamaterial technology is used to develop small antennas with high efficiency. Radio frequency (RF) transmission properties of human tissues have been investigated in several papers, such as [15, 16]. Compact wearable metamaterial and fractal antennas have been presented in papers in recent years as referred to in [17–41]. New wearable printed metamaterial antennas with high efficiency are presented in this chapter. The bandwidth of the metamaterial antenna with SRR and metallic strips is around 50% for voltage standing wave ratio (VSWR) better than 2.3:1. Computed and measured results of metamaterial antennas on the human body are presented in this chapter.

12.1.2 PRINTED WEARABLE DIPOLE ANTENNAS WITH SRRs

A microstrip dipole antenna with SRR is shown in Figure 12.1. The microstrip loaded dipole antenna with SRR provides horizontal polarization. The slot antenna provides

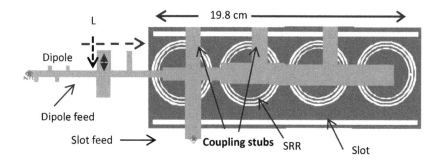

FIGURE 12.1 Printed antenna with split ring resonators.

vertical polarization. The resonant frequency of the antenna with SRR is 400 MHz. The resonant frequency of the antenna without SRR is 10% higher. The antennas shown in Figure 12.1 consist of two layers. The dipole feed network is printed on the first layer. The radiating dipole with SRR is printed on the second layer. The thickness of each layer is 0.8 mm. The dipole and the slot antenna create dual polarized antenna. The computed S11 parameters are presented in Figure 12.2.

The length of the dual polarized antenna with SRR shown in Figure 12.1 is 19.8 cm. The length of the dual polarized antenna without SRR shown in Figure 12.3 is 21 cm. The ring width is 1.4 mm; the spacing between the rings is 1.4 mm. The antennas have been analyzed by using Keysight Advanced Design System (Figure 12.7) software. The matching stubs' locations and dimensions have been optimized to get the best VSWR results. The length of the stub L in Figure 12.1 and Figure 12.3 is 10 mm. The locations and number of the coupling stubs may vary the antenna axial ratio from 0 to 30 dB. The number of coupling stubs may be minimized. The number of coupling stubs in Figure 12.1 is three. The antenna axial ratio value may also be adjusted by varying the slot feed location. The dimensions of the antenna shown in Figure 12.3 are presented in [1]. The bandwidth of the antenna shown in Figure 12.3 is around 10% for VSWR better than 2:1. The antenna beam width is 100°. The antenna gain is around 2 dBi. The computed S11 parameters are presented in Figure 12.4. Figure 12.5 presents the antenna's measured S11 parameters. There is a

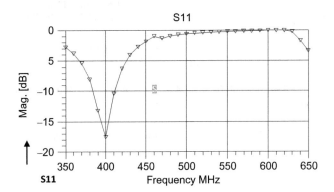

FIGURE 12.2 Antenna with split ring resonators, computed S11.

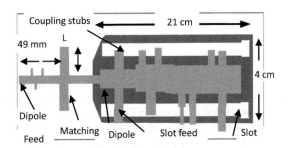

FIGURE 12.3 Dual polarized microstrip antenna.

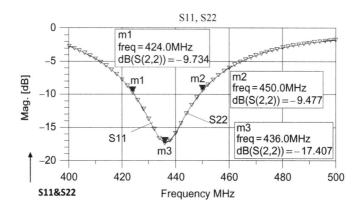

FIGURE 12.4 Computed S11 and S22 results, antenna without split ring resonator.

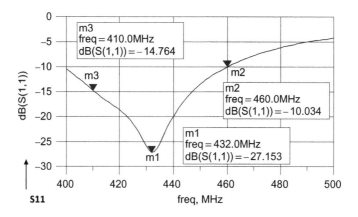

FIGURE 12.5 Measured S11 of the antenna without split ring resonator.

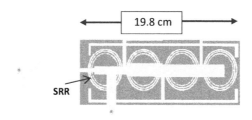

FIGURE 12.6 Antenna with split ring resonator with two resonant frequencies.

good agreement between measured and computed results. The antenna presented in Figure 12.1 has been modified as shown in Figure 12.6. The location and the dimension of the coupling stubs have been modified to get two resonant frequencies as presented in Figure 12.7. The first resonant frequency is 370 MHz and is lower by 20% than the resonant frequency of the antenna without the SRR.

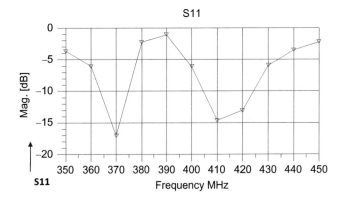

FIGURE 12.7   S11 for antenna with two resonant frequencies.

Metallic strips have been added to the antenna with SRR as presented in Figure 12.8. The computed S11 parameter of the antenna with metallic strips is presented in Figure 12.9. The antenna bandwidth is around 50% for VSWR better than 3:1. The computed radiation pattern is shown in Figure 12.10. The 3D computed radiation pattern is shown in Figure 12.11. Directivity and gain of the antenna with SRR is around 5 dBi, see Figure 12.12. The directivity of the antenna without SRR is around 2 dBi.

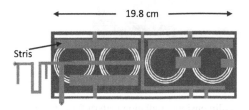

FIGURE 12.8   Antenna with split ring resonator and metallic strips.

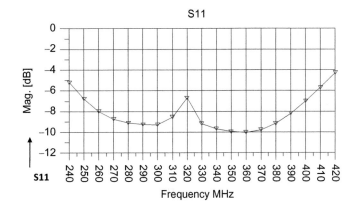

FIGURE 12.9   S11 for antenna with split ring resonator and metallic strips.

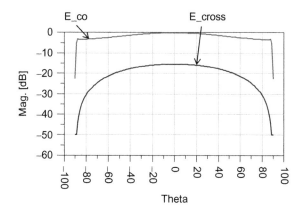

FIGURE 12.10 Radiation pattern for antenna with split ring resonator.

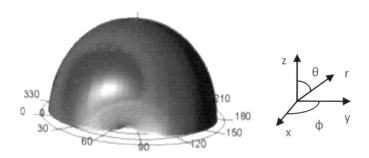

FIGURE 12.11 Three-dimensional radiation pattern for antenna with split ring resonator.

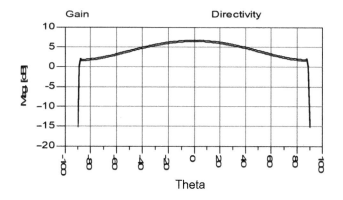

FIGURE 12.12 Directivity of the antenna with split ring resonator.

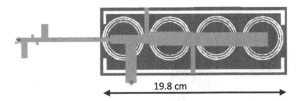

19.8 cm

FIGURE 12.13   Antenna with split ring resonator with two coupling stubs.

The length of the antennas with SRR is smaller by 5% than the antennas without SRR. Moreover, the resonant frequency of the antennas with SRR is lower by 5–10%.

The feed network of the antenna presented in Figure 12.8 has been optimized to yield VSWR better than 2:1 in frequency range of 250–440 MHz. Optimization of the number of the coupling stubs and the distance between the coupling stubs may be used to tune the antenna's resonant frequency. An optimized antenna with two coupling stubs has two resonant frequencies. The first resonant frequency is 370 MHz and the second resonant frequency is 420 MHz. An antenna with SRR with two coupling stubs is presented in Figure 12.13.

The computed S11 parameter of the antenna with two coupling stubs is presented in Figure 12.14. The 3D radiation pattern for antenna with SRR and two coupling stubs is shown in Figure 12.15.

The antenna with metallic strips has been optimized to yield wider bandwidth as shown in Figure 12.16. The computed S11 parameter of the modified antenna with metallic strips is presented in Figure 12.17. The antenna bandwidth is around 50% for VSWR better than 2.3:1.

12.1.3 FOLDED DIPOLE METAMATERIAL ANTENNA WITH SRR

The length of the antenna shown in Figure 12.3 may be reduced from 21 to 7 cm by folding the printed dipole as shown in Figure 12.18. Tuning bars are located along the feed line to tune the antenna to the desired frequency. The antenna bandwidth is around

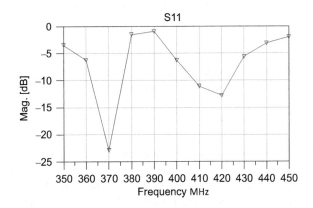

FIGURE 12.14   S11 for antenna with split ring resonator with two coupling stubs.

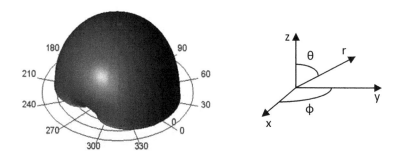

FIGURE 12.15 Three-dimensional radiation pattern for antenna with split ring resonator and two coupling stubs.

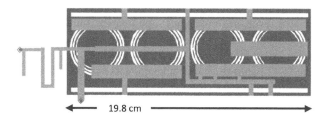

FIGURE 12.16 Wideband antenna with split ring resonator and metallic strips.

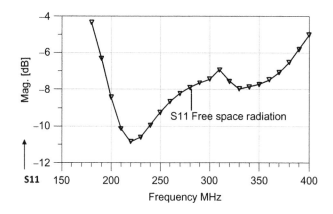

FIGURE 12.17 S11 for antenna with split ring resonator and metallic strips.

10% for VSWR better than 2:1 as shown in Figure 12.19. The antenna beam width is around 100°. The antenna gain is around 2 dBi. The size of the antenna with SRR shown in Figure 12.6 may be reduced by folding the printed dipole as shown in Figure 12.20. The dimensions of the folded dual polarized antenna with SRR presented in Figure 12.20 are $11 \times 11 \times 0.16$ cm. Figure 12.21 presents the antenna's

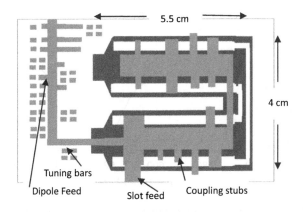

FIGURE 12.18   Folded dipole antenna, $7 \times 5 \times 0.16$ cm.

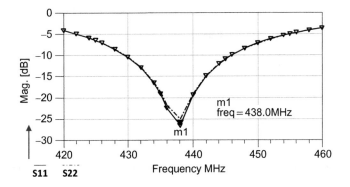

FIGURE 12.19   Folded antenna, computed S11 and S22 results.

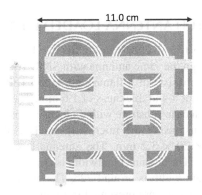

FIGURE 12.20   Folded dual polarized antenna with split ring resonator.

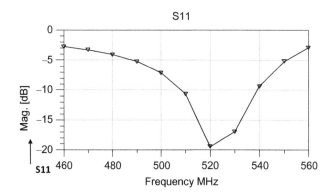

FIGURE 12.21   Folded antenna with split ring resonator, computed S11.

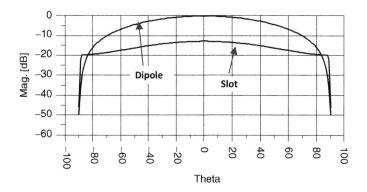

FIGURE 12.22   Radiation pattern of the folded antenna with split ring resonator.

computed S11 parameters. The antenna bandwidth is 10% for VSWR better than 2:1. The computed radiation pattern of the folded antenna with SRR is shown in Figure 12.22.

12.2 STACKED PATCH ANTENNA LOADED WITH SRR

As a first step, a microstrip stacked patch antenna [1–3] has been designed. The second step was to design the same antenna with SRR. The antenna consists of two layers. The first layer consists of a flame retardant-4 (FR-4) dielectric substrate with dielectric constant of 4 and thickness of 1.6 mm. The second layer consists of an RT-Duroid 5880 dielectric substrate with dielectric constant of 2.2 and thickness of 1.6 mm. The dimensions of the microstrip stacked patch antenna shown in Figure 12.23 are 33 × 20 × 3.2 mm. The antenna has been analyzed by using Keysight ADS software. The antenna bandwidth is around 5% for VSWR better than 2.5:1. The antenna beam width is around 72°. The antenna gain is around 7 dBi. The computed S11 parameters are presented in Figure 12.24. The radiation pattern of the microstrip stacked patch is shown in Figure 12.25. The antenna with SRR is

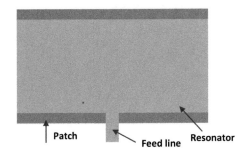

FIGURE 12.23 A microstrip stacked patch antenna.

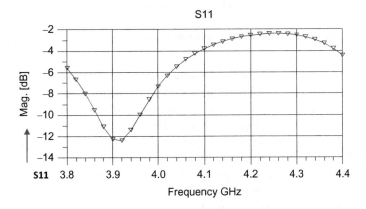

FIGURE 12.24 Computed S11 of the microstrip stacked patch.

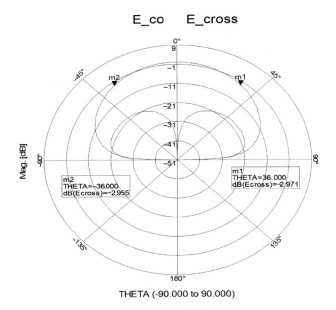

FIGURE 12.25 Radiation pattern of the microstrip stacked patch.

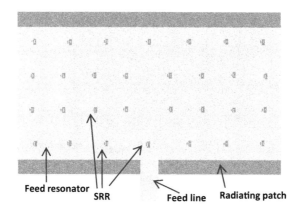

FIGURE 12.26 Printed antenna with split ring resonators.

shown in Figure 12.26. This antenna has the same structure as the antenna shown in
Figure 12.23. The ring width is 0.2 mm; the spacing between the rings is 0.25 mm.
Twenty-eight SRRs are placed on the radiating element. There is a good agreement
between measured and computed results. The measured S11 parameters of the
antenna with SRR are presented in Figure 12.27. The antenna bandwidth is around
12% for VSWR better than 2.5:1. By adding an air space of 4 mm between the
antenna layers the VSWR was improved to 2:1. The antenna gain is around 9–10 dBi.
The antenna efficiency is around 95%. The antenna's computed radiation pattern is
shown in Figure 12.28. The patch antenna with SRR performs as a loaded patch
antenna. The effective area of a patch antenna with SRR is higher than the effective

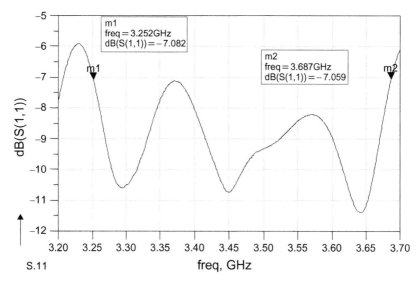

FIGURE 12.27 Patch with split ring resonators, measured S11.

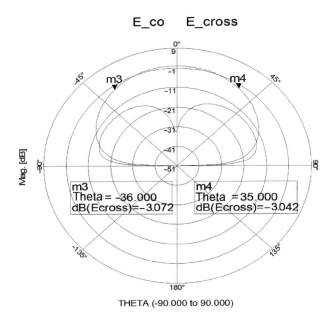

FIGURE 12.28   Radiation pattern for patch with split ring resonator.

area of a patch antenna without SRR. The resonant frequency of a patch antenna with SRR is lower by 10% than the resonant frequency of a patch antenna without SRR.

The antenna beam width is around 70°. The gain and directivity of the stacked patch antenna with SRR is higher by 2–3 dB than patch the antenna without SRR.

12.3   PATCH ANTENNA LOADED WITH SRRS

A patch antenna with SRRs has been designed. The antenna is printed on a RT-Duroid 5880 dielectric substrate with dielectric constant of 2.2 and thickness of 1.6 mm. The dimensions of the microstrip patch antenna shown in Figure 12.29 are 36 × 20 × 1.6 mm. The antenna bandwidth is around 5% for S11 lower than −9.5 dB. However, the antenna bandwidth is around 10% for VSWR better than 3:1. The antenna beam width is around 72°. The antenna gain is around 7.8 dBi. The directivity of the antenna is 8. The antenna gain is 6.03. The antenna efficiency is 77.25%. The measured S11 parameters are presented in Figure 12.30. The gain and directivity of the patch antenna with SRR is higher by 2.5 dB than the patch antenna without SRR.

12.4   METAMATERIAL ANTENNA CHARACTERISTICS IN VICINITY TO THE HUMAN BODY

The antenna's input impedance variation as a function of distance from the human body had been computed by using the structure presented in Figure 12.31. Electrical properties of human body tissues are listed in Table 12.1 see [15]. The antenna's location on a human body may be considered by calculating S11 for different dielectric

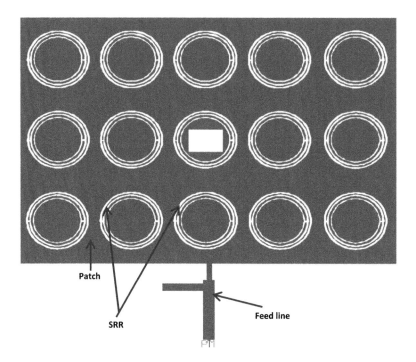

FIGURE 12.29   Patch antenna with split ring resonators.

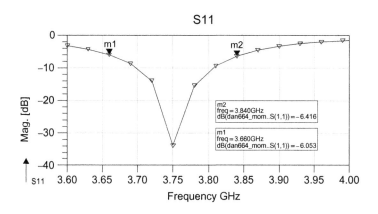

FIGURE 12.30   Patch with split ring resonators, computed S11.

constants of the body. The variation of the dielectric constant of the body from 43 at the stomach to 63 at the colon zone shifts the antenna's resonant frequency by to 2%. The antenna was placed inside a belt with thickness of between 1 and 4 mm with dielectric constant from 2 to 4. The air spacing between the belt and the patient's shirt varied from 0 to 8 mm. The dielectric constant of the patient's shirt varied from 2 to 4.

Figure 12.32 presents S11 results of the antenna with SRR shown in Figure 12.13 on the human body. The antenna's resonant frequency is shifted by 3%. Figure 12.33

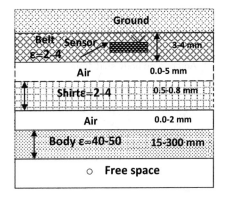

FIGURE 12.31   Wearable antenna environment.

TABLE 12.1
Summary of electrical properties of human body tissues

Tissue	Property	434 (MHz)	600 (MHz)	1000 (MHz)
Prostate	σ	0.75	0.90	1.00
	ε	50.53	47.4	47.0
Skin	σ	0.57	0.6	0.65
	ε	41.6	40.43	40.25
Stomach	σ	0.67	0.73	0.79
	ε	42.9	41.41	39.05
Colon, muscle	σ	0.98	1.06	1.30
	ε	63.6	61.9	60.00
Lung	σ	0.27	0.27	0.27
	ε	38.4	38.4	38.4

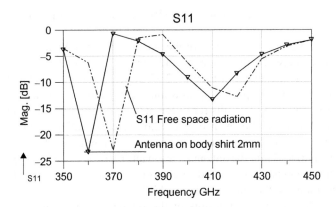

FIGURE 12.32   S11 of the antenna with split ring resonator on the human body.

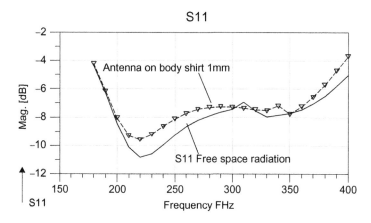

FIGURE 12.33 Antenna with split ring resonator, S11 results on the human body.

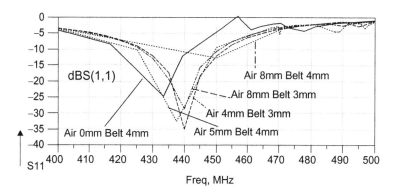

FIGURE 12.34 S11 results of the antenna shown in Figure 12.3 on the human body.

presents S11 results of the antenna with SRR and metallic strips, shown in Figure 12.16, on the human body. The antenna's resonant frequency is shifted by 1%.

Figure 12.4 presents S11 results (of the antenna shown in Figure 12.3) for different air spacing between the antennas and human body, belt thickness and shirt thickness. Results presented in Figure 12.33 indicate that the antenna has VSWR better than 2.5:1 for air spacing up to 8 mm between the antennas and the body. Figure 12.35 presents S11 results for different positions relative to the human body for the folded antenna shown in Figure 12.6. An explanation of Figure 12.35 is given in Table 12.2. If the air spacing between the antennas and the human body is increased from 0 to 5 mm the antenna's resonant frequency is shifted by 5%. A tunable wearable antenna may be used to control the antenna's resonant frequency at different positions on the human body, see [25].

Figure 12.36 presents S11 results of the folded antenna with SRR, shown in Figure 12.8, on the human body. The antenna's resonant frequency is shifted by 2%.

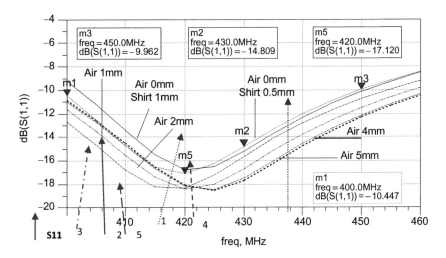

FIGURE 12.35 S11 results for different locations relative to the human body for the antenna shown in Figure 12.6.

TABLE 12.2
Explanation of Figure 12.35

Picture #	Line type	Sensor position
1	Dot	Shirt thickness 0.5 mm
2	Line	Shirt thickness 1 mm
3	Dash dot	Air spacing 2 mm
4	Dash	Air spacing 4 mm
5	Long dash	Air spacing 1 mm
6	Big dots	Air spacing 5 mm

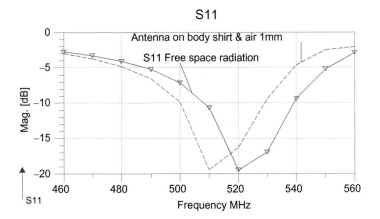

FIGURE 12.36 Folded antenna with split ring resonator, S11 results on the body.

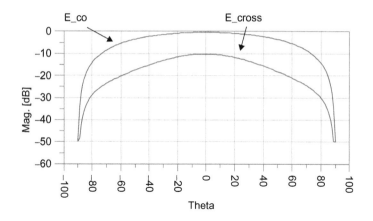

FIGURE 12.37 Radiation pattern of the folded antenna with split ring resonator on human body.

The radiation pattern of the folded antenna with SRR on the human body is presented in Figure 12.37.

12.5 METAMATERIAL WEARABLE ANTENNAS

The proposed wearable metamaterial antennas may be placed inside a belt as shown in Figure 12.38. Three to four antennas may be placed in a belt and attached to the patient's stomach. More antennas may be attached to the patient's back to improve the level of the received signal from different locations on the human body. The cable from each

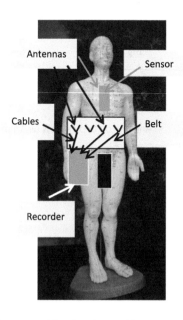

FIGURE 12.38 Medical system with printed wearable antenna.

antenna is connected to a recorder. The received signal is transferred via a SP8T switch to the receiver. The antennas receive a signal that is transmitted from various positions on the human body. The medical system selects the signal with the highest power.

In several systems, the distance separating the transmitting and receiving antennas is in the near-field zone. In these cases, the electric field intensity decays rapidly with distance. The near-fields only transfer energy to close distances from the antenna and do not radiate energy to far distances. The radiated power is trapped in the region near to the antenna. In the near-field zone the receiving and transmitting antennas are magnetically coupled. The inductive coupling value between two antennas is measured by their mutual inductance. In these systems we must consider only the near-field electromagnetic coupling.

In Figures 12.39 to 12.42 several photos of printed antennas for medical applications are shown. The dimensions of the folded dipole antenna are $7 \times 6 \times 0.16$ cm. The dimensions of the compact folded dipole presented in [5] and shown in Figure 12.39 are $5 \times 5 \times 0.5$ cm. The antennas' electrical characteristics on the human body have been measured by using a phantom. The phantom has been designed to represent the human body's electrical properties as presented in [5]. The tested antenna was attached to the phantom during the measurement of the antennas' electrical parameters.

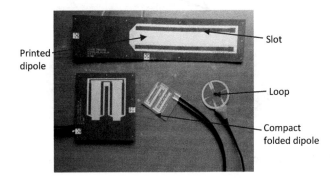

FIGURE 12.39 Microstrip antennas for medical applications.

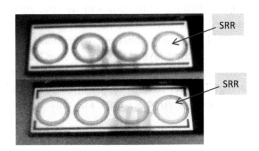

FIGURE 12.40 Metamaterial antennas for medical applications.

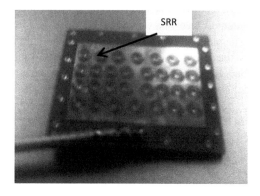

FIGURE 12.41   Metamaterial patch antenna with split ring resonator.

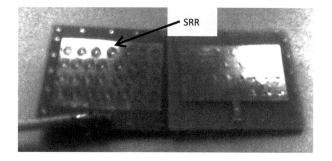

FIGURE 12.42   Metamaterial stacked patch antenna with split ring resonator.

12.6 WIDEBAND STACKED PATCH WITH SRR

A wideband microstrip stacked patch antenna with air spacing [1–3] was designed. The antenna was designed with SRR. The antenna consists of two layers. The first layer consists of an FR-4 dielectric substrate with dielectric constant of 4 and thickness of 1.6 mm. The second layer consists of an RT-Duroid 5880 dielectric substrate with dielectric constant of 2.2 and thickness of 1.6 mm. The layers are separated by air spacing. The dimensions of the microstrip stacked patch antenna shown in Figure 12.43 are 33 × 20 × 3.2 mm. The antenna has been analyzed by

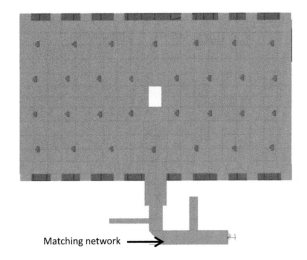

FIGURE 12.43 Wideband stacked patch antenna with split ring resonator.

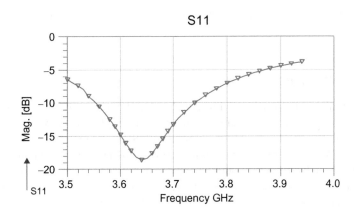

FIGURE 12.44 Wideband stacked antenna with split ring resonator, S11 results.

using Keysight ADS software. The antenna bandwidth is around 10% for VSWR better than 2.0:1. The antenna beam width is around 72°. The antenna gain is around 9–10 dBi. The antenna efficiency is around 95%. The computed S11 parameters are presented in Figure 12.44. The radiation pattern of the stacked patch is shown in Figure 12.45. There is a good agreement between measured and computed results.

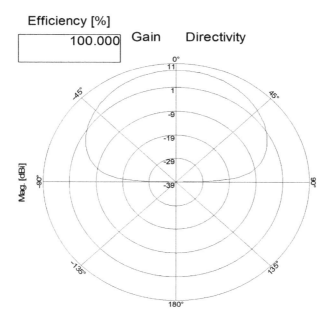

FIGURE 12.45   Radiation pattern of the stacked antenna with split ring resonator.

12.7 CONCLUSION

Metamaterial technology is used to develop small antennas with high efficiency. A new class of printed metamaterial antennas with high efficiency is presented in this chapter. The bandwidth of the antenna with SRR and metallic strips is around 50% for VSWR better than 2.3:1. Optimization of the number of the coupling stubs and the distance between the coupling stubs may be used to tune the antenna's resonant frequency and number of resonant frequencies. The length of the antennas with SRR is smaller by 5% than the antennas without SRR. Moreover, the resonant frequency of the antennas with SRR is lower by 5% to 10% than the antennas without SRR. The gain and directivity of the patch antenna with SRR is higher by 2–3 dB than the patch antenna without SRR. The resonant frequency of the antenna with SRR on a human body is shifted by 3%.

REFERENCES

[1] A. Sabban, *Low Visibility Antennas for Communication Systems*, Taylor & Francis Group, New York, 2015.
[2] A. Sabban, "New Wideband Printed Antennas for Medical Applications," *IEEE Transactions on Antennas and Propagation*, 61(1), 84–91, January 2013.
[3] A. Sabban, "A New Wideband Stacked Microstrip Antenna," in *IEEE Antenna and Propagation Symposium*, Houston, TX, June 1983.

[4] A. Sabban, *Microstrip Antenna Arrays*; N. Nasimuddin (Ed.), *Microstrip Antennas*, InTech, 2011, pp. 361–384, ISBN: 978-953-307-247-0, Riijeka, Croatia. http://www.intechopen.com/articles/show/title/microstrip-antenna-arrays.

[5] A. Sabban and K. C. Gupta, "Characterization of Radiation Loss from Microstrip Discontinuities Using a Multiport Network Modeling Approach," *IEEE Transactions on Microwave Theory and Techniques*, 39(4), 705–712, April 1991.

[6] J. B. Pendry, A. J. Holden, W. J. Stewart, and I. Youngs, "Extremely Low Frequency Plasmons in Metallic Mesostructures," *Physical Review Letters*, 76, 4773–4776, 1996.

[7] J. B. Pendry, A. J. Holden, D. J. Robbins, and W. J. Stewart, "Magnetism from Conductors and Enhanced Nonlinear Phenomena," *IEEE Transactions on Microwave Theory and Techniques*, 47, 2075–2084, 1999

[8] R. Marqués, F. Mesa, J. Martel, and F. Medina, "Comparative Analysis of Edge and Broadside Coupled Split Ring Resonators for Metamaterial Design. Theory and Experiment," *IEEE Transactions on Antennas and Propagations*, 51, 2572–2581, 2003.

[9] R. Marqués, J. D. Baena, J. Martel, F. Medina, F. Falcone, M. Sorolla, and F. Martin, "Novel Small Resonant Electromagnetic Particles for Metamaterial and Filter Design," in *Proceedings of the ICEAA'03*, Torino, Italy, 2003, pp. 439–442.

[10] R. Marqués, J. Martel, F. Mesa, and F. Medina, "Left-Handed-Media Simulation and Transmission of EM Waves in Resonator-Subwavelength Split-Ring-Loaded Metallic Waveguides." *Physics Review Letter*, 89, 183901, 2002.

[11] J. D. Baena, R. Marqués, J. Martel, and F. Medina, "Experimental Results on Metamaterial Simulation Using SRR-Loaded Waveguides," in Proceedings of the IEEE-AP/S International Symposium on Antennas and Propagation, Columbus, OH, June 2003, pp. 106–109.

[12] R. Marqués, J. Martel, F. Mesa, and F. Medina, "A New 2-D Isotropic Left-Handed Metamaterial Design: Theory and Experiment," *Microwave and Optical Technology Letters*, 35, 405–408, 2002

[13] R. A. Shelby, D. R. Smith, S. C. Nemat-Nasser, and S. Schultz, "Microwave Transmission through a Two-Dimensional, Isotropic, Left-Handed Metamaterial," *Applied Physics Letter*, 78, 489–491, 2001

[14] J. Zhu and G. V. Eleftheriades, "A Compact Transmission-Line Metamaterial Antenna with Extended Bandwidth," *IEEE Antennas and Wireless Propagation Letters*, 8, 295–298, 2009.

[15] L. C. Chirwa, P. A. Hammond, S. Roy, and D. R. S. Cumming, "Electromagnetic Radiation from Ingested Sources in the Human Intestine between 150 MHz and 1.2 GHz," *IEEE Transaction on Biomedical Engineering*, 50,(4), 484–492, April 2003.

[16] D. Werber, A. Schwentner, and E. M. Biebl, "Investigation of RF Transmission Properties of Human Tissues," *Advances in Radio Sciences*, 4, 357–360, 2006.

[17] B. Gupta, S. Sankaralingam, and S. Dhar, "Development of Wearable and Implantable Antennas in the Last Decade," in *2010 Mediterranean Microwave Symposium (MMS'2010)*, Cyprus, 2010, pp. 251–267.

[18] T. Thalmann, Z. Popovic, B. M. Notaros, and J. R. Mosig, "Investigation and Design of a Multi-Band Wearable Antenna," in *3rd European Conference on Antennas and Propagation, EuCAP 2009*, Berlin, Germany, 2009, pp. 462–465.

[19] P. Salonen, Y. Rahmat-Samii, and M. Kivikoski, "Wearable Antennas in the Vicinity of Human Body," in *IEEE Antennas and Propagation Society International Symposium, 2004*, Monterey, CA, 2004, Vol. 1, pp. 467–470.

[20] T. Kellomaki, J. Heikkinen, and M. Kivikoski, "Wearable Antennas for FM Reception," in *First European Conference on Antennas and Propagation, EuCAP 2006*, Nice, France, November 2006, pp. 1–6.

[21] A. Sabban, "Wideband Printed Antennas for Medical Applications," in *APMC 2009 Conference*, Singapore, December 2009.

[22] A. Alomainy, A. Sani et al., "Transient Characteristics of Wearable Antennas and Radio Propagation Channels for Ultrawideband B ody-centric Wireless Communication," *IEEE Transactions on Antennas and Propagation*, 57(4), 875–884, April 2009.

[23] M. Klemm and G. Troester, "Textile UWB Antenna for Wireless Body Area Networks," *IEEE Transactions on Antennas and Propagation*, 54(11), 3192–3197, November 2006.

[24] P. M. Izdebski, H. Rajagoplan, and Y. Rahmat-Sami, " Conformal Ingestible Capsule Antenna: A Novel Chandelier Meandered Design," *IEEE Transactions on Antennas and Propagation*, 57(4), 900–909, April 2009.

[25] A. Sabban, "Wideband Tunable Printed Antennas for Medical Applications," in *IEEE Antenna and Propagation Symposium*, Chicago IL, July 2012.

[26] B. B. Mandelbrot, *The Fractal Geometry of Nature*. W.H. Freeman and Company, New York, 1983.

[27] B.B. Mandelbrot, "How Long Is the Coast of Britain? Statistical Self-Similarity and Fractional Dimension," *Science*, 156, 636–638, 1967.

[28] F. J. Falkoner, *The Geometry of Fractal Sets*, Cambridge University Press, Cambridge, UK, 1990.

[29] C. A. Balanis, *Antenna Theory Analysis and Design*, 2nd edn., John Wiley & Sons, Inc., Hoboken, NJ, 1997

[30] K. Virga and Y. Rahmat-Samii, "Low-Profile Enhanced-Bandwidth PIFA Antennas for Wireless Communications Packaging," *IEEE Transactions on Microwave Theory and Techniques*, 45(10), 1879–1888, October 1997.

[31] M. I. Skolnik, *Introduction to Radar Systems*. McGraw Hill, London, UK, 1981.

[32] Patent US 5087515, Patent US 4976828, Patent US 4763127, Patent US 4600642, Patent US 3952307, Patent US 3725927.

[33] European Patent Application EP 1317018 A2, November 27, 2002.

[34] T. Chio and K. Wong, "Design of Compact Microstrip Antennas with a Slotted Ground Plane," in *IEEE-APS Symposium*, Boston, MA, July 8–12, 2001.

[35] R. C. Hansen, "Fundamental Limitations on Antennas," *Proceedings of the IEEE*, 69(2), 170–182, February 1981

[36] D. Pozar, *The Analysis and Design of Microstrip Antennas and Arrays*, IEEE Press, Piscataway, NJ, ISSN: 08855-1331.

[37] J. F. Zurcher and F. E. Gardiol, *Broadband Patch Antennas*. Artech House, Boston, MA, 1995.

[38] I. Minin, "Microwave and Millimeter Wave Technologies from Photonic Bandgap Devices to Antenna and Applications," in *Fractal Antenna Applications* by Mi. V. Rusu and R. Baican, eds., Intech, March 2010, Chapter 16, pp. 351–382, ISBN: 978-953-7619-66-4. Rijeka Croatia.

[39] M.V. Rusu, M. Hirvonen, H. Rahimi, P. Enoksson, C. Rusu, N. Pesonen, O. Vermesan, and H. Rustad, "Minkowski Fractal Microstrip Antenna for RFID Tags," in *Proceedings of the EuMW2008 Symposium*, Amsterdam, the Netherlands, October 2008.

[40] H. Rahimi, M. Rusu, P. Enoksson, D. Sandström, and C. Rusu, "Small Patch Antenna Based on Fractal Design for Wireless Sensors," in *MME07, 18th Workshop on Micromachining, Micromechanics, and Microsystems*, Portugal, September 16–18, 2007.

[41] J. R. James, P. S. Hall, and C. Wood, *Microstrip Antenna Theory and Design*. Peregrinus on behalf of the Institution of Electrical Engineers, London, 1981.

13 Wearable Compact Fractal Antennas for 5G and Medical Systems

Albert Sabban

CONTENTS

INTRODUCTION

Compact wearable printed antennas suffer from low efficiency. Fractal technology may be employed to develop compact efficient wearable antennas. In this chapter, several wearable fractal antennas are presented. The fractal antennas were designed to be worn in the vicinity of the human body. A fractal antenna is an antenna that uses antenna design with similar fractal segments to maximize the antenna effective area. Fractal antennas are also referred to as multilevel structures with space-filling curves.

The key aspect lies in a repetition of a motif over two or more scale sizes or "iterations." Fractal antennas are very compact, multiband or wideband, and have useful applications in cellular telephone and microwave communication systems. The development of compact fractal printed antennas for 5G communication, Internet of Things (IOT) and medical systems is presented in this chapter. These fractal antennas are compact and efficient. S-band, C-band and X-band fractal printed antennas were developed and fabricated for 5G communication, IOT and medical systems. Space-filling technique and Hilbert curves are employed to design compact fractal antennas. The antenna bandwidth is around 10% with voltage standing wave ratio better than 3:1. The antenna gain is around 8 dBi with 90% efficiency.

13.1 INTRODUCTION TO FRACTAL PRINTED ANTENNAS

Printed wearable antennas for communication and medical systems were presented in papers and books [1–18]. Compact wearable printed antennas suffer from low efficiency. Fractal technology may be employed to develop compact efficient wearable antennas. The electrical properties of human body tissues are presented in [7, 8]. In this chapter, several wearable fractal antennas are presented. The fractal antennas were designed to be worn in the vicinity of the human body. A fractal antenna is an antenna that uses antenna design with similar fractal segments to maximize the antenna effective area. Fractal antennas were presented in books, papers and patents, see [18–32]. Fractal antennas are also referred to as a multilevel structure with space-filling curves (SFCs). The key aspect lies in a repetition of a motif over two or more scale sizes or "iterations." Fractal antennas are very compact, multiband or wideband, and have useful applications in cellular telephone and microwave communication systems.

13.1.1 FRACTAL STRUCTURES

A curve, with endpoints, is represented by a continuous function whose domain is the unit interval [0, 1]. The curve may lie in a plane or in a three-dimensional space. A fractal curve is a densely self-intersecting curve that passes through every point of the unit square. A fractal curve is a continuous mapping from the unit interval to the unit square.

In mathematics, a SFC is a curve whose range contains the entire two-dimensional unit square. Most SFCs are constructed iteratively as a limit of a sequence of piecewise linear continuous curves, each one closely approximating the space-filling limit. In SFCs, where two subcurves intersect (in the technical sense), there is self-contact without self-crossing. A SFC can be (everywhere) self-crossing if its approximation curves are self-crossing. A SFC's approximations can be self-avoiding, as presented in Figure 13.1. In three dimensions, self-avoiding approximation curves can even contain joined ends. SFCs are special cases of fractal constructions. No differentiable SFC can exist.

The term "fractal curve" was introduced by B. Mandelbrot [18, 19] to describe a family of geometrical objects that are not defined in standard Euclidean geometry. Fractals are geometric shapes that repeat themselves over a variety of scale sizes. One of the key properties of a fractal curve is self-similarity. A self-similar object is unchanged after increasing or shrinking its size. An example of a repetitive geometry is the Koch curve, presented in 1904 by Helge von Koch; it is shown in Figure 13.1b.

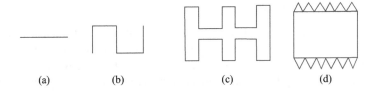

FIGURE 13.1 (a) Line, (b) motif of bended line, (c) bended line fractal structure, (d) fractal structure.

Koch generated the geometry by using a segment of a straight line and raising an equilateral triangle over its middle third. Repeating the process of erecting equilateral triangles over the middle thirds of straight lines led to what is presented in Figure 13.2a. Iterating the process infinitely many times results in a "curve" of infinite length. This geometry is continuous everywhere but is nowhere differentiable. If the Koch process is applied to an equilateral triangle, then, after many iterations, the iterations converge to the Koch snowflake shown in Figure 13.2b. This process can be applied to several geometries as shown in Figure 13.3 and Figure 13.4. Many variations of these geometries are presented in several papers.

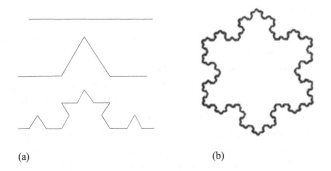

FIGURE 13.2 (a) Koch fractal structures, (b) Koch snowflakes.

FIGURE 13.3 Folded fractal structures.

FIGURE 13.4 Variations of Koch fractal structures.

13.1.2 FRACTAL ANTENNAS

Fractal geometries may be applied to the design of antennas and antenna arrays. The advantages of printed circuit technology and printed antennas enhance the design of fractal printed antennas and microwave components. The effective area of a fractal antenna is significantly higher than the effective area of a regular printed antenna. Fractal antennas may operate with good performance at several different frequencies simultaneously. Fractal antennas are compact multiband antennas. The directivity of fractal antennas is usually higher than the directivity of a regular printed antenna. The number of elements in a fractal antenna array may be reduced by a quarter of the number of elements in a regular array. A fractal antenna could be considered a non-uniform distribution of radiating elements. Each of the elements contributes to the total radiated power density at a given point with a given amplitude and phase. By spatially superposing these line radiators we can study the properties of a fractal antenna array.

The features of small antenna are:

- A large input reactance (either capacitive or inductive) that usually must be compensated with an external matching network.
- A small radiating resistance.
- Small bandwidth and low efficiency.

This means that it is highly challenging to design a resonant antenna in a small space in terms of the wavelength at resonance.

The use of microstrip antennas is well known in mobile telephony handsets [22]. The planar inverted-F antenna (PIFA) configuration is popular in mobile communication systems. The advantages of PIFAs are their low profile, their low fabrication costs and an easy integration within the system structure. One of the miniaturization techniques used in this antenna system is based on SFCs. In some specific cases of antenna configuration systems, the antenna shape may be described as a multilevel structure. A multilevel technique has already been proposed to reduce the physical dimensions of microstrip antennas. The presented integrated multiservice antenna system for a communication system comprises the following parts and features.

The antenna includes a conducting strip or wire shaped by a SFC, composed of at least 200 connected segments forming a substantial right angle with each adjacent segment smaller than a 100th of the free-space operating wavelength. The important reduction in size of such an antennas system is obtained by using space-filling geometries. A SFC can be described as a curve that is large in terms of physical length but small in terms of the area in which the curve can be included. A SFC can be fitted over a flat or curved surface, and due to the angles between segments, the physical length of the curve is always larger than that of any straight line that can be fitted in the same area (surface).

Additionally, to properly shape the structure of a miniature antenna, the segments of the SFCs must be shorter than a tenth of the free-space operating wavelength. The antenna is fed with a two-conductors structure, such as a coaxial cable, with one of the conductors connected to the lower tip of the multilevel structure and the other conductor connected to the metallic structure of the system which acts as a ground plane. This antenna type is 20% smaller than the typical size of a conventional

(a)   (b)

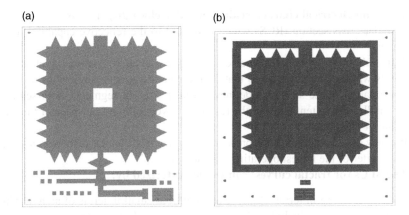

FIGURE 13.5   (a) Patch with space-filling perimeter of the conducting sheet, (b) microstrip patch with space-filling perimeter of the conducting sheet.

external quarter-wave whip antenna. This feature together with the small profile of the antenna, which can be printed in a low-cost dielectric substrate, allows a simple and compact integration of the antenna structure.

Reducing the size of the radiating elements can be achieved by using a PIFA configuration, consisting of connecting two parallel conducting sheets, separated either by air or a dielectric, of magnetic or magnetodielectric material. The sheets are connected through a conducting strip near one of the sheet's corners that is orthogonally mounted to both sheets.

The antenna is fed through a coaxial cable, having its outer conductor connected to the first sheet; the second sheet is coupled either by direct contact or is capacitive to inner conductor of the coaxial cable.

In Figures 13.5a and 13.5b are presented two examples of space-filling perimeter of the conducting sheet to achieve an optimized miniaturization of the antenna.

13.2 ANTI-RADAR FRACTALS AND/OR MULTILEVEL CHAFF DISPERSERS

Chaff was one of the forms of countermeasure employed against radar. It usually consists of a large number of electromagnetic dispersers and reflectors, which are normally arranged in the form of strips of metal foil packed in a bundle. Chaff is usually employed to foil or to confuse surveillance and tracking radar.

13.2.1 Geometry of Dispersers

Here are presented new geometries of the dispersers or reflectors that improve the properties of radar chaff [23]. Some of the geometries presented here of the dispersers or reflectors are related to some forms of antennas. Multilevel and fractal structure antennas are distinguished from regular antennas by their reduced size and multiband behavior, as has been expounded in patent publications [24].

The main electrical characteristic of a radar chaff disperser is:

Its radar cross-section (RCS) which is related to the reflective capability of the disperser.

A fractal curve for a chaff disperser is defined as a curve comprising at least ten segments which are connected so that each element forms an angle with its neighbors, no pair of these segments defines a longer straight segment, these segments being smaller than a tenth part of the resonant wavelength in free space of the entire structure of the disperser.

In many of the configurations presented, the size of the entire disperser is smaller than a quarter of the lowest operating wavelength.

The SFCs (or fractal curves) can be characterized by:

1. They are long in terms of physical length but small in terms of area in which the curve can be included. The dispersers with a fractal form are long electrically but can be included in a very small surface area. This means it is possible to obtain a smaller packaging and a denser chaff cloud using this technique.
2. Frequency response: Their complex geometry provides a spectrally richer signature when compared with known state-of-the-art rectilinear dispersers.

The fractal structure properties of the disperser not only introduce an advantage in terms of reflected radar signal response, but also in terms of the aerodynamic profile of the dispersers. It is known that a surface offers greater resistance to air than a line or a one-dimensional form.

Therefore, giving a fractal form to the dispersers with a dimension greater than unity ($D > 1$), increases resistance to the air and improves the time of suspension.

13.3 DEFINITION OF MULTILEVEL FRACTAL STRUCTURE

Multilevel structures are a geometry related to fractal structures. In the case of radar chaff, a multilevel structure is defined as a structure which includes a set of polygons, which are characterized in having the same number of sides, wherein these polygons are electro- magnetically coupled either by means of capacitive coupling or by means of an ohmic contact. The region of contact between the directly connected polygons is smaller than 50% of the perimeter of polygons in at least 75% of the polygons that constitute the defined multilevel structure.

A multilevel structure provides both:

- A reduction in the size of dispersers and an enhancement of their frequency response.
- Can resonate in a non-harmonic way and can even cover simultaneously and with the same relative bandwidth at least a portion of numerous bands.

The fractal structure (SFC) is preferred when a reduction in size is required, while multilevel structures are preferred when it is required that the most important considerations be given to the spectral response of radar chaff.

The main advantages of configuring the form of the chaff dispersers are:

1. The dispersers are small; consequently, more dispersers can be encapsulated in the same cartridge, rocket or launch vehicle.

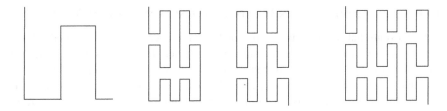

FIGURE 13.6   Fractal curves which can be used to configure a chaff disperser.

FIGURE 13.7   Hilbert fractal curves.

2. The dispersers are also lighter. Therefore, they can remain floating in the air longer than conventional chaff.
3. Due to the smaller size of the chaff dispersers, the launching devices (cartridges, rockets, etc.) can be smaller than state-of-the-art chaff systems and still provide the same RCS.
4. Due to the lighter weight of the chaff dispersers, the launching devices can shoot the packages of chaff farther from the launching devices and locations.
5. Chaff composed of multilevel and fractal structures provides larger RCS at longer wavelengths than conventional chaff dispersers of the same size.
6. The dispersers with long wavelengths can be configured and printed on light dielectric supports having a non-aerodynamic form and opposing a greater resistance to the air and thereby having a longer time of suspension.
7. The dispersers provide a better frequency response than state-of-the-art dispersers. In the following images, such size compression structures based on fractal curves are presented in Figure 13.6.

Figure 13.7 shows several examples of Hilbert fractal curves (with increasing iteration order) which can be used to configure the chaff disperser.

13.4 ADVANCED ANTENNA SYSTEM

The main advantage of an advanced antenna system lies in the multiband and multiservice performance of the antenna. This enables convenient and easy connection of a simple antenna for most communication systems and applications. The main advantages addressed by advanced antennas feature similar parameters (input impedance, radiation pattern) at several bands maintaining their performance, compared with conventional antennas. Fractal-shapes permit compact antennas of reduced dimensions compared to other conventional antennas. Multilevel antennas introduced a higher flexibility to the design of multiservice antennas for real applications, extending the theoretical capabilities of ideal fractal antennas to practical, commercial antennas.

13.4.1 Comparison between Euclidean Antennas and Fractal Antenna

Most conventional antennas are Euclidean design/geometry, where the closed antenna area is directly proportional to the antenna perimeter. Thus, for example, when the length of a Euclidean square is increased by a factor of three, the enclosed area of the antenna is increased by a factor of nine. The gain, directivity, impedance and efficiency of Euclidean antennas are a function of the antenna's size to wavelength ratio.

Euclidean antennas are typically desired to operate within a narrow range (e.g., 10%–40%) around a central frequency fc which in turn dictates the size of the antenna (e.g., half or quarter wavelength). When the size of a Euclidean antenna is made much smaller than the operating wavelength (λ), it becomes very inefficient because the antenna's radiation resistance decreases and becomes less than its ohmic resistance (i.e., it does not couple electromagnetic excitations efficiently to free space). Instead, it stores energy reactively within its vicinity (reactive impedance Xc). These aspects of Euclidean antennas work together to make it difficult for small Euclidean antennas to couple or match to feeding or excitation circuitry and cause them to have a high Quality (Q) factor (lower bandwidth). Q factor may be defined as approximately the ratio of input reactance X_{in} to radiation resistance R_r, $Q = X_{in}/R_r$.

The Q factor may also be defined as the ratio of average stored electric energies (or magnetic energies stored) to the average radiated power. Q can be shown to be inversely proportional to bandwidth.

Thus, small Euclidean antennas have very small bandwidth, which is of course undesirable (matching network may be needed). Many known Euclidean antennas are based upon closed-loop shapes.

Unfortunately, when small in size, such loop-shaped antennas are undesirable because, as discussed above, the radiation resistance decreases significantly when the antenna size is decreased. This is because the physical area (A) contained within the loop-shaped antenna's contour is related to the loop perimeter.

Radiation resistance R_r of a circular loop-shaped Euclidean antenna is defined by R_r, as given in Equation 13.1, k is a constant.

$$R_r = \eta\pi(2/3)(KA/\lambda)^2 = 20\pi^2\left(\frac{C}{\lambda}\right) \tag{13.1}$$

Since the resistance R_c is only proportional to perimeter (C), then for C < 1, the resistance R_c is greater than the radiation resistance R_r and the antenna is highly inefficient. This is generally true for any small circular Euclidean antenna. A small-sized antenna will exhibit a relatively large ohmic resistance and a relatively small radiation resistance R_r. This low efficiency limits the use of small antennas.

Fractal geometry is a non-Euclidean geometry which can be used to overcome the problems of small Euclidean antennas. Radiation resistance Rr of a fractal antenna decreases as a small power of the perimeter (C) compression. A fractal loop or island always has a substantially higher radiation resistance than a small Euclidean loop antenna of equal size. Fractal geometry may be grouped into:

- Random fractals, which may be called chaotic or Brownian fractals.

- Deterministic or exact fractals. In deterministic fractal geometry, a self-similar structure results from the repetition of a design or motif (generator) with self-similarity and structure at all scales. In deterministic or exact self-similarity, fractal antennas may be constructed through recursive or iterative means. In other words, fractal structures are often composed of many copies of themselves at different scales, thereby allowing them to defy the classic antenna performance constraint, which is size to wavelength ratio.

13.4.2 MULTILEVEL AND SPACE-FILLING GROUND PLANES FOR MINIATURE ANTENNAS

A new family of antenna ground planes of reduced size and enhanced performance based on an innovative set of geometries are presented in this section.

These new geometries are known as multilevel and space-filling structures, which had been previously used in the design of multiband and miniature antennas.

One of the key issues of the present antenna system is considering the ground plane of an antenna as an integral part of the antenna that mainly contributes to its radiation and impedance performance (impedance level, resonant frequency and bandwidth).

Multilevel and space-filling structures are used in the ground plane of the antenna to obtain a better return loss or voltage standing wave ratio (VSWR), a better bandwidth, a multiband behavior or a combination of all these effects. The technique can be seen as a way to reduce the size of the ground plane and therefore the size of the overall antenna. The key point of the present antenna system is shaping the ground plane of an antenna in such a way that the combined effect of the ground plane and the radiating element enhances the performance and characteristics of the whole antenna device, either in terms of bandwidth, VSWR, multiband, efficiency, size or gain.

13.4.3 MULTILEVEL GEOMETRY

The resulting geometry is no longer a solid, conventional ground plane, but a ground plane with a multilevel or space-filling geometry, at least in a portion of the ground plane.

A multilevel geometry for a ground plane consists of a conducting structure including a set of polygons, featuring the same number of sides, electromagnetically coupled either by means of a capacitive coupling or ohmic contact. The contact region between directly connected polygons is narrower than 50% of the perimeter of polygons in at least 75% of polygons defining the conducting ground plane. In this definition of multilevel geometry, circles and ellipses are included as well, since they can be understood as polygons with an infinite number of sides.

13.4.4 SFC

A SFC is a curve that is large in terms of physical length but small in terms of the area in which the curve can be included.

It is a curve composed of at least ten segments which are connected in such a way that each segment forms an angle with its neighbors, that is, no pair of adjacent segments define a larger straight segment, and wherein the curve can be optionally periodic along a fixed straight direction of space if, and only if, the period is defined by a non-periodic curve composed of at least ten connected segments and no pair of adjacent and connected segments defines a straight longer segment. A SFC can be fitted over a flat or curved surface, and due to the angles between segments, the physical length of the curve is always larger than that of any straight line that can be fitted in the same area (surface).

Additionally, to properly shape the ground plane, the segments of the SFC curves included in the ground plane must be shorter than a tenth of the free-space operating wavelength.

Figure 13.8 shows several examples of fractal geometries which can be used as SFCs. Figure 13.9 shows several examples of Hilbert fractal curves which can be used as SFCs.

The curves shown in Figure 13.9 are some examples of such SFCs.

Due to the special geometry of multilevel and space-filling structures, the current distributes over the ground plane in such a way that it enhances the antenna's performance and features in terms of:

- Reduced size compared to antennas with a solid ground plane.
- Enhanced bandwidth compared to antennas with a solid ground plane.
- Multi-frequency performance.
- Better VSWR feature at the operating band or bands.
- Better radiation efficiency.
- Enhanced gain.

Figure 13.10a shows a patch antenna above an example of a new ground plane structure formed by both multilevel and space-filling geometries. Figure 13.10b shows a monopole antenna above a ground plane structure formed by both multilevel and space-filling geometries.

Figure 13.11 shows several examples of different contour-shaped multilevel ground planes, such as rectangular 13.11a, 13.11b and circular ground plane 13.11c.

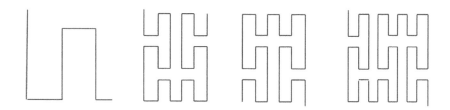

FIGURE 13.8 Fractal curves which can be used as space-filling curves.

FIGURE 13.9 Hilbert Fractal curves which can be used as space-filling curves.

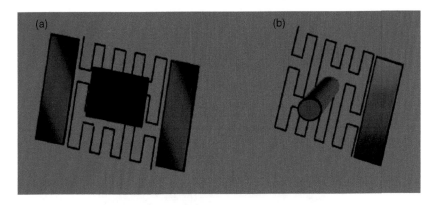

FIGURE 13.10   (a) Patch antenna above a new ground plane structure, (b) monopole antenna above a ground plane structure formed by both multilevel and space-filling geometries.

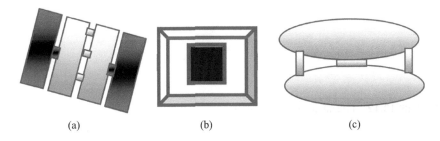

(a)   (b)   (c)

FIGURE 13.11   Examples of different contour-shaped multilevel ground planes. (a) Rectangular ground planes, (b) multilevel rectangular ground plane, (c) circular ground planes.

13.5   WEARABLE FRACTAL ANTENNAS FOR 5G AND IOT APPLICATIONS

In this chapter, several designs of wearable fractal printed antennas are presented for communication and Internet of Things (IOT) applications. These fractal antennas are compact and efficient. The antenna gain is around 8 dBi with 90% efficiency.

13.5.1   A WEARABLE 2.5 GHz FRACTAL ANTENNA FOR WIRELESS COMMUNICATION

A new fractal microstrip antenna was designed as presented in Figure 13.12. The antenna was printed on a Duroid substrate 0.8 mm thick with 2.2 dielectric constant. The antenna dimensions are 5.2 × 48.8 × 0.08 cm. The antenna was designed by using Advanced Design System (ADS) software.

The antenna bandwidth is around 2% around 2.5 GHz for VSWR better than 3:1. The antenna bandwidth may be improved to 5%, for VSWR better than 2:1, by adding a second layer above the resonator. A patch radiator is printed on the second layer as presented in Figure 13.13. The radiator was printed on flame retardant-4 (FR-4)

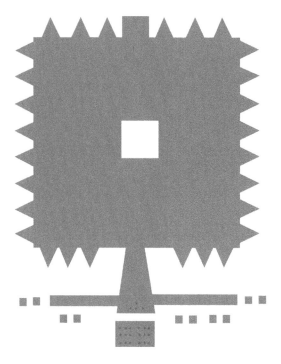

FIGURE 13.12   Fractal antenna resonators.

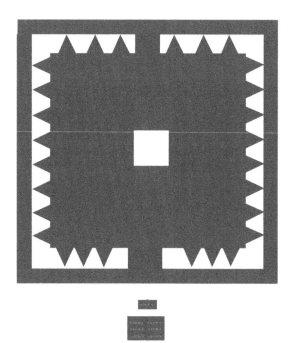

FIGURE 13.13   Fractal antenna patch radiator.

substrate 0.8 mm thick with 4.5 dielectric constant. The electromagnetic fields radiated by the resonator are electromagnetic coupled to the patch radiator. The patch radiator dimensions are 45.2 × 48.8 × 0.08 cm. The stacked fractal antenna structure is shown in Figure 13.14. The spacing between the two layers may be varied to get wider bandwidth. The stacked fractal antenna S11 parameters with 8 mm air spacing between the layers is presented in Figure 13.15. The S11 parameter of the fractal stacked patch antenna with 10 mm air spacing is given in Figure 13.16. The antenna bandwidth is improved to 5% for VSWR better than 2:1. The fractal stacked patch antenna radiation pattern is shown in Figure 13.17. The antenna beam width is around 76°, with 8 dBi

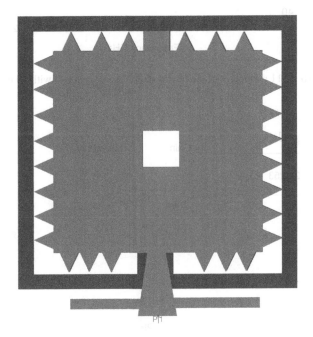

FIGURE 13.14   Fractal stacked patch antenna structure.

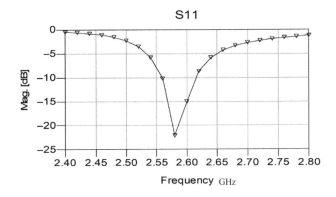

FIGURE 13.15   S11 parameter of the fractal stacked patch antenna with 8 mm air spacing.

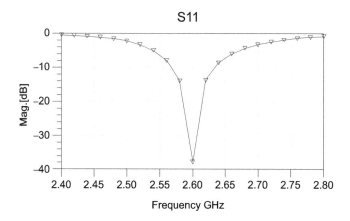

FIGURE 13.16   S11 parameter of the fractal stacked patch antenna with 10 mm air spacing.

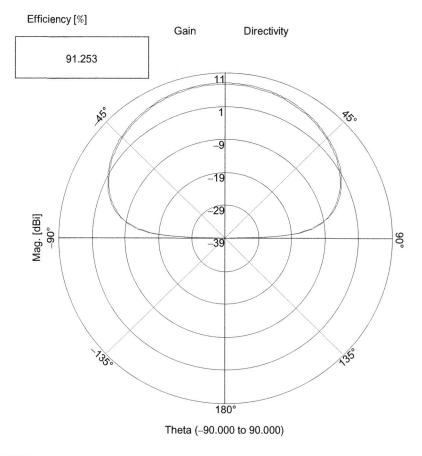

FIGURE 13.17   Fractal stacked patch antenna radiation pattern with 10 mm air spacing.

gain and 91% efficiency. A photo of the stacked fractal patch antenna is shown in Figure 13.18. The antenna resonators are shown in Figure 13.18a. The antenna radiator is shown in Figure 13.18b. A modified version of the antenna is shown in Figure 13.19. The S11 parameter of the modified fractal stacked patch antenna with 8 mm air spacing is given in Figure 13.20. The antenna bandwidth is around 10% for VSWR better than 3:1. The fractal stacked patch antenna radiation pattern is shown in Figure 13.21. The antenna beam width is around 76°, with 8 dBi gain and 91.82% efficiency.

(a)   (b)

FIGURE 13.18   Fractal stacked patch antenna. (a) Resonator, (b) radiator.

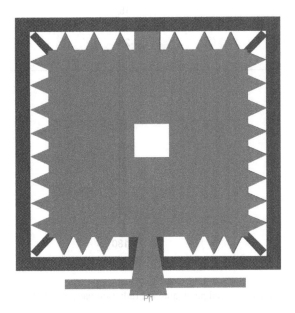

FIGURE 13.19   A modified fractal stacked patch antenna structure.

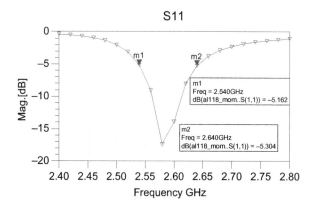

FIGURE 13.20 S11 parameter of the modified fractal stacked antenna with 8 mm air spacing.

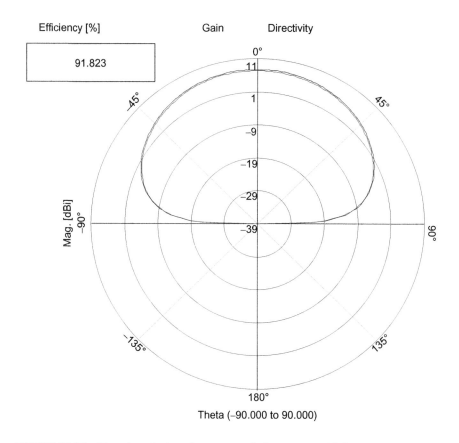

FIGURE 13.21 Fractal stacked patch antenna radiation pattern with 8 mm air spacing.

13.5.2 New Stacked Patch 2.5 GHz Fractal Printed Antennas

A new fractal microstrip antenna was designed as presented in Figure 13.22. The antenna was printed on Duroid substrate 0.8 mm thick with 2.2 dielectric constant. The antenna dimensions are 45.8 × 39.1 × 0.08 cm. The antenna was designed by using ADS software. The antenna resonator is shown in Figure 13.22a. The antenna resonator photo is shown in Figure 13.22b. The antenna radiator is shown in Figure 13.23.

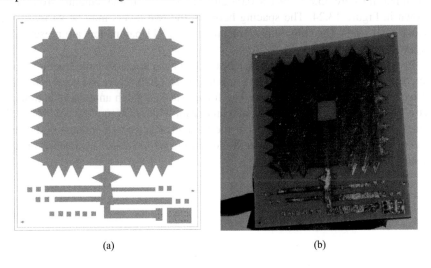

(a)   (b)

FIGURE 13.22   Resonator of a fractal stacked patch antenna. (a) Layout, (b) resonator photo.

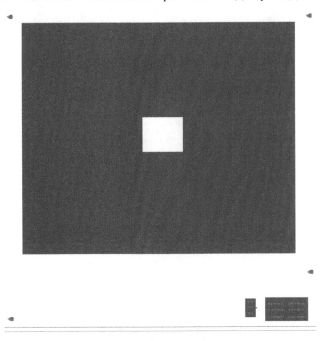

FIGURE 13.23   Radiator of a fractal stacked patch antenna.

The antenna resonator bandwidth is around 2% around 2.52 GHz for VSWR better than 3:1. The antenna bandwidth may be improved to 6%, for VSWR better than 3:1, by adding a second layer above the resonator. A patch radiator is printed on the second layer as presented in Figure 13.24. The radiator was printed on FR-4 substrate 0.8 mm thick with 4.5 dielectric constant. The electromagnetic fields radiated by the resonator are electromagnetic coupled to the patch radiator. The patch radiator dimensions are $45.8 \times 39.1 \times 0.08$ cm. The stacked fractal antenna structure is shown in Figure 13.24. The spacing between the two layers may be varied to get wider bandwidth. The single layer fractal antenna S11 parameters are presented in Figure 13.25.

A comparison of computed and measured S11 parameters of the fractal stacked patch antenna with no air spacing is given in Figure 13.26. There is a good agreement of measured and computed results. The fractal stacked patch antenna radiation pattern is shown in Figure 13.27. The antenna beam width is around 82°, with 7.5 dBi gain and 97.2% efficiency. A photo of the fractal stacked patch antenna is shown in Figure 13.28. The antenna resonator is shown in Figure 13.28a. The antenna radiator is shown in Figure 13.28b.

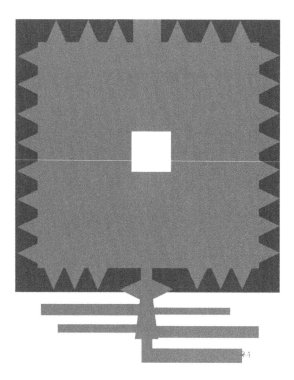

FIGURE 13.24 Layout of the fractal stacked patch antenna.

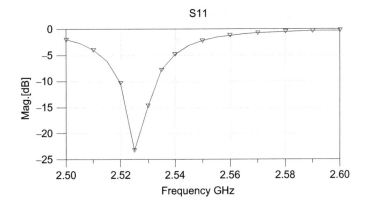

FIGURE 13.25 Computed S11 parameter of the single layer fractal antenna.

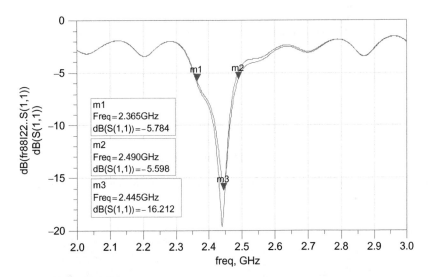

FIGURE 13.26 Measured and computed S11 parameters of the fractal stacked patch antenna with no air spacing between the layers.

13.6 X-BAND WEARABLE FRACTAL PRINTED ANTENNAS FOR 5G AND IOT APPLICATIONS

A new fractal microstrip antenna was designed as presented in Figure 13.29. The antenna was printed on Duroid substrate 0.8 mm thick with 2.2 dielectric constant. The antenna dimensions are $17.2 \times 21.8 \times 0.08$ cm. The antenna was designed by using ADS software.

The antenna resonator bandwidth is around 3% around 7. 2 GHz for VSWR better than 2:1. The antenna bandwidth is improved to 22%, for VSWR better than 3:1, by adding a second layer above the resonator. A patch radiator is printed on the second layer as presented in Figure 13.30. The radiator was printed on FR-4 substrate 0.8 mm thick

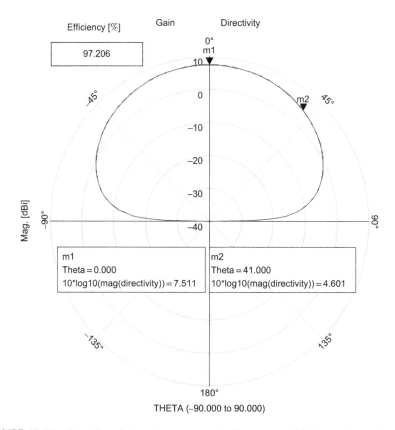

FIGURE 13.27 Fractal stacked patch antenna radiation pattern with 8 mm air spacing.

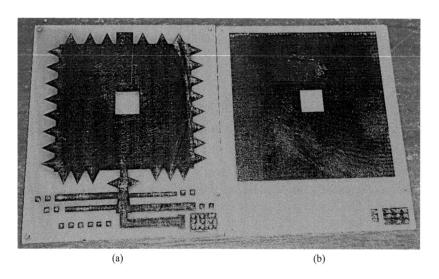

FIGURE 13.28 Fractal stacked patch antenna. (a) Resonator, (b) radiator.

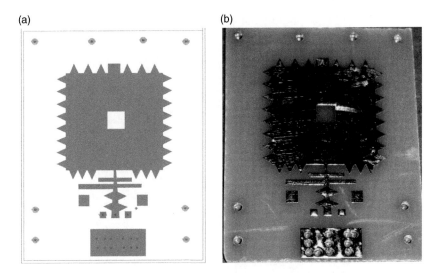

FIGURE 13.29 (a) Resonator of the 8 GHz fractal stacked patch antenna, (b) 8 GHz fractal resonator photo.

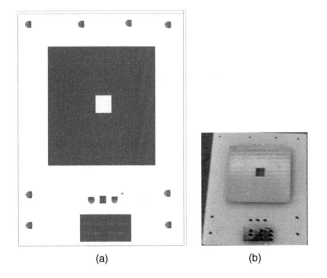

FIGURE 13.30 (a) Radiator of the 8 GHz fractal stacked patch antenna, (b) radiator photo.

with 4.5 dielectric constant. The electromagnetic fields radiated by the resonator are electromagnetic coupled to the patch radiator. The patch radiator dimensions are $17.2 \times 21.8 \times 0.08$ cm. The stacked fractal antenna structure is shown in Figure 13.31. The spacing between the two layers may be varied to get wider bandwidth. The single layer fractal antenna S11 parameters are presented in Figure 13.32. The computed S11 parameters of the fractal stacked patch antenna with 2 mm air spacing are given in Figure 13.33. The fractal stacked patch antenna radiation pattern at 7.5 GHz is shown in Figure 13.34. The antenna beam width is around 82°, with 7.8 dBi gain and

FIGURE 13.31 Layout of the 8 GHz fractal stacked patch antenna.

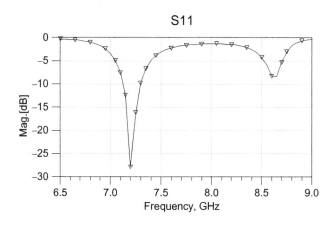

FIGURE 13.32 Computed S11 of the 8 GHz fractal stacked patch antenna.

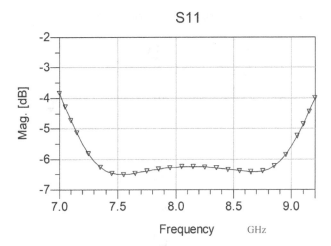

FIGURE 13.33 Computed S11 of the 8 GHz fractal stacked patch antenna with 2 mm air spacing.

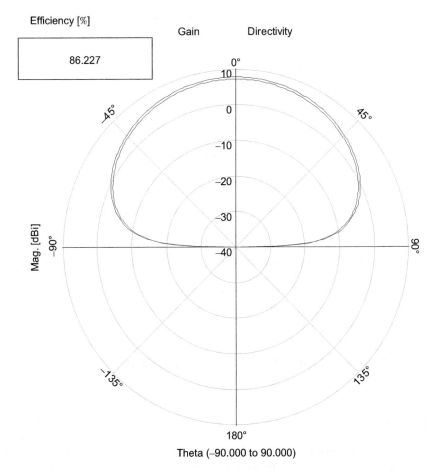

FIGURE 13.34 Fractal stacked patch antenna radiation pattern with 2 mm air spacing at 7.5 GHz.

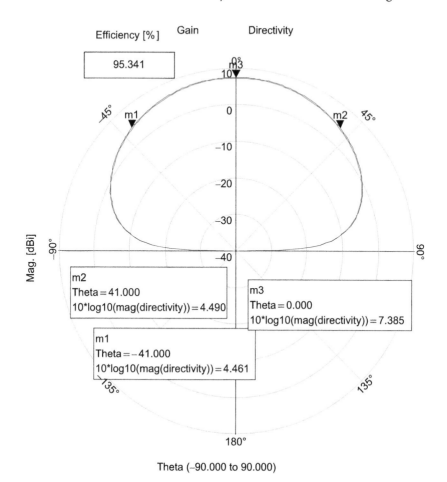

FIGURE 13.35 Fractal stacked patch antenna radiation pattern with 2 mm air spacing at 8 GHz.

82.2% efficiency. The fractal stacked patch antenna radiation pattern at 8 GHz is shown in Figure 13.35. The antenna beam width is around 82°, with 7.5 dBi gain and 95.3% efficiency. A photo of the fractal stacked patch antenna is shown in Figure 13.36. The antenna resonator is shown in Figure 13.36a. The antenna radiator is shown in Figure 13.36b.

13.7 WEARABLE STACKED PATCH 7.4 GHZ FRACTAL ANTENNA

A new fractal microstrip antenna was designed as presented in Figure 13.37. The antenna was printed on Duroid substrate 0.8 mm thick with 2.2 dielectric constant. The antenna dimensions are 18 × 12 × 0.08 cm. The antenna was designed by using ADS software.

The antenna resonator bandwidth is around 2% around 7. 4 GHz for VSWR better than 2:1. The antenna bandwidth is improved to 10%, for VSWR better than 3:1, by

(a)   (b)

FIGURE 13.36   A photo of the fractal stacked patch antenna. (a) Resonator, (b) radiator.

adding a second layer above the resonator. A patch radiator is printed on the second layer as presented in Figure 13.38. The radiator was printed on FR-4 substrate 0.8 mm thick with 4.5 dielectric constant. The electromagnetic fields radiated by the resonator are electromagnetic coupled to the patch radiator. The fractal stacked patch dimensions are 18 × 12 × 0.08 cm. The stacked fractal antenna structure is shown in Figure 13.39. The spacing between the two layers may be varied to get wider bandwidth. The computed S11 parameters of the fractal stacked patch antenna with 3 mm air spacing are given in Figure 13.40. The fractal stacked patch antenna radiation pattern at 7.5 GHz is shown in Figure 13.41. The antenna beam width is around 86°, with 7.9 dBi gain and 89.7% efficiency.

A modified antenna structure is presented in Figure 13.42. The antenna matching network has been modified and S11 at 7.45 GHz is −23.5 dB as shown in Figure 13.43. The computed S11 parameter of the fractal stacked patch antenna with 3 mm air spacing is given in Figure 13.43. The fractal stacked patch antenna bandwidth is around 9% for VSWR better than 3:1. The fractal stacked patch antenna radiation pattern at 7.5 GHz is shown in Figure 13.44. The antenna beam width is around 85°, with 7.8 dBi gain and 86.2% efficiency.

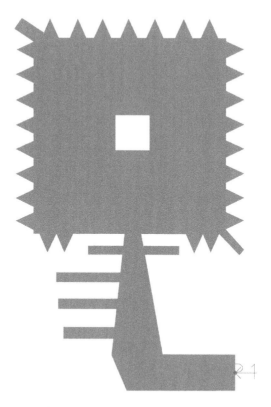

FIGURE 13.37   Layout of the 7.4 GHz fractal resonator.

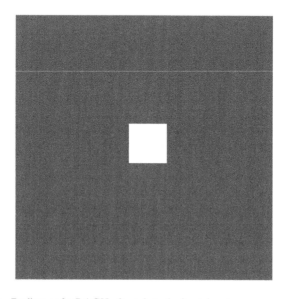

FIGURE 13.38   Radiator of a 7.4 GHz fractal stacked patch antenna.

FIGURE 13.39   Layout of the 8 GHz fractal stacked patch antenna.

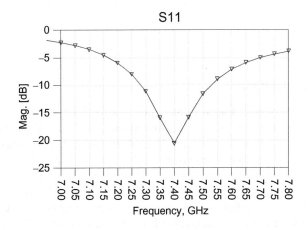

FIGURE 13.40   Computed S11 of the 7.4 GHz modified fractal antenna with 3 mm air spacing.

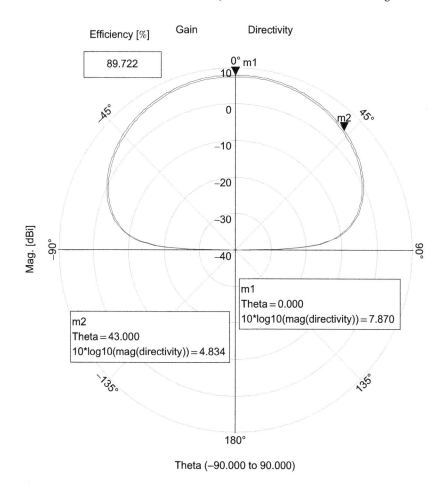

Efficiency [%] Gain Directivity

89.722

m1
Theta = 0.000
10*log10(mag(directivity)) = 7.870

m2
Theta = 43.000
10*log10(mag(directivity)) = 4.834

Mag. [dBi]

Theta (−90.000 to 90.000)

FIGURE 13.41 Fractal stacked patch antenna radiation pattern with 3 mm air spacing at 7.5 GHz.

13.8 CONCLUSION

Development of compact fractal printed antennas for 5G communication, IOT and medical systems have been presented in this chapter. These fractal antennas are compact and efficient. S-band, C-band and X-band fractal printed antennas were developed and fabricated for 5G communication, IOT and medical systems. Space-filling technique and Hilbert curves were employed to design compact fractal antennas. The antenna bandwidth is around 10% with VSWR better than 3:1. The antenna gain is around 8 dBi with 90% efficiency. There is a good agreement between computed and measured results. The fractal stacked patch antenna beam width at 7.5 GHz is around 85°, with 7.8 dBi gain and 86.2% efficiency.

FIGURE 13.42   Layout of the modified 7.4 GHz fractal stacked patch antenna.

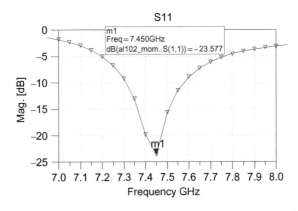

FIGURE 13.43   Computed S11 of the 7.4 GHz modified fractal antenna with 3 mm air spacing.

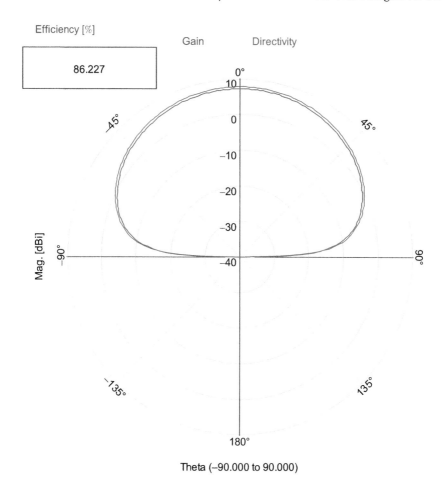

FIGURE 13.44 Modified fractal patch antenna radiation pattern with 3 mm air spacing at 7.5 GHz.

REFERENCES

[1] Sabban, A., *Low Visibility Antennas for Communication Systems*, Taylor & Francis Group, New York, 2015.
[2] Sabban, A., A New Wideband Stacked Microstrip Antenna, IEEE Antenna and Propagation Symp., Houston, Texas, USA, June 1983.
[3] Sabban, A., Microstrip Antenna Arrays, *Microstrip Antennas*, Nasimuddin Nasimuddin (Ed.), InTech, pp. 361–384, 2011, ISBN: 978-953-307-247-0, Rijeka, Croatia. http://www.intechopen.com/articles/show/title/microstrip-antenna-arrays.
[4] Sabban, A., New Wideband Printed Antennas for Medical Applications, *IEEE Journal, Transactions on Antennas and Propagation*, 61(1), 84–91, January 2013.
[5] James, J.R., Hall, P.S., and Wood, C., *Microstrip Antenna Theory and Design*, The Institution of Engineering and Technology, IET, London, UK, 1981.

[6] Sabban, A. and Gupta, K.C., Characterization of Radiation Loss from Microstrip Discontinuities Using a Multiport Network Modeling Approach, *IEEE Transactions on MTT*, 39(4), 705–712, April 1991.

[7] Chirwa, L.C., Hammond, P.A., Roy, S., and Cumming, D.R.S., Electromagnetic Radiation from Ingested Sources in the Human Intestine between 150 MHz and 1.2 GHz, *IEEE Transaction on Biomedical Engineering*, 50(4), 484–492, April 2003.

[8] Werber, D., Schwentner, A., and Biebl, E. M., Investigation of RF Transmission Properties of Human Tissues, *Advanced Radio Science*, 4, 357–360, 2006.

[9] Gupta, B., Sankaralingam, S., and Dhar, S., Development of Wearable and Implantable Antennas in the Last Decade, Microwave Symposium (MMS), 2010 Mediterranean 2010, Cyprus, 251–267, 2010.

[10] Thalmann, T., Popovic, Z., Notaros, B.M, and Mosig, J.R., Investigation and design of a multi-band wearable antenna, 3rd European Conference on Antennas and Propagation, EuCAP 2009, Berlin, Germany, 462–465.

[11] Salonen, P., Rahmat-Samii, Y., and Kivikoski, M., Wearable antennas in the vicinity of human body, IEEE Antennas and Propagation Society Symposium, Monterey, USA, 2004, Vol. 1, 467–470.

[12] Kellomaki, T., Heikkinen, J., and Kivikoski, M., Wearable antennas for FM reception, First European Conference on Antennas and Propagation, EuCAP 2006, Nice, France, 2006, 1–6.

[13] Sabban, A., Wideband printed antennas for medical applications, APMC 2009 Conference, Singapore, Dec. 2009.

[14] Alomainy, A., Sani, A. et al., Transient Characteristics of Wearable Antennas and Radio Propagation Channels for Ultrawideband Body-centric Wireless Communication" *IEEE Transactions on Antennas and Propagation*, 57(4), 875–884, April 2009.

[15] Klemm, M. and Troester, G., Textile UWB Antenna for Wireless Body Area Networks, *IEEE Transactions on Antennas and Propagation*, 54(11), 3192–3197, Nov. 2006.

[16] Izdebski, P.M., Rajagoplan, H., and Rahmat-Sami, ,Y., Conformal Ingestible Capsule Antenna: A Novel Chandelier Meandered Design, *IEEE Transactions on Antennas and Propagation*, 57(4), 900–909, April 2009.

[17] Sabban, A., Wideband Tunable Printed Antennas for Medical Applications, IEEE Antenna and Propagation Symp, Chicago, IL, USA, July 2012.

[18] Mandelbrot, B.B., The Fractal Geometry of Nature. W.H Freeman and Company, New York, 1983.

[19] Mandelbrot, B.B., How Long is the Coast of Britain? Statistical Self-Similarity and Fractional Dimension, *Science* 156, 636–638, 1967.

[20] Falkoner, F.J., *The Geometry of Fractal Sets*. Cambridge University Press, 1990.

[21] Balanis, C.A., *Antenna Theory Analysis and Design*, Second Edition. John Wiley & Sons, Inc., 1997.

[22] Virga, K. and Rahmat-Samii, Y., Low-Profile Enhanced-Bandwidth PIFA Antennas for Wireless Communications Packaging, *IEEE Transactions on Microwave Theory and Techniques*, 45(10), 1879–1888, October 1997.

[23] Skolnik M.I, *Introduction to Radar Systems*, McGraw Hill, London, 1981.

[24] Patent US 5087515, Patent US 4976828, Patent US 4763127, Patent US 4600642, Patent US 3952307, Patent US 3725927.

[25] European Patent Application EP 1317018 A2 / 27.11.2002

[26] Chiou, T. and Wong, K., Design of Compact Microstrip Antennas with a Slotted Ground Plane, IEEE-APS Symposium, Boston, 8–12, July 2001.

[27] Hansen R.C., Fundamental limitations on Antennas, *Proceedings of IEEE*, 69(2), 170–182, February 1981

[28] Pozar, D., *The Analysis and Design of Microstrip Antennas and Arrays*, IEEE Press, Piscataway, NJ, 08855-1331.

[29] Zurcher, J.F. and Gardiol, F.E., *Broadband Patch Antennas*, Artech House, 1995.

[30] Minin, I., Fractal Antenna Applications, Mircea V. Rusu and Roman Baican, *Microwave and Millimeter Wave Technologies from Photonic Bandgap Devices to Antenna and Applications*, Intech, ISBN 978-953-7619-66-4, Chapter 16, 351–382, March 2010.

[31] Rusu, M.V., Hirvonen, M., Rahimi, H., Enoksson, P., Rusu, C., Pesonen, N., Vermesan, O., and Rustad, H., Minkowski Fractal Microstrip Antenna for RFID Tags, Proc. EuMW2008 Symposium, Amsterdam, October 2008.

[32] Rahimi, H., Rusu, M., Enoksson, P, Sandström, D., and Rusu, C., Small Patch Antenna Based on Fractal Design for Wireless Sensors, MME07, 18th Workshop on Micromachining, Micromechanics, and Microsystems, Portugal, 16–18, Sept. 2007

14 Reconfigurable Wearable Antennas

Sema Dumanli

CONTENTS

14.1 INTRODUCTION

Various wearable sensors are being introduced to the market; they monitor diverse physiological parameters for fitness tracking of healthy individuals or the diagnosis, therapy, and rehabilitation of patients [1–3]. In the future, it is envisaged that each person is going to be wearing multiple sensors on their body that are part of a body area network (BAN). Some common challenges for wearable sensors can be listed as: achieving robust and quality sensing, high-level integration hence low-cost production, longer battery lifetime, and user convenience [4]. Although wearable sensor design is highly multidisciplinary, it can be argued that the careful design of the wireless links is essential to the solution of the previously mentioned goals. These links within the BANs are classified under three categories: in-body, on-body, and off-body [5]. The links that a wearable sensor forms with an implantable device are called in-body links. On-body links are the links created between two wearable devices and, finally, the links that a wearable sensor form with an off-body device, such as an access point, are called off-body links.

This chapter discusses antenna solutions for wearable devices; therefore, all of these propagation links are going to be taken into consideration. The network architecture should be designed so that the challenges previously listed for wearable sensor design are addressed (e.g., low cost, efficiency and user acceptance). In addition, the network should operate and coexist with other networks in similar frequency bands [6]. Here, a highly reconfigurable network architecture that opportunistically selects the best possible radiation characteristic for each link is considered. Reconfiguration is proposed in order to improve the reliability and enhance the battery lifetime, however, this functionality should not interfere with user convenience and cost. Therefore, the antennas proposed should support radiation pattern diversity along with other requirements.

Antenna design is one of the most important elements of the optimum wearable device, which should operate reliably for a long time without restricting user activity and causing any behavior modification. In order to increase the battery lifetime, the energy efficiency of the device should be improved. Considering the fact that the energy consumed during radio frequency (RF) transmission is a high percentage of the overall consumption, decreasing the number of retransmissions and improving the link budget by having higher antenna gain or pattern diversity can directly be translated into longer battery life.

A convenient form factor is also related to the antenna since it is one of the largest elements of the device alongside the battery. Flexible or small-sized rigid antennas are required so that the sensor can be incorporated seamlessly into clothing. When an antenna is located near lossy human tissues (i.e., worn by a person), its frequency response changes and radiation efficiency degrades. In order to avoid these, a body phantom should be included in the design process and the antenna should be electromagnetically isolated from the human body as much as possible with a ground plane. If the antenna is designed to be immune to these near-field effects of the body, the overall system efficiency is going to be maintained. In addition to being efficient and immune to detuning, the antenna can be reconfigurable in order to further improve system performance. The optimum radiation pattern will change depending on the link to be formed. Directional antennas were proven to perform the best for off-body links [7], whereas a radiation pattern with a null in the vertical plane according to the human body surface is best for on-body links [8]. Vertical polarization is better in launching surface waves [9]. In-body links are trickier since the propagation medium is extremely lossy. Therefore, as well as pointing the radiation toward the implant, one should minimize near-field losses either by increasing the separation between the antenna and the human body or by utilizing a bolus layer [10, 11].

The wearable reconfigurable antenna design challenge has been taken on by several researchers [12]. The proposed solutions can be grouped into two categories:

a. Flexible and comparatively larger antennas that can be printed on clothing.
b. Miniaturized rigid antennas that can be incorporated into clothing.

To the author's best knowledge, antennas that can achieve reconfigurability on flexible substrates are limited to five proposals [12–17]. A felt antenna operating at 2.45 GHz that can switch between four different radiation patterns has been proposed in [12]. The antenna has four different radiators each connected to its feed through a pin diode. As one pin diode is activated, the radiation of its corresponding radiator becomes dominant in the resultant radiation pattern. The patterns created are directive patterns targeting $\Phi = 0°$, 90°, 180°, and 270°. One can argue that all these patterns are more suitable for **off-body links**. The size of the antenna is 88 mm × 88 mm × 2 mm. [13] described another 2.45 GHz felt antenna with an overall size of 100 mm × 100 mm × 3 mm. It switches between a monopolar and a directive radiation pattern. This is a critical property as it has the potential to cover both **on-body and off-body links**. In this chapter, however, the pin diodes were not realized. Hence, the efficiency figures of 38% and 45% for each mode are calculated excluding the switching stage. In [14], an antenna designed to operate at 6 GHz has been proposed. Due to the high operating frequency, its dimensions are 30 mm × 60 mm × 1.5 mm. It was printed on a mixed fabric made of polyester (66.2%) and cotton (33.8%). The antenna switches between three directive radiation patterns, two of them creating only a tilt of 30° in θ. All the patterns created are more suitable for **off-body links**. The prototype excludes the switching stage. In [15] an ultra-wideband (UWB) wearable antenna is prototyped using different techniques, such as conductive thread embroidery or thin laser-cut sheet of copper on cotton or denim substrate. Its size is relatively small: 42 mm × 80 mm × 1 mm. It achieves three different radiation patterns that steer on the H plane. The antenna does not have a strong directive mode, which is suitable for **on-body links**. Finally, [16, 17] described a flexible antenna operating at MedRadio band (401–406 MHz). This is an electrically small antenna with low gain values. It can switch between two radiation modes, however, the change in the maximum radiation direction is only 30°. All of these fabric antennas suffer from low efficiency values; however, they cover quite a wide range of prototyping techniques some of which overcome the problem of repeatability and durability [15]. Remember that the connection between the flexible and rigid structures remained a question mark. For example, the switching has been realized only in [12] where no repeatability analysis was performed.

On the other hand, the rigid reconfigurable wearable antennas being realized use more established manufacturing techniques that are more repeatable. They are expected to have greater efficiencies with smaller conductor losses associated with the radiator. However, for the rigid substrates, antenna sizes in the order of 100 mm are out of the question. Hence, the reconfigurable rigid wearable antennas in the literature are not more common than their flexible counterparts.

In [18], a dual port antenna with dimensions of 68 mm × 68 mm × 6.35 mm has been proposed. One of the ports excites a monopolar radiation pattern while the other one excites a directional one which can support both **on-body and off-body** links. The correlation between the ports is insignificantly small at the operating frequency of 2.45 GHz providing excellent diversity values. The efficiency of the antenna is greater than 92% for both modes, which is rare for wearable antennas. The antenna has two ports and the switching is excluded from the work. Another fine example of a rigid on-body antenna supporting both **on-body and off-body** links was presented in [19]. It is a circular antenna with a diameter of 48 mm and a thickness of 3.2 mm. Its performance is remarkable. Another rigid on-body antenna suitable for use in the 2.4 GHz industrial, scientific and medical (ISM) band has been proposed in [20]. The proposal is 120 mm × 155 mm × >6 mm. Although the size is too large to be used in practice, the antenna is unique in comprising a high impedance surface as its ground plane. It has two modes both of which have the same θ (43°) and opposite Φ (0°, 180°). Considering the tilt angle, both modes are suitable for **off-body** links.

When it comes to in-body links, reconfiguration efforts focus on reconfiguring the antenna's pattern and operating frequency together. UWB, Medical Implant Communication System (MICS), 2.4 or 5.8 ISM band have previously been used for in-body links; hence, reconfigurable antennas in the literature combine these frequencies with 2.4 or 5.8 GHz ISM bands intended for either on-body or off-body links. [21] described a dual band antenna which has an **in-body** mode at the MICS band and a directive off-body mode operating at 2.45 GHz. The size of the antenna, 40 mm × 40 mm × 3.2 mm, is comparable to other 2.4 GHz ISM band counterparts. [22, 23] proposed to use the 2.45 GHz ISM band for in-body communications. [22] described an antenna creating an off-body mode at the upper UWB band. It has a circular shape with a diameter of 35 mm and a very low profile of 0.76 mm thickness. On the other hand, [23] utilized the 5.8 GHz ISM band for on-body communications. The in-body mode has circular polarization while the on-body mode has a monopolar radiation pattern. The overall dimensions of the antenna are 29 mm × 31.3 mm × 5.8 mm. Despite the common assumption of operating the in-body mode at lower frequency bands, [24] proposed an antenna with an on-body mode at 2.45 GHz and an in-body mode at 5.8 GHz. The on-body mode has a monopolar radiation pattern and the in-body mode directs its energy toward the human body. The thick profile (15.5 mm × 10.5 mm × 28 mm) can be counted as a drawback, especially for a wearable antenna. Note that none of the proposed reconfigurable wearable antennas comprising an in-body mode is flexible.

The author has proposed several rigid and flexible pattern reconfigurable antennas that are comparable to the literature. Antennas that provide diversity for off-body links [25, 26], for off-body and on-body links [27–30], and off-body and in-body modes [31] are intended for various applications, including smart glasses and wristwatches charged with far-field wireless power transfer. The details of these antennas are going to be provided in Sections 14.2.1, 14.2.2, and 14.2.3.

14.2 EXAMPLE ANTENNAS

Examples provided here were simulated on Ansys human body phantom, male, 4 mm accuracy consideration, in all cases [11], as seen in Figure 14.1. The Specific Absorption Rate (SAR) of all the antennas discussed here are kept well under the 1.6 W/kg limitation averaged over 1 g of tissue [32].

The examples that will be provided here have been designed to create a complete BAN in which the sensors form wireless connections; thus, making use of the reconfigurability of antennas to maximize system performance. The envisaged BAN can be seen in Figure 14.2.

14.3 PROVIDING DIVERSITY FOR OFF-BODY LINKS

14.3.1 The Design of a Pattern Reconfigurable Antenna Suitable for Smart Glasses

A novel pattern reconfigurable wearable antenna suitable for a smart glasses application is designed to operate at the 2.4 GHz ISM band [26]. The antenna consists

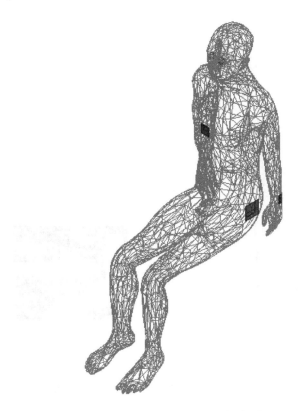

FIGURE 14.1   Wearables simulated on Ansys human body phantom: sitting position, male, 4 mm accuracy.

of two slots that are placed perpendicular to each other and fed with a coplanar waveguide transmission line placed at the corner as seen in Figure 14.3(a–b). The switches that are located directly on the slots are manipulated to activate different polarizations; hence, they provide polarization diversity. The gain patterns of each mode were simulated in a vacuum and their corresponding frequency responses can be seen in Figure 14.3(c–e). The antenna is prototyped on the glass as seen in Figure 14.3(b) and the simulations are validated through measurements. The antenna has been shown to provide two distinct patterns with a correlation coefficient of less than 0.1. Note that a pattern reconfigurable glass antenna has not been proposed before.

14.3.2 A Wrist Wearable Dual Port Dual Band Stacked Patch Antenna for Wireless Information and Power Transmission

A novel antenna with orthogonally polarized dual ports, one of which is dual band, is proposed [25]. It has a stacked patch configuration creating a dual band dual port operation unlike the heretofore aim of achieving a wideband operation.

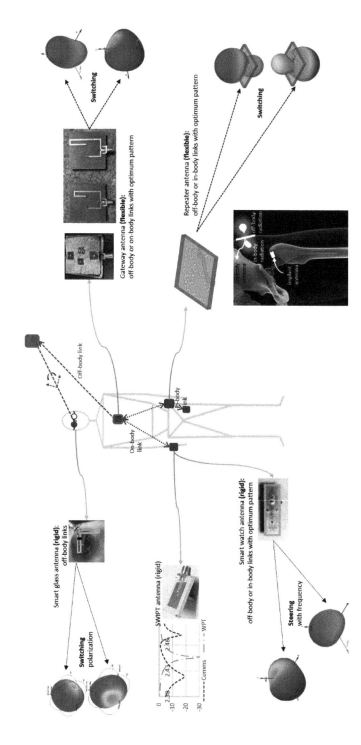

FIGURE 14.2 Reconfigurable wearable antennas previously proposed by the author presented at the system level.

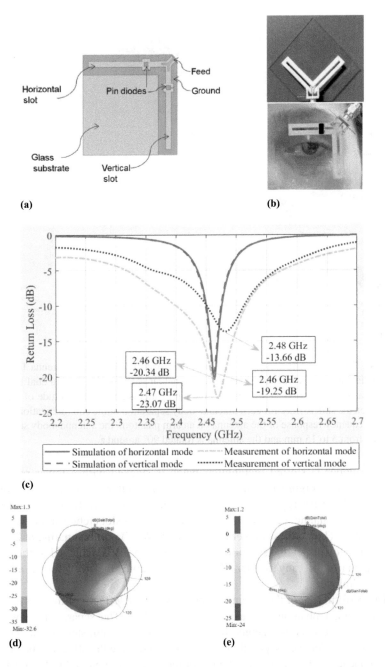

FIGURE 14.3 Pattern reconfigurable glass antenna suitable for smart glass applications providing polarization diversity of off-body links. (a) Antenna model including the pin diode pads, (b) antenna prototype without the pin diodes, (c) the simulated and measured frequency response of the antenna for each off-body mode, (d) simulated gain pattern of the horizontal slot while the vertical slot is shorted at 2.45 GHz, and (e) simulated gain pattern of the vertical antenna while the horizontal slot is shorted at 2.45 GHz.

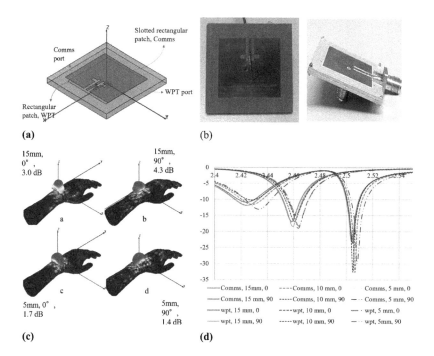

FIGURE 14.4 A wrist wearable dual port dual band stacked patch antenna for wireless information and power transmission. (a) A rectangular patch stacked with a dual band U slot loaded patch, (b) prototyped antenna, (c) the gain patterns and the magnitude of the current density, on the skin surface for each antenna–body separation of the communications port, and (d) the frequency response of the antenna on an arm phantom, the antenna–body separation is changed from 5 to 15 mm and the antenna is rotated $90°$ around z.

Here an upper U slotted rectangular patch is stacked with a lower rectangular patch as seen in Figure 14.4(a–b). The data port excites the upper patch covering Bluetooth Low Energy (BLE) advertisement channels while having a notch at around 22nd Channel (2.45 GHz) generated by the U slot etched at the edge of the upper patch. The wireless power transmission (WPT) port resonating at 2.45 GHz excites the lower patch. Isolation of more than 35 dB is achieved at the ports of the antenna before any further filtering, eliminating the need for an additional discrete diplexer.

The performance of the antenna is analyzed on the left arm of the Ansys computational human phantom as seen in Figure 14.4(c). The separation between the antenna and the wrist is changed from 5 to 15 mm in 5 mm steps while the arm is rotated for $90°$ for each separation in order to interrogate the effects of polarization. The frequency response of the antenna for these cases are plotted in Figure 14.4(d). It is observed that the detuning is more severe for WPT if there is no rotation, and more severe for the communications port if there is a $90°$ rotation. We can conclude that, for a wrist wearable antenna, if the polarization of the off-body mode is aligned with the arm, the detuning is stronger.

14.4 SWITCHING BETWEEN ON-BODY AND OFF-BODY LINKS

14.4.1 PATTERN DIVERSITY ANTENNA FOR ON-BODY AND OFF-BODY WIRELESS BAN (WBAN) LINKS

A novel pattern reconfigurable flexible antenna is presented that can switch between two different radiation patterns. For the on-body and the off-body links, the TM_{00} and the TM_{01} modes (transverse magnetic modes) of a rectangular patch antenna are excited, respectively. The TM_{00} mode generates a quasi-omnidirectional pattern with a simulated directivity of 2 dB in the horizontal plane while the TM_{01} mode generates a directional pattern with a maximum directivity value of 4.5 dB. Both modes are matched at the 2.45 GHz ISM band. Finally, using numerical and physical phantoms and male subjects, the antenna has been shown to perform well near human bodies: insignificant detuning was observed and the degradation in the radiation efficiency was measured to be 17% in the worst-case scenario of locating the antenna on the body with 0 mm separation. Note that the size of the antenna is no bigger than a conventional half-wavelength patch antenna.

The TM_{00} mode of a shorted rectangular patch antenna creates an omnidirectional radiation pattern. Although the TM_{00} mode of a shorted rectangular patch has not been investigated extensively, it is analogous to the widely studied TM_{01} mode of shorted ring patch antennas. The TM_m mode describes the generation of the omnidirectional mode in non-circular patches. The application of it to BANs was discussed in [33] and it was compared with a rectangular patch and a monopole near the body surface. The shorted patch was shown to be superior for on-body links. Here, an antenna capable of generating this omnidirectional mode and a directional mode is designed.

As seen in Figure 14.5(a–b), it comprises a square patch and a full square ground plane underneath. The radiating patch is printed with a copper plate. The patch and the ground plane are separated with 4.8-mm-thick polyethylene foam by Emerson & Cuming. The ground plane is shorted to the patch at symmetrical shorting pins. The shorting pin positions and their radii are optimized through simulations. Beneath the ground plane, a feeding network is etched on a 1.6-mm-thick flame retardant-4 (FR-4) substrate with 35-μm-thick copper. The ground plane of the patch is used as the ground plane of the microstrip line feeding network, as well. The network excites the patch at two symmetrical points while these feeding points are isolated from the ground plane with two circular slots. The feed line comprises a T junction and its branches of unequal length are connected to the aforementioned symmetrical feed points. Being unequal, the lengths of the branches cause a phase difference, which can be changed by altering the length of one branch by utilizing a switching mechanism as seen in Figure 14.5(a). Two prototypes with altered branches were manufactured to test both modes. In addition to that, the antenna is prototyped on a flexible substrate, replacing both the polyethylene foam and FR-4 with felt. Copper is replaced with silver fabric and silver ink in an attempt to compare the two as seen in Figure 14.5(c–d).

Figure 14.5(e–f) shows the E field distribution in vector form created by each mode: TM_{00} and TM_{01}. Feeding the excitation points with a 0° phase difference will

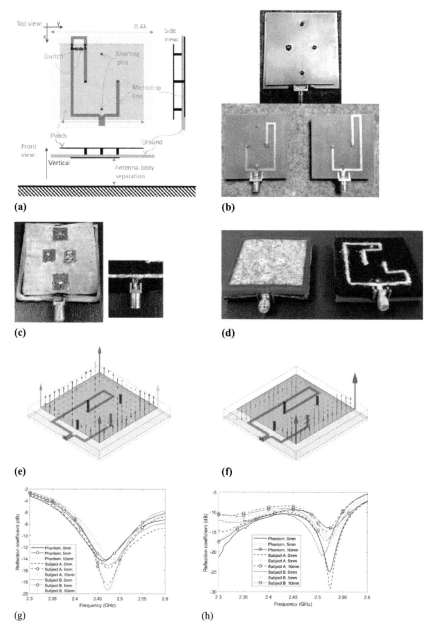

FIGURE 14.5   Pattern diversity antenna for on-body and off-body wireless body area network links [28]. (a) The proposed dual mode antenna which comprises a switching mechanism. (b) The prototyped on-body and off-body antennas, rigid version. (c) The prototyped on-body and off-body antennas, flexible version with silver fabric on felt. (d) The prototyped on-body and off-body antennas, flexible version with silver paint on felt. (e) Vector E field distributions of TM_{00}, on-body mode. (f) Vector E field distributions of TM_{01}, off-body mode. (g) $|S11|$ (dB) vs. frequency (GHz) of the on-body mode. (h) $|S11|$ (dB) vs. frequency (GHz) of the off-body mode.

(*Continued*)

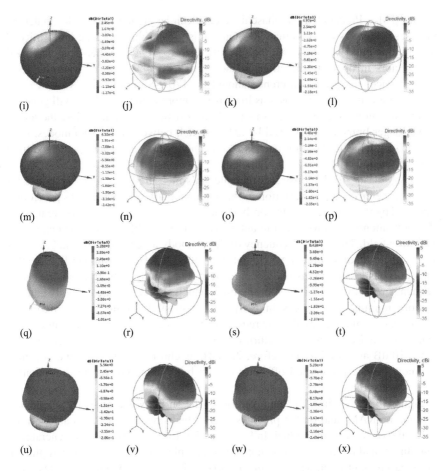

FIGURE 14.5 (CONTINUED) (i) On-body antenna in vacuum, simulation. (j) On-body antenna on RAM, meas, max. dir. of 4 dBi. (k) On-body antenna on the phantom with 0 mm antenna–body separation, simulation. (l) On-body antenna on the phantom with 0 mm antenna–body separation, Meas, max. dir. of 5.5 dBi. (m) On-body antenna on the phantom with 5 mm antenna–body separation, simulation. (n) On-body antenna on the phantom with 5 mm antenna–body separation, meas, max. dir. of 5 dBi. (o) On-body antenna on the phantom with 10 mm antenna–body separation, simulation. (p) On-body antenna on the phantom with 10 mm antenna–body separation, meas, max. dir. of 4 dBi. (q) Off-body antenna in vacuum, simulation. (r) Off-body antenna on RAM, meas, max. dir. of 8 dBi. (s) Off-body antenna on the phantom with 0 mm antenna–body separation, simulation. (t) Off-body antenna on the phantom with 0 mm antenna–body separation, meas, max. dir. of 7.6 dBi. (u) Off-body antenna on the phantom with 5 mm antenna–body separation, simulation. (v) Off-body antenna on the phantom with 5 mm antenna–body separation, meas, max. dir. of 7 dBi. (w) Off-body antenna on the phantom with 10 mm antenna–body separation, simulation. (x) Off-body antenna on the phantom with 10 mm antenna–body separation, meas, max. dir. of 6.5 dBi. Meas, max. dir. is measured maximum directivity.

activate TM_{00}. Practically, the phase difference can be greater than $0°$ for matching purposes as long as it is less than $90°$. Once it is greater than $90°$, the uniformity of the radiation pattern along φ will be severely distorted. On the other hand, having E field vectors in opposite directions at the excitation points will activate the TM_{01} mode. Hence, the phase difference should be $180°$. Similarly, this value can be decreased for matching purposes as long as it is more than $90°$, which will generate a more directional radiation pattern suitable for off-body operation. Here the length of the longer branch is decreased by $\lambda/4$ to switch between these two modes. Note that the length of the shorter branch is 28 mm and the longer branch is changed from 73 to 85 mm.

Figure 14.5(g–h) demonstrates the antenna's measured frequency response for different scenarios to the variation in the antenna–body separations. As previously predicted by the simulations, the on-body antenna's return loss is not as great as the off-body antenna's; however, both antennas are fairly stable against the changes in their near-field. The center frequency of the off-body antenna remains at 2.52 GHz for the phantom and Subject A, while it shifts to 2.51 GHz for Subject B. Note that Subject B has less fat in his body composition. For the on-body antenna, the center frequency remains at 2.47 most of the time, which shows that the on-body antenna is more immune to changes in body composition. In terms of the frequency response, the antenna satisfies the requirements of a BAN antenna and can be used near lossy environments. The return losses of both antennas for all scenarios are above 10 dB at 2.5 GHz. Therefore, 2.5 GHz is chosen as the frequency of measurement for the prototyped off-body and on-body antennas during the radiation pattern measurements.

Figure 14.5(l–x) show the gain patterns of the on-body and off-body antennas in vacuum, on RAM, and on the previously described single-layer flat phantom mimicking the chest of an individual. In Figure 14.5(l) the on-body antenna generates an omnidirectional radiation pattern in its horizontal plane at 2.5 GHz. The complimentary measurement is shown in Figure 14.5(j). Although the antenna generates a null at $\theta = 0°$ as predicted, the pattern is distorted. This is due to the fact that the antenna is measured on a block of RAM backed with a metal rotating post carrying the antenna. Therefore, the antenna does not act in a way identical to the way it acts in vacuum. When the on-body antenna is positioned on the phantom, the agreement between the simulations and the measurements gets much better. Comparing Figure 14.5(l) and (k), it can be observed that once the antenna is positioned on the body phantom, its radiation is shifted in the vertical direction and the energy will be emitted mostly in the positive half-space. This is favorable considering the SAR. Finally, Figure 14.5(k-p) show that the null at $\theta = 0°$ is maintained in all cases. In vacuum, the off-body antenna generates a directional pattern at 2.5 GHz and the measured half-power beam width is approximately $100°$, see Figure 14.5(q) and (r). Comparing these figures, the agreement between the simulation in vacuum and the measurement on RAM is better compared to the on-body antenna case since the off-body antenna's radiation in the lower half-space is minimal. When the off-body antenna is located on the phantom, the pattern goes through slight changes as predicted by the simulations. However, its directional nature is maintained, making it suitable for off-body communications. In all measured results, the effect of the feed connector is visible, which is positioned along the negative x-axis of the radiation pattern. This distortion leads to a slight increase in the cross-polarization levels of the off-body antenna. When the off-body antenna and the on-body antenna are simulated and measured on the phantom with different separations from 0 to 10 mm with 5-mm increments, the radiation patterns remain directional and omnidirectional, respectively.

Both antennas show little backward radiation, which restricts the absorption by the phantom. This is because the proposed antennas incorporate a full ground plane below their main radiator.

Note that this antenna currently is the smallest flexible antenna providing diversity for on-body and off-body links to the author's knowledge.

14.4.2 A RADIATION PATTERN DIVERSITY ANTENNA OPERATING AT THE 2.4 GHz ISM BAND

A novel rigid antenna meant to be worn on the wrist, operating in the 2.4 GHz ISM band, whose radiation pattern can be steered as the frequency of operation changes has been proposed [29]. This is achieved by merging the different radiation modes of the antenna together within a single operating band as seen in Figure 14.6(a–b), in contrast to the conventional procedure of covering the whole band with a single radiation mode. Using this method, a single antenna can provide radiation pattern diversity across the operating band without an external switching mechanism. The proposed antenna is a probe-fed patch shorted at four symmetrical points as seen in Figure 14.6(c–d). It is analyzed in terms of its on-body and off-body propagation performance as well as its sensitivity to antenna–body spacing. The antenna is shown to perform well up to 5 mm spacing. It has been proved that the antenna provides 9 dB advantage on average for the on-body link compared to the case where its off-body radiation mode is used to connect to the on-body sensor and vice versa. Moreover, the antenna is benchmarked against a monopole and a patch antenna in a residential setting. The performance of the antenna and subsequently the benefits of the pattern-switching technique are successfully quantified. The holistic method includes both antenna measurements and channel simulation with ray tracing. The results are verified against real-world measurements [30].

By changing the frequency within the ISM band, the maximum radiation direction is changed. If BLE standards are used, the first ten channels (Channel 37, Channels 0–8) can be used for the azimuth mode and the remaining 30 channels (Channels 9–39) can be used for the elevation mode. If ZigBee is preferred, Channels 11–14 can be used for the azimuth mode while Channels 15–26 will be more suitable for the elevation mode. Note that the Zigbee Channels 1–10 are located in the 915 MHz ISM band. The number of channels can be adjusted according to the needs of the system by changing the operating frequency of each mode. Looking into the interference rejection aspect of the application, approximately 10 dB difference is observed between the modes at $\theta = 0°$. If there is an interferer arriving at the antenna from the $\pm z$ direction, it is going to be attenuated by the antenna. Likewise, approximately 5 and 10 dB difference are observed between the modes at $\theta = 90°$ and $\theta = -90°$, respectively. Note that these numbers are subject to the specific channels chosen.

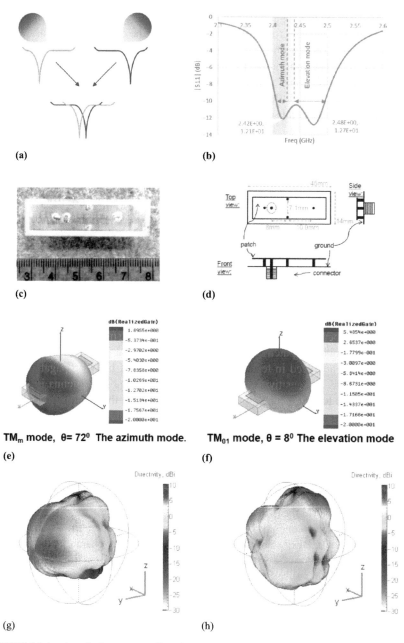

FIGURE 14.6 A radiation pattern diversity antenna operating at the 2.4 GHz industrial, scientific and medical band [29, 30]. (a) Merging the different radiation modes of the antenna together within a single operating band. (b) TM_m mode is excited along with the dominant TM_{01} mode in 2.4 GHz industrial, scientific and medical band. (c) Shorted patch antenna prototyped on Rogers RT/Duroid 6006. (d) Antenna model. (e) Simulated gain pattern of the on-body mode: TM_m. (f) Simulated gain pattern of the off-body mode: TM_{01}. (g) Measured gain pattern of the on-body mode, TM_m. (h) Measured gain pattern of the off-body mode, TM_{01}.

(Continued)

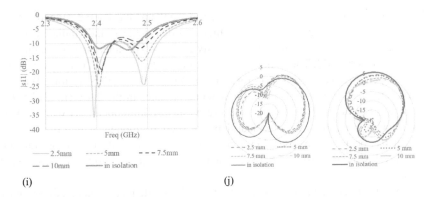

FIGURE 14.6 (CONTINUED) (i) Effect of antenna–body separation on the return loss. (j) Effect of antenna–body separation on the gain (on-body at 2.4 GHz, off-body at 2.48 GHz).

For the first selected frequency of 2.4 GHz, 2.6 dB maximum gain is observed. The maximum gain direction is $\theta = 72°$, $\Phi = 0°$ with minimal radiation in the vertical direction with respect to the body surface as seen in Figure 14.6(e–h). This stops energy from being wasted in the body or transmitted away from the body. It directs the energy toward the other on-body antennas along the body. Moreover, it has vertical polarization which is favored over horizontal polarization for on-body communications due to the excitation of the surface waves. On the other hand, at the second selected frequency 2.48 GHz, the radiation pattern is optimized for connecting to an off-body gateway with 5.8 dB maximum gain at $\theta = 8°$, $\Phi = 0°$.

The effect of antenna–body spacing has also been analyzed. The antenna–body separation is changed with 2.5 mm steps from 2.5 to 10 mm and the frequency characteristics demonstrating the two different modes and their frequency coverage for these antenna–body separations can be seen in Figure 14.6(i). As the antenna–body separation is decreased, the modes are shifted away from each other. The efficiency of the antenna degrades as the separation between the ground plane and the phantom decreases as seen in Figure 14.6(j). This is expected due to the lossy nature of the tissues in the phantom. However, in all cases, the efficiency is greater than 50%. We can see that the degradation in efficiency once the antenna is located on the phantom is greater for the off-body mode. This is due to the off-body mode having a higher electric field and lower magnetic field in the near-field of the antenna. Note that magnetic near fields are less susceptible to dissipation on the human body since human tissues have no magnetic losses ($\mu_r'' = 0$).

This antenna provides a unique middle ground between size and reconfigurability. The approach can be applied to other radiators.

14.5 SWITCHING BETWEEN IN-BODY AND OFF-BODY LINKS

The wireless link between an implant and an off-body gateway may be difficult to secure due to the fact that electromagnetic waves quickly attenuate as they propagate through human tissues. Depending on the depth of the implant within the body, the signal strength may be quite weak by the time the waves reach the skin. In order to address this problem, a digitally assisted repeater antenna has been designed [31] to be located outside the patient's body, which can detect the signals radiated by the implant and relay those signals to the off-body gateway. The antenna is based on U.S. Patent US10149636B2. The radiation pattern of the antenna is switched between two modes depending on the link it is forming: in-body link or off-body link. With an overall size of 30 × 30 × 3.15 mm, the antenna operates in the 2.4 GHz ISM band. The repeater is aimed to be used to secure wireless communications with a smart deep implant. Therefore, for a typical depth of such an implant, which is 4 cm, the repeater has been shown to enable a decrease of more than 40 dB in the transmit power level while the distance between the implant and the off-body gateway is kept constant.

Two slots are carved into the opposite faces of a single rectangular shallow cavity and both slots are fed with the same stripline which meanders in between the slots as seen in Figure 14.7(a). The slots are activated in an alternating way by means of two switches located in the center of each slot. When a switch is turned on, the slot on the opposite side is activated. This is a novel and a simple way of achieving radiation pattern diversity in a cavity-backed slot antenna. Having a shorted slot on the opposite wall of the cavity does not affect the activated slot since the shorted slot operates at a higher frequency and its polarization is orthogonal to the active slot.

The frequency response of each slot while the other one is shorted with a switch is plotted in Figure 14.7(b). The generated radiation patterns are plotted in 3D in Figure 14.7(c–d). Note that these radiation patterns are generated while the antenna is located in vacuum. Surely, the response will change once the antenna is located near lossy body tissue. Further analysis has been performed in order to demonstrate the performance of the antenna as a repeater. In a realistic scenario, the repeater is expected to be located near the human body to collect data from the implant. Locating the antenna near the lossy tissues deteriorates the performance of the antenna, especially during the in-body mode of operation. Therefore, the antenna's reflection coefficient is monitored while the separation between the skin and the antenna is changed. As seen in Figure 14.7(b), 40 mm has been found to be an acceptable separation with minimal detuning. Figure 14.7(e–f) shows the E field distribution in the vertical plane within a transmission scenario where a 4 cm deep implantable antenna operating at 2.4 GHz is excited and the repeater is located 4 cm away from the skin. These demonstrate how the energy is focused toward the desired direction.

The antenna is meant for prototyping on a felt substrate, and expected simulated efficiencies are 90% for the off-body mode and 72% for the in-body mode with 4 cm separation. The antenna can be located closer to the human body by replacing the gap with a bolus layer which is still work under progress for this particular project.

This antenna currently is the only flexible repeater antenna that can accommodate for the in-body and off-body communications to be formed at the same frequency band.

14.6 CONCLUSION

A challenging aspect of antenna design in wearable devices is the dynamic nature of the human body leading to dynamic propagation channels. Antenna requirements are

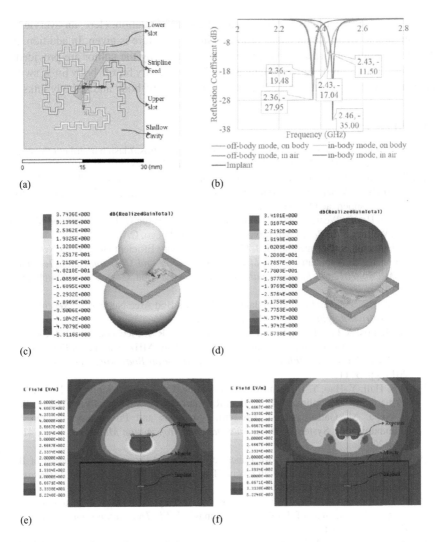

FIGURE 14.7 A digitally assisted repeater antenna for implant communications. (a) Antenna diagram, slots miniaturized according to second order Koch-fractal configuration. (b) The reflection coefficient of the in-body and off-body mode when the antenna is located in air and on human body with 4 cm separation. (c) The simulated gain pattern of the in-body mode when the antenna is located in vacuum. (d) The simulated gain pattern of the off-body mode when the antenna is located in vacuum. (e) E field distribution for in-body mode. (f) E field distribution for off-body mode.

different and sometimes contradictory for the connections established by different wearables placed at various points in the human body. In order to tackle this challenge, re-shaping the radiation pattern according to the direction of arrival of the wanted signal is a logical step forward. However, with small space available on the wearable devices and the limited energy supply, realizing these objectives are difficult since they require complex hardware and intensive calculation. Here, an overview of reconfigurable wearable antennas proposed in the literature is given. In addition, five different reconfigurable antennas previously proposed by the author are provided as examples. The examples cover antennas providing diversity for off-body links, switching between off-body and on-body links and between off-body and in-body links.

REFERENCES

[1] CodeBlue: Wireless Sensors for Medical Care, https://dash.harvard.edu/bitstream/handle/1/3191012/1242078272-bsn.pdf?sequence=2, accessed January 2020.

[2] The Human ++ Project, https://www.thehumanproject.org/, accessed January 2020.

[3] SPHERE – A Sensor Platform for HEalthcare in a Residential Environment, https://www.irc-sphere.ac.uk/, accessed January 2020.

[4] C. Van Hoof, "Wearable Wireless Sensor Technologies for Truly Personalized Medicine and Wellness," in *Radio Wireless Symposium*, San Diego, CA, 2015.

[5] S. Dumanli, S. Gormus, and I. J. Craddock, "Energy Efficient Body Area Networking for mHealth Applications," in *Medical Information and Communication Technology (ISMICT), 2012 6th International Symposium on*, La Jolla, CA, pp. 1–4, March 25–29, 2012.

[6] M. R. Yuce, "Wearable and Implantable Wireless Bod Area Networks," *Recent Patents on Electrical Engineering*, 2, 115–124, 2009.

[7] E. Mellios, A. Goulianos, S. Dumanli, G. S. Hilton, R. J. Piechocki, and I. J. Craddock, "Off-Body Channel Measurements at 2.4 GHz and 868 MHz in an Indoor Environment," in *Proceedings of the 9th International Conference on Body Area Networks*, London, 2014, pp. 5–11.

[8] P. S. Hall, Y. Hao, Y. I. Nechayev, A. Alomainy, C. C. Constantinou, C. Parini, M. R. Kamarudin, T. Z. Salim, D. T. M. Hee, R. Dubrovka, A. S. Owadally, S. Wei, A. Serra, P. Nepa, M. Gallo, and M. Bozzetti, "Antennas and Propagation for On-Body Communication Systems," *IEEE Antennas and Propagation Magazine*, 49(3), 41–58, June 2007.

[9] T. Alves, B. Poussot, and J. M. Laheurte, "Analytical Propagation Modeling of BAN Channels Based on the Creeping-Wave Theory," *IEEE Transactions on Antennas and Propagation*, 59(4), 1269–1274, 2011

[10] F. Merli, B. Fuchs, J. R. Mosig, and A. K. Skriverviky, "The Effect of Insulating Layers on the Performance of Implanted Antennas," *IEEE Transactions on Antennas and Propagation*, 59(1), 21–31, January 2011.

[11] W. Xia, K. Saito, M. Takahashi, and K. Ito, "Performances of an Implanted Cavity Slot Antenna Embedded in the Human Arm," *IEEE Transactions on Antennas and Propagation*, 57(4), 894–899, April 2009.

[12] M. I. Jais, M. F. Jamlos, M. Jusoh, T. Sabapathy, and M. R. Kamarudin, "2.45 GHz Beam-Steering Textile Antenna for WBAN Application," in *2013 IEEE Antennas and Propagation Society International Symposium (APSURSI)*, Orlando, FL, 2013, pp. 200–201.

[13] S. Yan and G. A. E. Vandenbosch, "Wearable Pattern Reconfigurable Patch Antenna," in *2016 IEEE International Symposium on Antennas and Propagation (APSURSI)*, Fajardo, 2016, pp. 1665–1666.

[14] S. Ha and C. W. Jung, "Reconfigurable Beam Steering Using a Microstrip Patch Antenna With a U-Slot for Wearable Fabric Applications," *IEEE Antennas and Wireless Propagation Letters*, 10, 1228–1231, 2011.

[15] A. da Conceição Andrade, I. P. Fonseca, S. F. Jilani, and A. Alomainy, "Reconfigurable Textile-Based Ultra-Wideband Antenna for Wearable Applications," *2016 10th European Conference on Antennas and Propagation (EuCAP)*, Davos, 2016, pp. 1–4.

[16] S. Kang and C. W. Jung, "Wearable Fabric Antenna on Upper Arm for MedRadio Band Applications with Reconfigurable Beam Capability," *Electronics Letters*, 51(17), 1314–1316, 2015.

[17] S. Kang and C. W. Jung, "Wearable fabric reconfigurable beam steering antenna for on/off-body communication system," in *2015 IEEE International Symposium on Antennas and Propagation & USNC/URSI National Radio Science Meeting*, Vancouver, BC, 2015, pp. 1211–1212.

[18] R. Masood, C. Person, and R. Sauleau, "A Dual-Mode, Dual-Port Pattern Diversity Antenna for 2.45-GHz WBAN," *IEEE Antennas and Wireless Propagation Letters*, 16, 1064–1067, 2017.

[19] X. Tong, C. Liu, X. Liu, H. Guo, and X. Yang, "Switchable ON-/OFF-Body Antenna for 2.45 GHz WBAN Applications," *IEEE Transactions on Antennas and Propagation*, 66(2), 967–971, February 2018.

[20] M. Li, S. Xiao, and B. Wang, "Pattern-Reconfigurable Antenna for On-Body Communication," *2013 IEEE MTT-S International Microwave Workshop Series on RF and Wireless Technologies for Biomedical and Healthcare Applications (IMWS-BIO)*, Singapore, 2013, pp. 1–3.

[21] K. Kwon, J. Tak, and J. Choi, "Design of a Dual-Band Antenna for Wearable Wireless Body Area Network Repeater Systems," in *2013 7th European Conference on Antennas and Propagation (EuCAP)*, Gothenburg, 2013, pp. 418–421.

[22] J. M. Felício, J. R. Costa and C. A. Fernandes, "Dual-Band Skin-Adhesive Repeater Antenna for Continuous Body Signals Monitoring," *IEEE Journal of Electromagnetics, RF and Microwaves in Medicine and Biology*, 2(1), 25–32, March 2018.

[23] K. Li, L. Xu, Z. Duan, Y. Tang, and Y. Bo, "Dual-Band and Dual-Polarized Repeater Antenna for Wearable Applications," in *2018 IEEE MTT-S International Wireless Symposium (IWS)*, Chengdu, 2018, pp. 1–3.

[24] J. Tak, K. Kwon, and J. Choi, "Design of a Dual Band Repeater Antenna for Medical Self-Monitoring Applications," in *2013 IEEE Antennas and Propagation Society International Symposium (APSURSI)*, Orlando, FL, 2013, pp. 2091–2092.

[25] S. Dumanli, "A Wrist Wearable Dual Port Dual Band Stacked Patch Antenna for Wireless Information and Power Transmission," in *2016 10th European Conference on Antennas and Propagation (EuCAP)*, Davos, 2016, pp. 1–5.

[26] E. Cil and S. Dumanli, "The Design of a Pattern Reconfigurable Antenna Suitable for Smart Glasses," in *2019 IEEE 30th International Symposium on Personal, Indoor and Mobile Radio Communications (PIMRC Workshops)*, Istanbul, Turkey, 2019, pp. 1–4.

[27] S. Dumanli, "On-Body Antenna with Reconfigurable Radiation Pattern," in *2014 IEEE MTT-S International Microwave Workshop Series on RF and Wireless Technologies for Biomedical and Healthcare Applications (IMWS-Bio2014)*, London, 2014, pp. 1–3.

[28] S. Dumanli, "Pattern Diversity Antenna for On-Body and Off-Body WBAN Links." *Turkish Journal of Electrical Engineering & Computer Sciences*, 26(5), 2395–2405, 2018.

[29] S. Dumanli, "A Radiation Pattern Diversity Antenna Operating at the 2.4 GHz ISM Band," in *2015 IEEE Radio and Wireless Symposium (RWS)*, San Diego, CA, 2015, pp. 102–104.

[30] S. Dumanli, L. Sayer, E. Mellios, X. Fafoutis, G. S. Hilton, and I. J. Craddock, "Off-Body Antenna Wireless Performance Evaluation in a Residential Environment," *IEEE Transactions on Antennas and Propagation*, 65(11), 6076–6084, November 2017.

[31] S. Dumanli, "A Digitally Assisted Repeater Antenna for Implant Communications," in *2017 11th European Conference on Antennas and Propagation (EUCAP)*, Paris, 2017, pp. 181–184.

[32] FCC's SAR Guidelines, *OET Bulletin* 65, [online] https://transition.fcc.gov/bureaus/oet/info/documents/bulletins/oet65/oet65.pdf, accessed on January 2019.

[33] G. A. Conway and W. G. Scanlon, "Antennas for Over-Body-Surface Communication at 2.45 GHz." *IEEE Transactions on Antennas & Propagation*, 57, 844–855, 2009.

15 Active Wearable Antennas for 5G and Medical Applications

Albert Sabban

CONTENTS

INTRODUCTION

Wearable printed active and tunable antennas are not widely presented in the literature. However, several types of microstrip antennas are widely presented in books and journals. Low profile compact active and tunable antennas are an important part of several communication, Internet of things (IOT) and medical systems. Printed active antennas possess attractive features such as low profile, flexibility, light weight, small volume and low production cost. A microstrip antenna's resonant frequency is altered due to environment condition, different antenna locations and different system modes of operation. Wideband active and tunable wearable antennas for 5G systems, IOT and medical systems are presented in this chapter. A voltage-controlled varactor is used to control the antenna's resonant frequency, as a function of the varactor bias voltage, at different locations on the human body. Amplifiers may be connected to the wearable antenna feed line to increase the system's dynamic range. Low profile compact tunable antennas are crucial in the development of wearable human biomedical systems.

Active wearable antennas may be used in receiving or transmitting channels for 5G, IOT and medical applications. In transmitting channels, a power amplifier is connected to the antenna. In receiving channels, a low noise amplifier is connected to the receiving antenna. A matching network matches the amplifiers to the antenna.

15.1 TUNABLE WEARABLE PRINTED ANTENNAS FOR WIRELESS COMMUNICATION SYSTEMS

The communication and biomedical industries have been in continuous growth in the past decade. Low profile compact tunable antennas are crucial in the development of wearable human biomedical systems, see [1-7]. Tunable antennas consist of a radiating element and a voltage-controlled diode (varactor). Varactor diodes are semiconductor devices that are voltage-controlled variable capacitance. The radiating element may be a microstrip patch antenna, dipole or loop antenna. The antenna's resonant frequency may be tuned by using a varactor to compensate for variations in antenna's resonant frequency at different locations.

15.2 VARACTORS BASIC THEORY

Varactor diodes are semiconductor devices that are used in many microwave systems where a voltage-controlled variable capacitance is required.

PN junction diodes exhibit a variable capacitance effect and PN diodes may be used as a voltage-controlled variable capacitance. However, special PN diodes are optimized and fabricated to give the required capacitance values. Varactor diodes normally enable much higher ranges of capacitance change to be achieved as a result of the PN diodes' optimized design. Varactor diodes are widely used in radio frequency (RF) devices. The circuit capacitance is varied by applying a controlled voltage. Varactor diodes are used in voltage-controlled oscillators (VCOs). Varactor diodes are also used in tunable filters and antennas.

15.2.1 VARACTOR DIODE BASICS

The varactor diode consists of a standard PN junction, see Figure 15.1. The diode is operated under reverse bias conditions and this gives rise to three regions and there is no conduction. The left and right ends of the diode are P and N regions, where current can be conducted. However, around the junction is the depletion region where no current carriers are available. As a result, current can be carried in the P and N regions, but the depletion region is an insulator. This is similar to a capacitor structure. It has conductive plates separated by an insulating dielectric. The capacitance of a capacitor

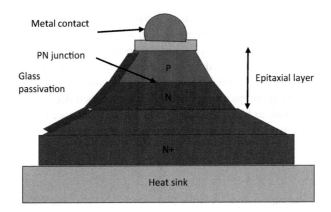

Metal contact

PN junction

Glass passivation

P

N

Epitaxial layer

N+

Heat sink

FIGURE 15.1 Internal structure of a varactor.

depends on the plate area, the dielectric constant of the insulator between the plates and the distance between the two plates. In the case of the varactor diode, it is possible to increase and decrease the width of the depletion region by changing the level of the reverse bias. This has the effect of changing the distance between the plates of the capacitor. However, to be able to use varactor diodes to their best advantage it is necessary to understand features of varactor diodes, including the capacitance ratio, Q, gamma, reverse voltage and the like.

A varactor provides an electrically controllable capacitance, which can be used in tuned circuits. It is small and inexpensive. Its disadvantages compared to a manually controlled variable capacitor are a lower Q, nonlinearity, lower voltage rating and a more limited range. The varactor model is shown in Figure 15.2.

Any PN junction has a junction capacitance that is a function of the voltage across the junction. The electric field in the depletion layer that is set up by the ionized donors and acceptors is responsible for the voltage difference that balances the applied voltage. A higher reverse bias widens the depletion layer, uncovering more fixed charge and raising the junction potential. The capacitance of the junction is $C = Q(V)/V$, and the *incremental capacitance* is $c = \Delta Q(V)/\Delta V$. The capacitance to be used in the formula for the resonant frequency is the incremental capacitance, where it is assumed that the incremental voltage ΔV is small compared to V. Finite voltages give rise to

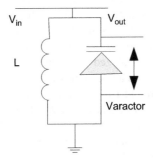

V_{in} V_{out}

L

Varactor

FIGURE 15.2 Varactor model.

nonlinearities. The capacitance decreases as the reverse bias increases, according to the relation $C = C_o/(1 + V/V_o)^n$, where C_o and V_o are constants. The diode forward voltage is approximately V_o. The exponent n depends on how the doping density of the semiconductors is a function of the distance away from the junction. For a graded junction (linear variation), $n = 0.33$. For an abrupt junction (constant doping density), $n = 0.5$. If the density jumps abruptly at the junction, then decreases (called hyper-abrupt), n can be made as high as $n = 2$. The varactor capacitance is given in Equation 15.1. The circuit frequency f_r may be calculated by using Equation 15.2.

$$C = \frac{A\varepsilon}{d} \qquad (15.1)$$

Where C is the capacitance, A is the plate area, d is the diode thickness.

$$f_r = \frac{1}{2\pi\sqrt{LC}} \qquad (15.2)$$

15.2.2 TYPES OF VARACTORS

Abrupt and hyperabrupt type: When the changeover p-n junction is abrupt then it is called abrupt varactor. When change is very abrupt, it is called hyperabrupt varac-tor. Varactors are used in oscillators to sweep for different frequencies.

Gallium arsenide (GaAs) varactor diodes: The semiconductor material used is GaAs. They are used for frequencies from 18 GHz up to and beyond 600 GHz.

15.3 DUAL POLARIZED TUNABLE DIPOLE ANTENNA

A compact tunable microstrip dipole antenna has been designed to provide horizontal polarization. The antenna consists of two layers. The substrate thickness affects the antenna bandwidth. The first layer is a 0.8 mm thick dielectric substrate. The second layer is a Duroid 5880 dielectric substrate, 0.8 mm thick. The printed slot antenna provides a vertical polarization. The printed dipole and the slot antenna provide dual orthogonal polarizations. The dimensions of the dual polarized antenna are $26 \times 6 \times 0.16$ cm. Tunable compact folded dual polarized antennas have also been designed. The dimensions of the compact antennas are $5 \times 5 \times 0.05$ cm. Varactors are connected to the antenna feed lines as shown in Figure 15.3. The voltage-controlled varactors are used to control the antenna's resonant frequency. The varactor bias voltage may be varied automatically to set the antenna's resonant frequency at different locations on the human body. The antenna may be used as a wearable antenna on a human body. The antenna may be attached to the patient's shirt on the patient's stomach or back zone. The antenna has been analyzed by using Agilent Advanced Design System (ADS) software. There is a good agreement between measured and computed results. The antenna bandwidth is around 10% for voltage standing wave ratio (VSWR) better than 2:1. The antenna beam width is around 100°. The antenna gain is around 2 dBi.

Figure 15.4 presents the measured S_{11} parameters without a varactor. Figure 15.5 presents the antenna S_{11} parameters as a function of different varactor capacitances. Figure 15.6 presents the tunable antenna's resonant frequency as a function of the varactor capacitance. The antenna's resonant frequency varies around 5% for

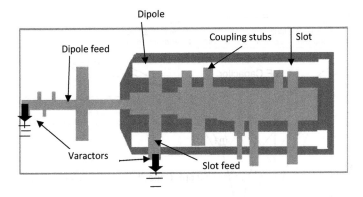

FIGURE 15.3   Dual polarized tunable antenna, 26 × 6 × 0.16 cm.

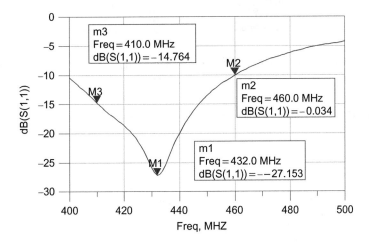

FIGURE 15.4   Measured S_{11} of the antenna on human body.

capacitances up to 2.5 pF. The antenna beam width is 100°. The antenna's cross-polarized field strength may be adjusted by varying the slot feed location.

15.4 WEARABLE TUNABLE ANTENNAS FOR 5G, INTERNET OF THINGS (IOT) AND MEDICAL APPLICATIONS

As presented in Chapter 9, the antenna's input impedance varies as a function of distance from the body. Properties of human body tissues are listed in Table 9.3, see [8–9]. Wearable antennas were presented in books and papers in the last decade [10–17]. A wearable antenna's electrical performance was computed by using ADS software [18]. The analyzed structure is presented in Figure 15.7. Wearable tunable antennas are not widely presented. In this section, several wearable tunable antennas are presented. Figure 15.8 presents S_{11} results for different belt and shirt thicknesses, and air spacing between the antennas and a human body. When an antenna's resonant frequency is shifted, the voltage across the varactor is varied to tune the antenna's resonant frequency.

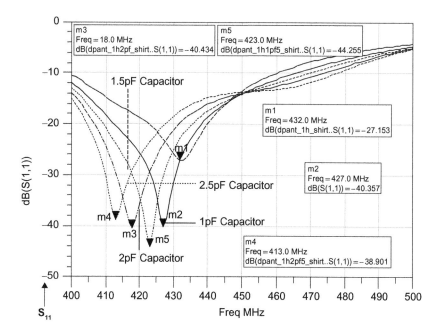

FIGURE 15.5   The tunable antenna S_{11} parameter as a function of varactor capacitance.

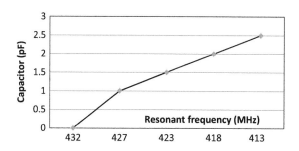

FIGURE 15.6   Antenna's resonant frequency as a function of varactor capacitance.

If the air spacing between the sensors and the human body is increased from 0 to 5 mm the antenna's resonant frequency is shifted by 5%. A voltage-controlled varactor is used to tune the antenna's resonant frequency due to different antenna locations on a human body. Figure 15.8 presents several compact tunable antennas for medical applications. A voltage-controlled varactor may also be used to tune a loop antenna's resonant frequency at different antenna locations on the body.

15.5 VARACTORS' ELECTRICAL CHARACTERISTICS

Tuning varactors are voltage variable capacitors designed to provide electronic tuning of microwave components. Varactors are manufactured on silicon and GaAs substrates. GaAs varactors offer higher Q and may be used at higher frequencies than

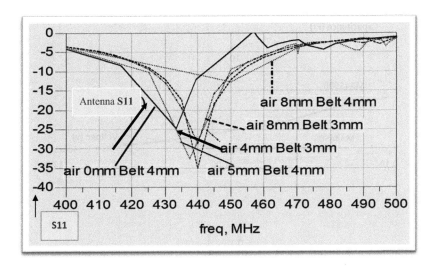

FIGURE 15.7   S11 of the antenna for different spacing relative to the human body.

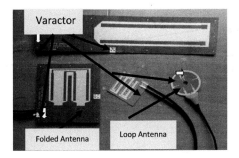

FIGURE 15.8   Tunable antennas for medical applications.

silicon varactors. Hyperabrupt varactors provide nearly linear variation of frequency with applied control voltage. However abrupt varactors provide inverse fourth root frequency dependence. MACOM offers several GaAs hyperabrupt varactors, such as the MA46 series. Figure 15.9 presents the capacitance versus bias voltage (C-V) curves of varactors MA46505 to MA46506.

Figure 15.10 presents the capacitance versus bias voltage curves of varactors MA46H070 to MA46H074.

15.6 MEASUREMENTS OF WEARABLE TUNABLE ANTENNAS

Figure 15.11 presents a compact tunable antenna with a varactor. A varactor was connected to the antenna feed line. The varactor bias voltage was varied from 0 to 9 V. Figure 15.12 presents measured S_{11} as a function of varactor bias voltage. The antenna's resonant frequency was shifted by 5% for bias voltage between 7 and 9 V. We may conclude that varactors may be used to compensate for variations in the antenna's resonant frequency at different locations on the human body.

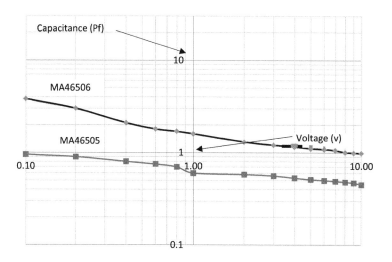

FIGURE 15.9 Varactor capacitance as a function of bias voltage.

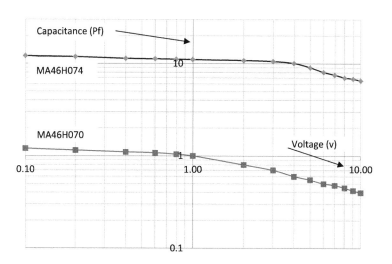

FIGURE 15.10 Capacitance versus bias voltage curves of varactors MA46H070 to MA46H074.

15.7 FOLDED DUAL POLARIZED TUNABLE ANTENNA FOR IOT AND MEDICAL APPLICATIONS

The dimensions of the folded dual polarized antenna presented in Figure 15.13 are 7 × 5 × 0.16 cm. The length and width of the coupling stubs in Figure 15.13 are 12 mm by 9 mm. Small tuning bars are located along the feed line to tune the antenna to the desired resonant frequency.

Figure 15.14 presents the antenna computed S11 and S22 parameters. The computed radiation pattern of the folded dipole is shown in Figure 15.15.

FIGURE 15.11   Tunable antenna with a varactor.

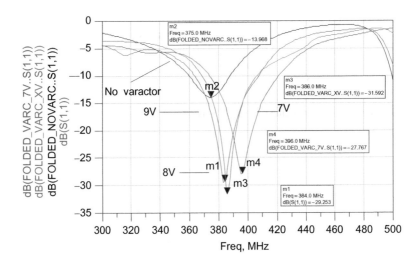

FIGURE 15.12   Measured S_{11} of the tunable antenna as a function of varactor bias voltage.

15.8 MEDICAL APPLICATIONS FOR WEARABLE TUNABLE ANTENNAS

Three to four tunable folded dipole or tunable loop antennas may be assembled in a belt and attached to a patient's stomach or back, as shown in Figure 15.16. The bias voltage to the varactors is supplied by recorder battery. The RF and direct current

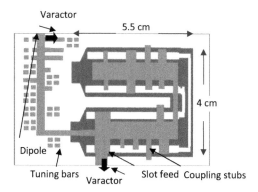

FIGURE 15.13 Tunable folded dual polarized antenna.

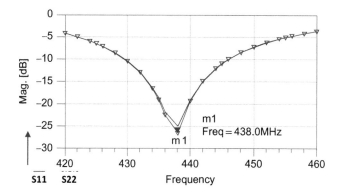

FIGURE 15.14 Folded antenna computed S11and S22 results.

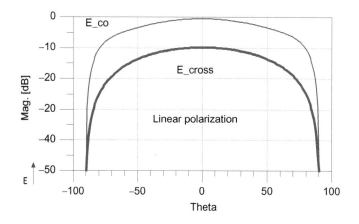

FIGURE 15.15 Folded antenna radiation pattern.

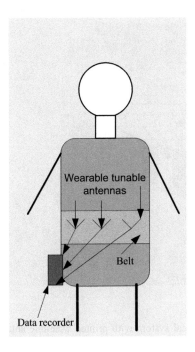

FIGURE 15.16 Tunable wearable system and antennas on human body.

(DC) cables from each antenna are connected to a recorder. The received signal is routed to a switching matrix. The signal with the highest level is selected during the medical test. The varactor's bias voltage may be varied to tune the antenna's resonant frequency. The antennas receive a signal that is transmitted from various positions in the human body. Tunable antennas may be attached to the patient's back in order to improve the level of the received signal from different locations in the human body. In several applications the distance separating the transmitting and receiving antennas is less than the far field distance, $2D^2/\lambda$, where D is the largest dimension of the source of the radiation. λ is the wavelength. In these applications, the amplitude of the electromagnetic field close to the antenna falls off rapidly with distance from the antenna. The electromagnetic fields do not radiate energy to infinite distances, but instead their energies remain trapped in the antenna near zone. The near fields transfer energy only to close distances from the receivers. In these applications we have to refer to the near field and not to the far field radiation. The receiving and transmitting antennas are magnetically coupled. Change in current flow through one wire induces a voltage across the ends of the other wire through electromagnetic induction. The proposed tunable wearable antennas may be placed on the patient's body as shown in Figure 15.17a. The patient in Figure 15.17b is wearing a wearable antenna. The antennas belt is attached to the patient's front or back. Figure 15.18 presents several compact tunable antennas for medical applications. A voltage-controlled varactor may also be used to tune the wearable antenna's resonant frequency at different antenna locations on the body.

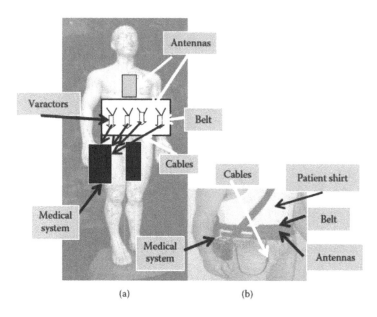

FIGURE 15.17 (a) Medical system with printed wearable antennas. (b) Patient wearing a wearable system and antennas.

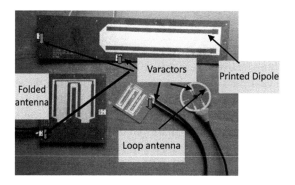

FIGURE 15.18 Tunable antennas for 5G, Internet of Things and medical applications.

15.9 ACTIVE WEARABLE ANTENNAS FOR 5G, IOT AND MEDICAL APPLICATIONS

Active wearable antennas may be used in receiving or transmitting channels for 5G, IOT and medical applications. In transmitting channels, a power amplifier is connected to the antenna. In receiving channels, a low noise amplifier (LNA) is connected to the receiving antenna. A matching network matches the amplifiers to the antenna.

15.9.1 BASIC CONCEPT OF ACTIVE ANTENNAS (AAs)

Active antennas (AAs) are devices combining a radiating element with active components. The radiating element is designed to provide the optimal load to the active elements. The integration of the antenna and the active components drastically reduces the complexity of the matching network. In the last decade, AAs were employed in wireless and in medical communication systems, see [19–32]. The current major applications of AAs are large electronically scanned arrays (phased arrays). Arrays of AAs, or active arrays, are well suited for mobile terminals requiring dynamic satellite tracking. The most common approach toward achieving fast-beam scanning is through the integration of monolithic microwave integrated circuit (MMIC) phase shifters, LNAs, and solid state power amplifiers with the antenna elements. In some cases, hybrid electro/mechanical arrays combining mechanical steering with electrical steering/shaping are considered. This architecture is often used to reduce the number of active control elements by limiting the electrical scanning in only one plane. This is often the case for mobile user terminals where azimuth scanning is performed by mechanical rotation and elevation agility is realized by a linear phased array. In the last decade, printed transmission lines and antennas replaced coaxial transmission lines and metallic radiators in phased array systems. Developments in MMIC technology and other fabrication processes allowed automated low cost production processing of phased arrays with a high integration level.

Phased arrays emerged as a new promising technology for radar and communication systems around 1970. Phased arrays replace mechanically scanned antennas. Phased arrays are much faster for beam switching than mechanically scanned antennas. Phased arrays significantly reduce the size, weight and power associated with a gimbal. Early phased array antennas were passive antennas. The front end of the antenna was composed of array elements with phase shifters. A passive manifold was employed for RF combining to form a single beam. At the output of the manifold was a switch with an LNA (receive channel) and power amplifier (transmit channel). Solid state electronic device capability was not developed enough to include active amplifiers at the front end of the array for each array element. With the LNA and amplifier behind the manifold the amount of RF loss was quite large causing inefficiency on transmit (power aperture) and limitations on sensitivity of the receiving. With the great progress of GaAs MMIC technology in the last 20 years, the dimensions of solid state devices have been minimized to the size of the array elements, enabling distributed phased arrays architectures. High power amplifiers (HPAs) and LNAs could be placed close to the front end and connected to each radiating element. This resulted in drastic power efficiency improvement and much higher receive sensitivity, since the only loss before the first LNA was the radiating element and a radome. The amplifiers were packaged in transmit/receive (T/R) modules with phase shifter and attenuator. This phased array architecture was called an "active" phased array.

15.9.2 ACTIVE WEARABLE RECEIVING LOOP ANTENNA

Figure 15.19 presents a basic receiver block diagram with an active antenna. In Figure 15.19 the LNA is an integral part of the antenna (Figure 15.20).

An enhancement-mode pseudo-morphic high-electron-mobility transistor (E-PHEMT) LNA was connected to a loop antenna. A receiving AA block diagram

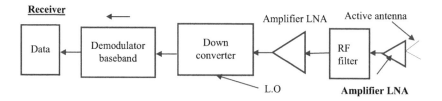

FIGURE 15.19   Receiver block diagram with active receiving antenna.

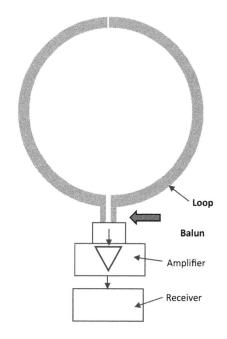

FIGURE 15.20   Active printed loop antenna layout.

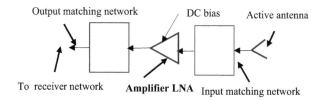

FIGURE 15.21   Receiving active antenna block diagram.

is presented in Figure 15.21. The radiating element is connected to the LNA via an input matching network. An output matching network connects the amplifier port to the receiver. A DC bias network supplies the required voltages to the amplifiers. The amplifier specification is listed in Table 15.1. The amplifier complex S parameters are listed in Table 15.2. The amplifier noise parameters are listed in Table 15.3.

TABLE 15.1
Low noise amplifier specification

Parameter	Specification	Remarks
Frequency range	0.4–3 GHz	
Gain	26 dB @ 0.4 GHz 18 dB @ 2 GHz	Vds = 3 V; Ids = 60 mA
Noise filter	0.4 dB @ 0.4 GHz 0.5 dB @ 2 GHz	Vds = 3 V; Ids = 60 mA
P1dB	18.9 dBm @ 0.4 GHz 19.1 dBm @ 2 GHz	Vds = 3 V; Ids = 60 mA
OIP3	32.1 dBm @ 0.4 GHz 33.6 dBm @ 2 GHz	Vds = 3 V; Ids = 60 mA
Max. input power	17 dBm	
Vgs	0.48 V	Vds = 3 V; Ids = 60 mA
Vds	3 V	
Ids	60 mA	
Supply voltage	±5 V	
Package	Surface Mount	
Operating temperature	−40°C to +80°C	
Storage temperature	−50°C to +100°C	

TABLE 15.2
Low noise amplifier S parameters

F-GHz	S11	S11°	S21	S21°	S12	S12°	S22	S22°
0.10	0.986	−17.17	25.43	168.9	0.008	88.22	0.55	−14.38
0.19	−31.76	0.964	24.13	158.9	0.016	74.88	0.54	−22.98
0.279	0.93	−45.77	22.97	149.5	0.021	65.77	0.51	−33.65
0.323	0.92	−53.39	22.45	145.3	0.026	62.38	0.49	−39.2
0.413	0.89	−65.72	20.98	137.27	0.03	57.9	0.46	−49.3
0.50	0.87	−77.1	19.54	130.3	0.034	53.03	0.43	−57.5
0.59	0.83	−87.12	18.08	124.14	0.038	48.18	0.40	−64.12
0.726	0.8	−100.8	16.22	115.7	0.042	42.06	0.36	−74.86
0.816	0.77	−108.8	15.07	110.75	0.044	39.53	0.34	−80.87
1.04	0.74	−126.2	12.74	100.13	0.049	33.69	0.29	−94.96
1.21	0.71	−137.6	11.25	92.91	0.051	30.05	0.26	−104
1.53	0.687	−154.2	9.29	82.06	0.055	26.08	0.22	−119
1.75	0.67	−164.1	8.24	75.31	0.058	23.14	0.20	−128.4
2.02	0.67	−174.6	7.27	67.82	0.06	20.88	0.18	−138.8

Loop antenna S11 parameter on human body is presented in Figure 15.22. A textile sleeve covers the loop antenna to match the loop to the antenna environment. The radiating loop antenna and the textile sleeve are attached to the human body. The antenna bandwidth is around 20% for VSWR better than 3:1. The active loop antenna S21 parameter, gain, on human body is presented in Figure 15.23.

TABLE 15.3
Noise parameters

F-GHz	NFMIN	N11X	N11Y	rn
0.5	0.079	0.3284	24.56	0.056
0.7	0.112	0.334	36.08	0.05
0.9	0.144	0.3396	47.4	0.045
1	0.16	0.3424	52.98	0.042
1.9	0.306	0.3682	100.93	0.029
2	0.322	0.3711	106.01	0.029
2.4	0.387	0.3829	125.79	0.029
3	0.484	0.401	153.93	0.036
3.9	0.629	0.429	−167.3	0.059
5	0.808	0.4645	−125.53	0.11
5.8	0.937	0.4912	−99.03	0.162
6	0.969	0.498	−92.92	0.177

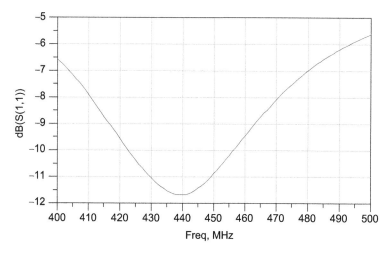

FIGURE 15.22   Loop antenna S11 parameter on human body.

The AA gain is 25 ± 2.5 dB for frequencies ranging from 350 to 580 MHz. The active loop antenna noise figure is presented in Figure 15.24. The active loop antenna noise figure is 0.7 ± 0.2 dB for frequencies ranging from 400 to 900 MHz.

15.9.3 COMPACT DUAL POLARIZED RECEIVING AA

A printed compact dual polarized antenna is shown in Figure 15.25. The compact antenna consists of two layers. The first layer consists of flame retardant-4 (FR-4) 0.25 mm dielectric substrate. The second layer consists of Kapton 0.25 mm dielectric substrate. In Figure 15.26 the LNA, presented in Section 15.9.2, is an integral part of the dual polarized receiving antenna. The active dual polarized antenna S21 parameter, gain, on a human body is presented in Figure 15.27. The AA gain is 25 ± 3 dB

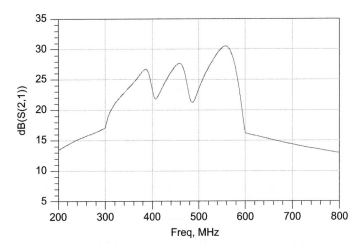

FIGURE 15.23   Active loop antenna S21 parameter, gain, on human body

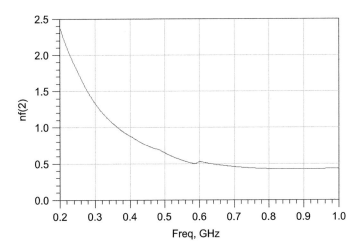

FIGURE 15.24   Active loop antenna noise figure.

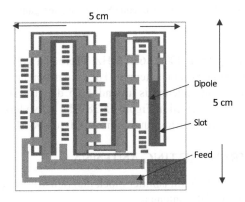

FIGURE 15.25   Printed compact dual polarized antenna.

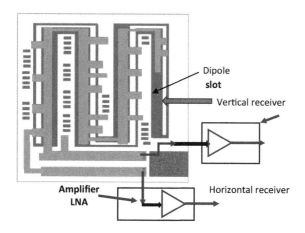

FIGURE 15.26   Active printed compact dual polarized receiving antenna layout.

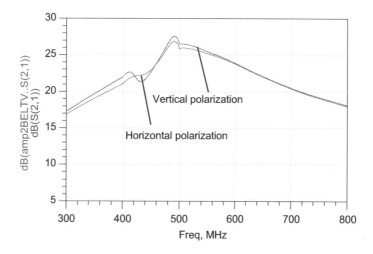

FIGURE 15.27   Active dual polarized antenna S21 parameter, gain, on human body.

for frequencies ranging from 400 to 650 MHz. There is a good match between the gain of the vertical and horizontal antennas. The gain difference between the gain of the vertical and horizontal antenna is around ±0.5 dB. The active dual polarized antenna noise figure is presented in Figure 15.28. The active loop antenna noise figure is 0.8 ± 0.4 dB for frequencies ranging from 400 to 900 MHz.

15.10 ACTIVE TRANSMITTING ANTENNA

Figure 15.29 presents a basic transmitter block diagram with an active antenna. The HPA is an integral part of the antenna.

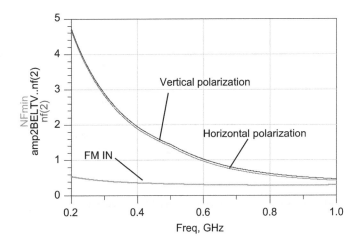

FIGURE 15.28 Active dual polarized antenna noise figure.

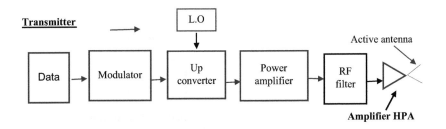

FIGURE 15.29 Transmitter block diagram with active transmitting antenna.

15.10.1 Compact Dual Polarized Active Transmitting Antenna

A printed compact dual polarized transmitting antenna is shown in Figure 15.30. The antenna's dimensions are $5 \times 5 \times 0.05$ cm.

The active transmitting dual polarized antenna layout is shown in Figure 15.31. In Figure 15.31 the HPA is an integral part of the antenna and is connected to the transmitting antenna. HPA specifications are listed in Table 15.4. The HPA is a MMIC GaAs, metal–semiconductor field-effect transistor (MESFET).

The transmitting AA block diagram is presented in Figure 15.32. The radiating element is connected to the HPA via an output HPA matching network. An HPA input matching network connects the amplifier port to the transmitter.

The HPA complex S parameters are listed in Table 15.5.

The active transmitting dual polarized antenna S11 parameter on a human body is presented in Figure 15.33. The active transmitting dual polarized antenna S21 parameter, gain, on a human body is presented in Figure 15.34. The active dual polarized antenna gain is 14 ± 3 dB for frequencies ranging from 380 to 600 MHz. The active transmitting dual polarized antenna output power is around 18 dBm.

The amplifier pin description is given and presented in Figure 15.35.

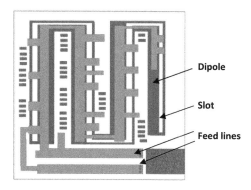

FIGURE 15.30   Transmitting printed compact dual polarized antenna.

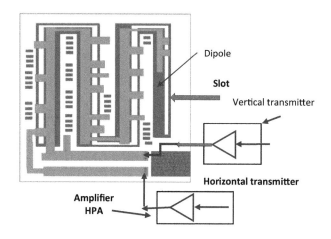

FIGURE 15.31   Transmitting active printed dual polarized antenna layout.

TABLE 15.4
High power amplifier specification

Parameter	Specification	Remarks
Frequency range	0.4–2.5 GHz	
Gain	15 dB @ 0.4 GHz	Vds = 5 V; Ids = 85 mA
	17.8 dB @ 2 GHz	
Noise filter	5.5 dB @ 0.4 GHz	Vds = 5 V; Ids = 85 mA
	5.5 dB @ 2 GHz	
P1dB	18.0 dBm @ 0.4 GHz	Vds = 5 V; Ids = 85 mA
	18.0 dBm @ 2 GHz	
OIP3	29 dBm @ 0.4 GHz	Vds = 5 V; Ids = 85 mA
	29 dBm @ 2 GHz	
Max. input power	10 dBm	
Vgs	0.48 V	Vds = 5 V; Ids = 85 mA
Vds	5 V	

(Continued)

TABLE 15.4 (*Continued*)
High power amplifier specification

Parameter	Specification	Remarks
Ids	mA85	
Supply voltage	±5 V	
Package	Surface Mount	
Operating temperature	−40°C to +80°C	
Storage temperature	−50°C to +100°C	

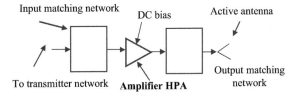

FIGURE 15.32   Transmitting active antenna block diagram.

TABLE 15.5
High power amplifier S parameters

F-GHz	S11 dB	S11°	S21 dB	S21°	S12 dB	S12°	S22 dB	S22°
0.20	0.065	−38.75	−3.09	−139.2	−47.56	157.63	−1.03	−74.66
0.28	−0.14	−60.8	7.46	163.7	−40.45	114	−3.6	−109.3
0.344	−1.1	−77	11.8	118.7	−37.9	78.4	−7.2	−131.6
0.4	−2.2	−88.5	13.8	85.24	−36.9	52.76	−11	−143.5
0.48	−3.7	−101.8	15.35	46.5	−36.6	25.4	−17	−143.1
0.52	−4.44	−107.5	15.8	30.2	−36.7	14.1	−19.4	−132.5
0.56	−5.1	−112.7	16.2	15.3	−36.8	3.64	−20.5	−118.9
0.64	−6.4	−122	16.8	−11.4	−37.2	−12.6	−19.2	−100.3
0.712	−7.4	−100.8	17.13	−32.7	−37.5	−25.4	−17.7	−100.6
0.8	−8.45	−137.6	17.5	−56.8	−38.1	−37.8	−16.3	−108.8
0.88	−9.2	−144.6	17.72	−77.1	−38.5	−49.4	−15.7	−119.7
1.04	−10.4	−158.6	18.1	−115.1	−39.6	−67.5	−15.3	−144.5
1.12	−10.8	−166.2	18.23	−133.3	−40.3	−75.8	−15.35	−157.5
1.24	−11.3	−178.7	18.37	−159.7	−41.3	−86.9	−15.9	−178.7
1.36	−11.8	167.4	18.4	174.4	−42.4	−91.4	−16.5	159.2
1.48	−12.2	151.2	18.4	149	−43.6	−94.9	−17.5	136.8
1.6	−12.8	134.3	18.3	123.3	−44.2	−93.4	−18.9	113.7
1.8	−14.3	101.2	17.9	83	−43	−86.3	−22	69.5
2	−16.5	61.8	17.3	43.5	−40.4	94.6	−27	6.42
2.16	−18.5	22.1	16.8	12.9	−38.	−105.5	−27.8	−70.2
2.28	−19.6	−14.9	16.3	−9.6	−37.2	−116.2	−25.1	−113.2
2.4	−19.4	−53.9	15.7	−31.8	−36	−128	−22.2	−147.2
2.56	−17.7	99.7	15	−60	−34.6	−145.6	−19.3	−179.4
2.7	−15.7	131	14.3	−84.3	−33.8	−160.3	−17.5	158.1
2.86	−13.7	159	13.5	−111.1	−33	−177.7	−16	134.7
3	−12.2	179.1	12.7	−134.1	−32.4	167.4	15.2	116.3

Pin	Function
1D	C bias
3R	F in
6R	F out
2,4,5,7,8G	Ground. Via holes.

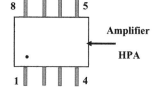

FIGURE 15.33 Active dual polarize antenna S11 parameter.

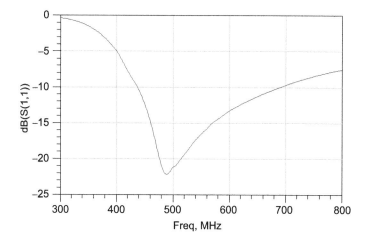

FIGURE 15.34 Active dual polarized antenna S21 parameter, gain, on human body.

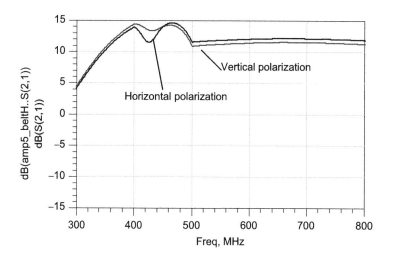

FIGURE 15.35 Amplifier pin description.

15.10.2 ACTIVE TRANSMITTING LOOP ANTENNA

The printed loop antenna is shown in Figure 15.36. The antenna's dimensions are $5 \times 5 \times 0.05$ cm. In Figure 15.36 the HPA is an integral part of the antenna. The HPA is a MMIC GaAs MESFET.

The active transmitting loop antenna S21 parameter, gain, on a human body is presented in Figure 15.37. The active antenna gain is 14 ± 3 dB for frequencies

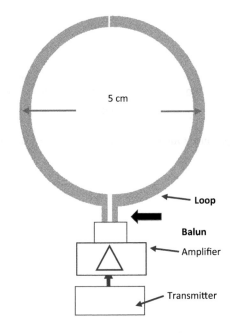

FIGURE 15.36 Active transmitting printed loop antenna layout.

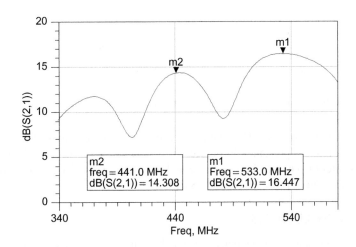

FIGURE 15.37 S21 parameter for an active transmitting loop antenna.

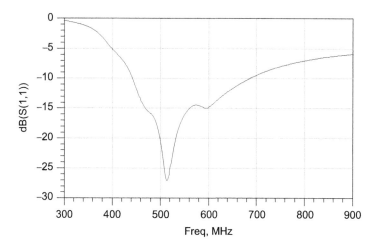

FIGURE 15.38   S11 parameter for an active transmitting loop antenna.

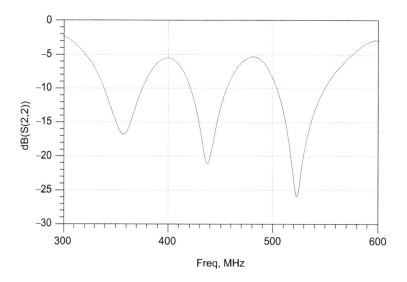

FIGURE 15.39   S22 parameter for an active transmitting loop antenna.

ranging from 400 to 600 MHz. The active transmitting loop antenna S11 parameter on a human body is presented in Figure 15.38. The active transmitting loop antenna S22 parameter on a human body is presented in Figure 15.39. The active transmitting loop antenna output power is around 18 dBm.

15.11 CONCLUSIONS

This chapter presents wideband active printed antennas with high efficiency for commercial and medical applications. The antenna dimensions may vary from 26 cm by 6 cm by 0.16 cm to 5 cm by 5 cm by 0.05 cm according to the medical system specification. The antennas' bandwidth is around 10% for VSWR better than 2:1. The antenna beam width is around 100°. The tunable antennas' gain varies from 0 to 2 dBi. If the air spacing between the dual polarized antenna and the human body is increased from 0 to 5 mm an antenna's resonant frequency is shifted by 5%. A varactor is employed to compensate variations in the antenna's resonant frequency at different locations on the human body.

Active wearable antennas may be used in receiving or transmitting channels. In transmitting channels a power amplifier is connected to the antenna. In receiving channels a LNA is connected to the receiving antenna. The active loop antenna gain is 25 ± 2.5 dB for frequencies ranging from 350 to 580 MHz. The active loop antenna noise figure is 0.7 ± 0.2 dB for frequencies ranging from 400 to 900 MHz. The active dual polarized antenna gain is 25 ± 3 dB for frequencies ranging from 400 to 650 MHz. The gain difference between the gain of the vertical and horizontal antenna is around ± 0.5 dB. The active dual polarized antenna noise figure is 0.8 ± 0.4 dB for frequencies ranging from 400 to 900 MHz. The active transmitting dual polarized antenna gain is 14 ± 3 dB for frequencies ranging from 380 to 600 MHz. The active transmitting dual polarized antenna output power is around 18 dBm.

REFERENCES

[1] Sabban, A., Microstrip Antenna Arrays, *Microstrip Antennas*, Nasimuddin Nasimuddin (Ed.), InTech, 361–384, 2011, ISBN: 978-953-307-247, Rijeka, Croatia. http://www.intechopen.com/articles/show/title/microstrip-antenna-arrays.

[2] Sabban, A. and Gupta, K.C., Characterization of Radiation Loss from Microstrip Discontinuities Using a Multiport Network Modeling Approach, *IEEE Transactions on MTT*, 39(4), 705–712, April 1991.

[3] Sabban, A., A New Wideband Stacked Microstrip Antenna, IEEE Antenna and Propagation Symp., Houston, Texas, USA, June 1983.

[4] Sabban, A., *Low Visibility Antennas for Communication Systems*, Taylor & Francis Group, New York, 2015.

[5] Kastner, R., Heyman, E., and Sabban, A., Spectral Domain Iterative Analysis of Single and Double-Layered Microstrip Antennas Using the Conjugate Gradient Algorithm, *IEEE Transactions on Antennas and Propagation*, 36(9), 1204–1212, Sept. 1988.

[6] Sabban, A., Wideband Microstrip Antenna Arrays, IEEE Antenna and Propagation Symposium MELCOM, Tel-Aviv,1981.

[7] James, J.R., Hall, P.S., and Wood, C., *Microstrip Antenna Theory and Design*, The Institution of Engineering and Technology, IET, London, UK, 1981.

[8] Chirwa, L.C., Hammond, P.A., Roy, S., and Cumming, D.R.S., Electromagnetic Radiation from Ingested Sources in the Human Intestine between 150 MHz and 1.2 GHz, *IEEE Transaction on Biomedical Engineering*, 50(4), 484–492, April 2003.

[9] Werber, D., Schwentner, A., and Biebl, E. M., Investigation of RF Transmission Properties of Human Tissues, *Advanced Radio Science*, 4, 357–360, 2006.

[10] Gupta, B., Sankaralingam, S., and Dhar, S., Development of Wearable and Implantable Antennas in the Last Decade, Microwave Symposium (MMS), 2010 Mediterranean 2010, Cyprus, 251–267, 2010.

[11] Thalmann, T., Popovic, Z., Notaros, B.M, and Mosig, J.R., Investigation and design of a multi-band wearable antenna, 3rd European Conference on Antennas and Propagation, EuCAP 2009, Berlin, Germany, 2009, 462–465.

[12] Salonen, P., Rahmat-Samii, Y., and Kivikoski, M., Wearable antennas in the vicinity of human body, IEEE Antennas and Propagation Society Symposium, Monterey, USA, 2004, Vol. 1, 467–470.

[13] Kellomaki, T., Heikkinen, J., and Kivikoski, M., Wearable antennas for FM reception, First European Conference on Antennas and Propagation, EuCAP 2006, Nice, France, 2006, 1–6

[14] Sabban, A., Wideband printed antennas for medical applications, APMC 2009 Conference, Singapore, December 2009.

[15] Lee, Y., *Antenna Circuit Design for RFID Applications*, Microchip Technology Inc, Microchip, AN, 710c.

[16] U.S. Patent, Inventors: Albert Sabban, Microstrip antenna arrays, (U.S. Patent US 1986/4,623,893), 1986, USA.

[17] U.S Patent, Inventors: Albert Sabban, Dual polarized dipole wearable antenna, U.S Patent number: 8203497, June 19, 2012, USA.

[18] ADS Software, Keysight, https://www.keysight.com/il/en/products/software/pathwave-design-software/pathwave-advanced-design-system.html.

[19] Wheeler, H.A., Small Antennas, *IEEE Transactions on Antennas and Propagation*, 23(4), 462–469, 1975.

[20] Sabban, A., Active Compact Wearable Body Area Networks for Wireless Communication, Medical and IOT Applications, *MDPI ASI, Applied System Innovation Journal*, 46, 1–20. December 2018.

[21] Sabban, A., New Wideband Passive and Active Wearable Antennas for Energy Harvesting Applications, *Journal of Sensor Technology*, 53–70, December 2017.

[22] Sabban, A., New Wideband Compact Wearable Slot Antennas for Medical and Sport Applications, *Journal of Sensor Technology*, 08, 18–34, December 2017.

[23] Sabban, A., *Novel Wearable Antennas for Communication and Medical Systems*, Taylor & Francis Group, October 2017.

[24] Sabban, A., *Wideband RF Technologies and Antenna in Microwave Frequencies*, Wiley Sons, New York, July 2016.

[25] Lin, J. and Itoh, T., Active Integrated Antennas, *IEEE Transactions on Microwave Theory and Techniques*, 42(12), 2186–2194, 1994.

[26] Mortazwi, A., Itoh, T., and Harvey, J., *Active Antennas and Quasi-Optical Arrays*, John Wiley & Sons, New York, 1998.

[27] Jacobsen, S. and Klemetsen, Ø., Improved Detectability in Medical Microwave Radio-Thermometers as Obtained by Active Antennas, *IEEE Transactions on Biomedical Engineering*, 55(12), 2778–2785, 2008.

[28] Jacobsen, S. and Klemetsen, Ø., Active Antennas in Medical Microwave Radiometry, *Electronics Letters*, 43(11), 606–608, 2007.

[29] Ellingson, S.W., Simonetti, J.H., and Patterson, C.D., Design and Evaluation of an Active Antenna for a 29–47 MHz Radio Telescope Array, *IEEE Transactions on Antennas and Propagation*, 55(3), 826–831, 2007.

[30] Segovia-Vargas, D., Castro-Galan, D., Garcia-Munoz, L.E., and Gonzalez-Posadas, V., Broadband Active Receiving Patch with Resistive Equalization, *IEEE Transactions on Microwave Theory and Techniques*, 56(1), 56–64, 2008.

[31] Rizzoli, V., Costanzo, A., and Spadoni, P., Computer-Aided Design of Ultra-Wideband Active Antennas by Means of a New Figure of Merit, *IEEE Microwave and Wireless Components Letters*, 18(4), 290–292, April 2008

[32] Yun, G., Compact Active Integrated Microstrip Antennas with Circular Polarisation Diversity, *IET Microwaves, Antennas and Propagation*, 2(1), 82–87, April 2008.

New Wideband Passive and Active Wearable Slot and Notch Antennas for Wireless and Medical 5G Communication Systems

Albert Sabban

CONTENTS

INTRODUCTION

This chapter describes the design of wideband wearable slot and notch antennas. Wideband efficient wearable antennas for communication systems are significant in the development of novel wearable 5G communication and Internet of things (IOT) systems. Slot antennas are low profile, compact and low cost and may be assembled in 5G and IOT communication systems. The dynamic range and the efficiency of a communication system may be improved by using active wearable antennas. An amplifier may be added to the wearable antenna. This chapter presents new compact ultra-wideband slot and notch antennas in frequencies ranging from 1 to 18 GHz. The slot and notch antennas were analyzed by using 3D full-wave software. The antenna bandwidth is from 50% to 100% with voltage standing wave ratio better than 3:1. The antenna gain is around 3 dBi with efficiency higher than 90%.

16.1 SLOT ANTENNAS BASIC THEORY

Slot antennas may be printed on a dielectric substrate or cut out of the surface they are to be mounted on [1–9]. Slot antennas may be excited by connecting a transmission line to the slot edges as shown in Figure 16.1(a). Slot antennas may be excited by a microstrip line as shown in Figure 16.1(b). The radiation pattern of slot antennas is determined by the size and shape of the slot in a radiating surface. Slot antennas may be used at frequencies between 200 MHz and up to 40 GHz.

The feed transmission line excites electric field distribution within the slot and currents that travel around the slot perimeter. If we replace the slot with metal and the ground plane with air, then we get a dipole antenna. The slot antenna is dual to the dipole antenna as shown in Figure 16.1(a). Babinet's principle [1] relates the electromagnetic fields of a slot antenna to its dual antenna. Babinet's principle relates the slot antenna to the dipole antenna. Babinet's principle states that the impedance of the slot antenna Z_s is related to the impedance of its dual antenna Zd by Equation 16.1. Where η is intrinsic impedance of free space and is equal to 120π Ω.

$$Z_s Z_d = \frac{\pi\eta^2}{4} \tag{16.1}$$

The impedance of 0.5λ long dipole is 73 Ω. By using Equation 16.1 the impedance of a 0.5λ long slot antenna is around 486 Ω. The second major result of Babinet's principle is that the fields of the dual antenna are almost the same as the slot antenna (the fields' components are interchanged, and called "duals"). That is, the fields of the slot antenna (given with a subscript s) are related to the fields of its complement (given with a subscript d) as written in Equation 16.2.

$$E_{\theta s} = H_{\theta d}$$

$$E_{\phi s} = H_{\phi d}$$

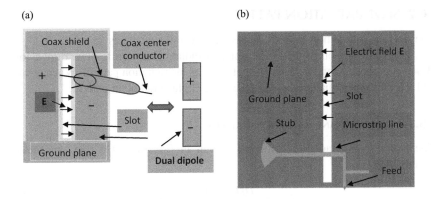

FIGURE 16.1 (a) Slot antenna and its dual dipole, (b) printed slot antenna and microstrip line feed.

$$H_{\theta s} = \frac{-E_{\theta d}}{\eta^2} \tag{16.2}$$

$$H_{\theta s} = \frac{-E_{\phi d}}{\eta^2}$$

The polarizations of the dual antennas are reversed. The dipole antenna in Figure 16.1 is vertically polarized and the slot antenna will be horizontally polarized. In the far fields the dipole's electromagnetic fields vary as the inverse of r and $\sin \theta$ as written in Equation 16.3.

$$
\begin{aligned}
E_r &= 0 \\
E_\theta &= j\eta_0 \frac{l\beta I_0 \sin \theta}{4\pi r} e^{j(\omega t - \beta r)} \\
H_\phi &= j\frac{l\beta I_0 \sin \theta}{4\pi r} e^{j(\omega t - \beta r)}
\end{aligned}
\tag{16.3}
$$

In the far fields the slot's electromagnetic fields vary as the inverse of r and $\sin \theta$ as written in Equation 16.4.

$$
\begin{aligned}
E_r &= 0 \\
H_{\theta s} &= -j\frac{l\beta I_0 \sin \theta}{4\pi r \eta_0} e^{j(\omega t - \beta r)} \\
E_{\varphi s} &= j\frac{l\beta I_0 \sin \theta}{4\pi r} e^{j(\omega t - \beta r)}
\end{aligned}
\tag{16.4}
$$

16.2 SLOT RADIATION PATTERN

The antenna's radiation pattern represents the radiated fields in space at a point $P(r, \theta, \varphi)$ as function of θ, φ. The antenna radiation pattern is three dimensional. When φ is constant and θ varies we get the E plane radiation pattern. When φ varies and θ is constant, usually $\theta = \pi/2$, we get the H plane radiation pattern.

16.2.1 SLOT E PLANE RADIATION PATTERN

The slot E plane radiation pattern is given in Equation 16.5 and presented in Figure 16.2.

$$\left|E_{\varphi s}\right| = j\frac{l\beta I_0 \sin\theta}{4\pi r} = A\sin\theta \qquad (16.5)$$

At a given point $P(r,\theta,\varphi)$ the slot E plane radiation pattern is given in Equation 16.6.

$$\left|E_{\varphi s}\right| = j\frac{l\beta I_0 \left|\sin\theta\right|}{4\pi r} = A\left|\sin\theta\right|$$
$$\text{Choose} \quad A = 1 \qquad (16.6)$$
$$\left|E_{\varphi s}\right| = \left|\sin\theta\right|$$

Slot E plane radiation pattern in spherical coordinate system is shown in Figure 16.3.

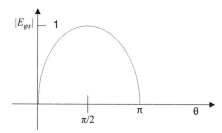

FIGURE 16.2   Slot E plane radiation pattern.

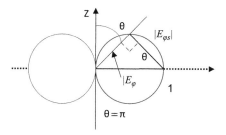

FIGURE 16.3   Slot E plane radiation pattern in spherical coordinate system.

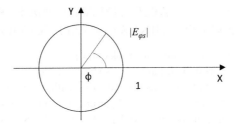

FIGURE 16.4   Slot H plane radiation pattern for $\theta = \pi/2$.

16.2.2 SLOT H PLANE RADIATION PATTERN

For $\theta = \pi/2$ the slot H plane radiation pattern is given in Equation 16.6 and presented in Figure 16.4. At a given point $P(r, \theta, \varphi)$ the slot H plane radiation pattern is given in Equation 16.7.

$$\left|E_{\varphi s}\right| = j \frac{l\beta I_0}{4\pi r} = A$$
$$\text{Choose} \quad A = 1 \tag{16.7}$$
$$\left|E_{\varphi s}\right| = 1$$

The slot H plane radiation pattern in xy plane is a circle with $r = 1$.

The radiation pattern of a vertical slot is omnidirectional. It radiates equal power in all azimuthal directions perpendicular to the axis of the antenna. A slot H plane radiation pattern in spherical coordinate system is shown in Figure 16.4.

16.3 SLOT ANTENNA IMPEDANCE

Antenna impedance determines the efficiency of transmitting and receiving energy in antennas. The dipole impedance is given in Equation 16.8. The slot impedance is given in Equation 16.9.

$$R_{rad} = \frac{2W_T}{I_0^2}$$
$$\text{For a dipole}: R_{rad} = \frac{80\pi^2 l^2}{\lambda^2} \tag{16.8}$$

$$Z_s = \frac{\pi\eta^2}{4Z_d} = \frac{\eta^2\lambda^2}{320\pi l^2} \tag{16.9}$$

By using Equation 16.9 the impedance of a 0.5λ long slot antenna is around 565 Ω.

16.4 A WIDEBAND WEARABLE SLOT ANTENNA FOR MEDICAL AND INTERNET OF THINGS (IOT) APPLICATIONS

Wearable antennas were presented in books and papers in the last decade [10–22]. A wideband wearable printed slot antenna is shown in Figure 16.5. The slot antenna is printed on a RT-Duroid 5880 dielectric substrate with dielectric constant of 2.2 and 1.2 mm thickness. The antenna's electrical parameters were calculated and optimized by using Advanced Design System (ADS) software [23]. Slot antennas theory and design are presented in [24]. The dimensions of the slot antenna shown in Figure 16.5 are $66 \times 60 \times 1.2$ mm. The slot antenna center frequency is 2.5 GHz. The computed S11 parameters are presented in Figure 16.6. The antenna bandwidth is around 50% for voltage standing wave ratio (VSWR) better than 2:1. The antenna bandwidth is around 70% for VSWR better than 3:1. Radiation pattern of the slot antenna is shown in Figure 16.7. The antenna beam width is around 90° at 2 GHz, as shown in Figure 16.7.

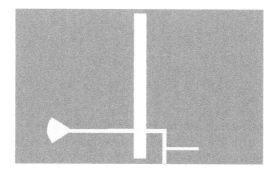

FIGURE 16.5   A wideband wearable printed slot antenna.

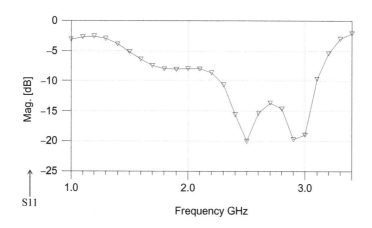

FIGURE 16.6   S11 of a wideband wearable printed slot antenna.

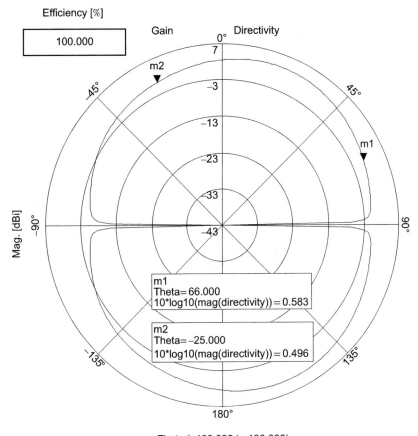

FIGURE 16.7   Radiation pattern of a wideband wearable printed slot antenna at 2 GHz.

The antenna gain is around 3 dBi. The slot antenna radiation pattern at 2.5 GHz is shown in Figure 16.8.

16.5 A WIDEBAND COMPACT T SHAPE WEARABLE PRINTED SLOT ANTENNA

A wideband compact T shape wearable printed slot antenna is shown in Figure 16.9. The slot antenna is printed on 5880 Duroid dielectric substrate with dielectric constant of 2.2 and 1.2 mm thickness. The antenna's electrical parameters were calculated and optimized by using full-wave electromagnetic software. The dimensions of the slot antenna shown in Figure 16.9 are 66 × 60 × 1.2 mm. The slot antenna center frequency is around 2.25 GHz. The computed S11 parameters are presented in Figure 16.10. The antenna bandwidth is around 57% for VSWR better than 2:1. The antenna bandwidth is around 90% for VSWR better than 3:1. The radiation pattern of the T shape slot antenna is shown in Figure 16.11. The antenna beam width is around 82° at 1.5 GHz, as shown in Figure 16.11. The antenna gain is around 3 dBi.

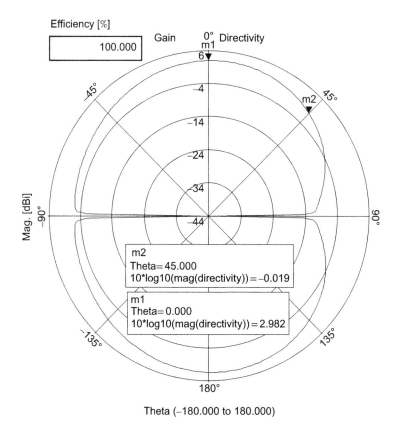

FIGURE 16.8   Radiation pattern of a wideband wearable printed slot antenna at 2.5 GHz.

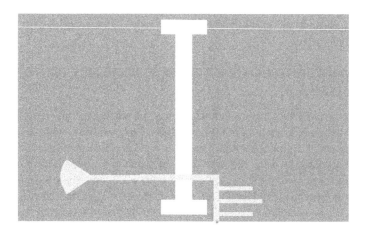

FIGURE 16.9   A wideband T shape wearable printed slot antenna.

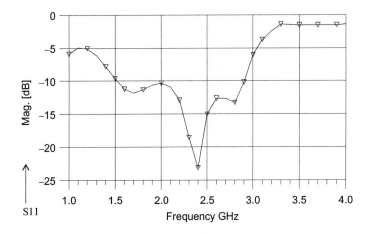

FIGURE 16.10   S11 of a wideband wearable printed slot antenna.

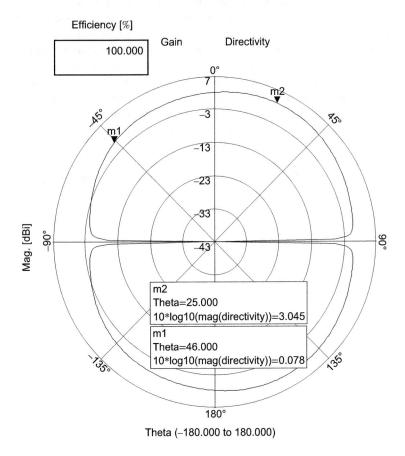

FIGURE 16.11   Radiation pattern of a wideband wearable printed slot antenna at 1.5 GHz.

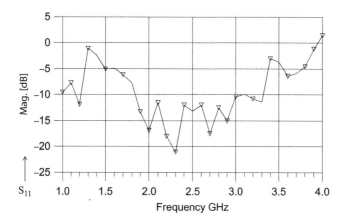

FIGURE 16.12   S11 of a wideband wearable printed slot antenna on human body.

The computed S11 parameters of the T shape slot on a human body are presented in Figure 16.12. The dielectric constant of human body tissue was taken as 45. The antenna was attached to a shirt with dielectric constant of 2.2 and 1 mm thickness. The antenna bandwidth is around 50% for VSWR better than 2:1. The antenna bandwidth is around 57% for VSWR better than 3:1. The antenna center frequency is shifted by 10%. The feed network of the antenna shown in Figure 16.9 was optimized to match the antenna to the human body environment, see Figure 16.13. The computed S11 parameters of the modified T shape slot on a human body are presented in Figure 16.14. The modified antenna VSWR is better than 3:1 for frequencies ranging from 0.8 to 3.9 GHz. The antenna gain at 1.5 GHz of the modified antenna is around 3 dBi. The radiation pattern of the modified T shape slot antenna at 1 GHz is shown in Figure 16.15. The radiation pattern of the modified T shape slot antenna at 1.5 GHz is shown in Figure 16.16.

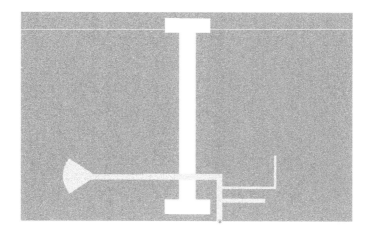

FIGURE 16.13   A modified wideband T shape wearable printed slot antenna.

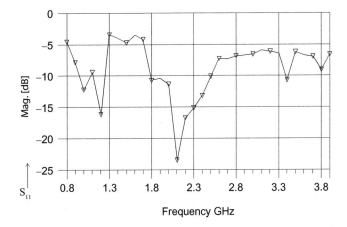

FIGURE 16.14   S11 of the modified T shape wearable slot antenna on human body.

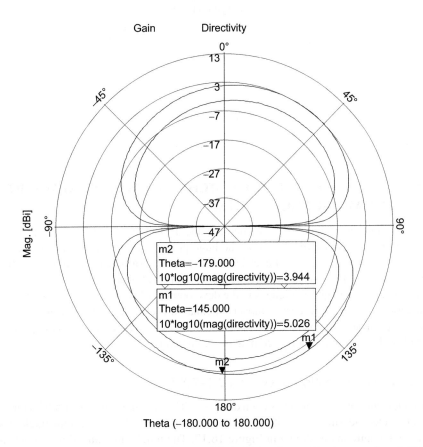

FIGURE 16.15   Radiation pattern of the modified wideband wearable slot antenna at 1 GHz.

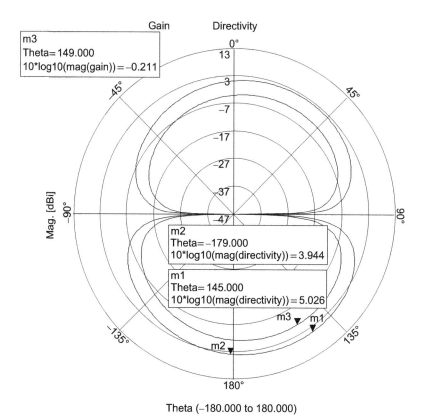

FIGURE 16.16   Radiation pattern of the modified wideband wearable slot antenna at 1.5 GHz.

16.6 WIDEBAND WEARABLE NOTCH ANTENNA FOR 5G AND IOT COMMUNICATION SYSTEMS

The wireless communication, biomedical and IOT industries have been in rapid growth in the past 10 years. Due to the huge progress in the development of communication systems in the last decade, the development of wideband communication systems is in continuous growth. However, development of wideband efficient antennas is one of the major challenges in the development of wideband wireless communication systems. Low cost compact antennas are crucial in the development of communication systems. Printed notch antennas and miniaturization techniques are employed to develop efficient compact notch antennas.

16.6.1 WIDEBAND NOTCH ANTENNA 2.1–7.8 GHz

A wideband notch antenna has been designed. The antenna is printed on RT-Duroid 5880 dielectric substrate with dielectric constant of 2.2 and 1.2 mm thickness. The notch antenna is shown in Figure 16.17. The notch antenna's dimensions are 116.4 × 71.4 mm. The antenna center frequency is 5 GHz. The antenna bandwidth is around 100% for S11 lower than −6.5 dB, as presented in Figure 16.18. The notch antenna VSWR is better than 3:1 for frequencies from 2.1 to 7.8 GHz. The antenna beam width is around 84°. The antenna gain is around 2.5 dBi. Figure 16.19 presents the radiation pattern of the wideband notch antenna at 3.5 GHz. Figure 16.19 presents the radiation pattern of the wideband notch antenna at 3 GHz.

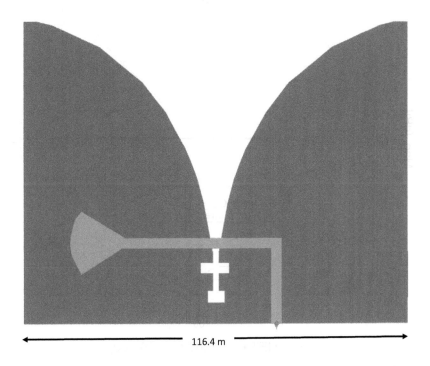

116.4 m

FIGURE 16.17 A wideband 2.1–7.8 GHz notch antenna.

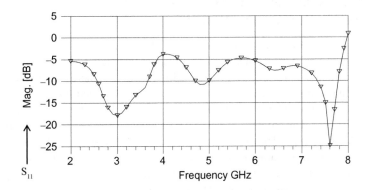

FIGURE 16.18 A wideband 2.1–7.8 GHz notch antenna, computed S11.

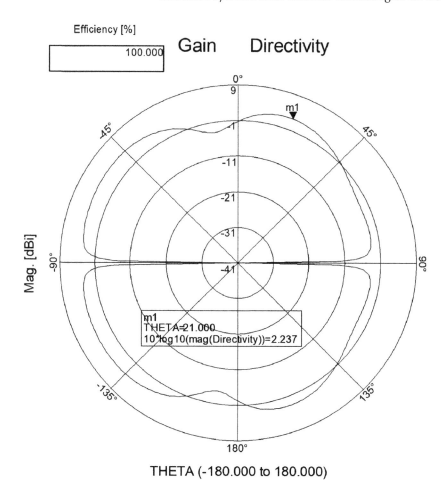

FIGURE 16.19   Radiation pattern of the wideband notch antenna at 3.5 GHz.

16.7 WEARABLE TUNABLE SLOT ANTENNAS FOR 5G AND IOT COMMUNICATION SYSTEMS

A wideband wearable tunable slot antenna is shown in Figure 16.21. Tunable slot antennas consist of a slot antenna and of a voltage-controlled diode (varactor) [6]. The antenna's resonant frequency may be tuned by using a varactor to compensate for variations in the antenna's resonant frequency at different locations. The slot antenna is printed on RT-Duroid 5880 dielectric substrate with dielectric constant of 2.2 and 1.2 mm thickness. The antenna's electrical parameters were calculated and optimized by using ADS software. The dimensions of the slot antenna shown in Figure 16.21 are $66 \times 60 \times 1.2$ mm. A varactor is connected to the slot feed line. The varactor bias voltage may be varied automatically to set the antenna's resonant frequency at different locations and environments. The slot antenna center frequency is 2.5 GHz. The S11 parameters for varactor capacitances ranging 0.1–1 pF are presented in

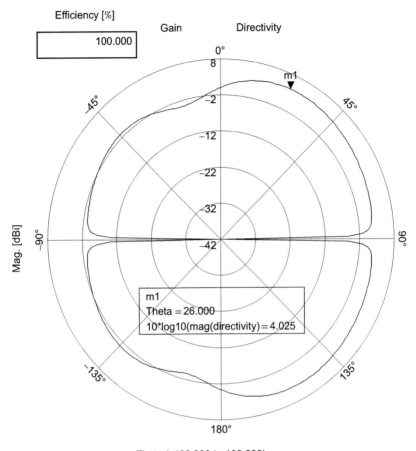

FIGURE 16.20   Radiation pattern of the wideband notch antenna at 3 GHz.

Figure 16.22. The antenna bandwidth is around 40% for VSWR better than 2:1. The antenna bandwidth is 60% for VSWR better than 3:1.

16.8 A WIDEBAND T SHAPE TUNABLE WEARABLE PRINTED SLOT ANTENNA

A wideband T shape wearable printed slot antenna is shown in Figure 16.23. The slot antenna is printed on RT-Duroid 5880 dielectric substrate with dielectric constant of 2.2 and 1.2 mm thickness. The antenna's electrical parameters were calculated and optimized by using ADS software. The dimensions of the slot antenna shown in Figure 16.22 are 66 × 60 × 1.2 mm. The slot antenna center frequency is around 2.25 GHz. The computed S11 parameters are presented in Figure 16.24. The antenna bandwidth is around 57% for VSWR better than 2:1. The antenna bandwidth is around 90% for VSWR better than 3:1. A varactor is connected to the slot feed line.

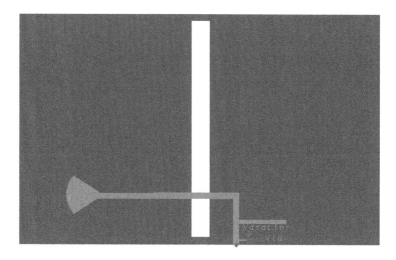

FIGURE 16.21   A wideband tunable wearable slot antenna.

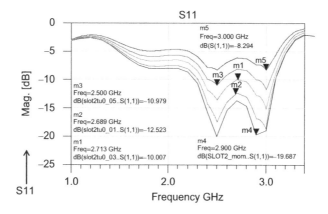

FIGURE 16.22   S11 of a wideband tunable wearable printed slot antenna.

The varactor bias voltage may be varied automatically to set the antenna's resonant frequency at different locations and environments. The S11 parameters for varactor capacitances ranging 0.1–1 pF are presented in Figure 16.25.

16.9 WEARABLE ACTIVE SLOT ANTENNAS FOR 5G COMMUNICATION AND IOT SYSTEMS

A wideband active wearable receiving slot antenna is shown in Figure 16.26. Active slot antennas are devices combining radiating elements with active components, such as amplifiers and diodes. The radiating element is designed to provide the optimal load to the active elements. The slot antenna is printed on RT-Duroid 5880 dielectric

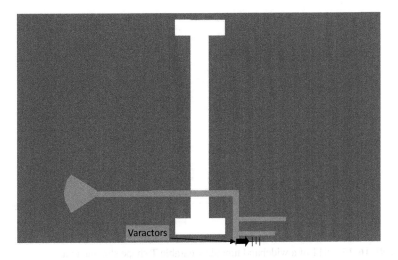

FIGURE 16.23   A wideband tunable wearable T shape slot antenna.

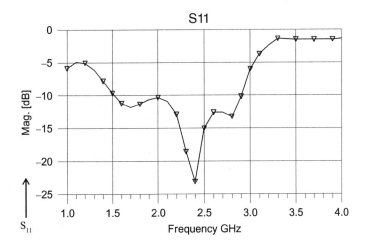

FIGURE 16.24   S11 of a wideband wearable T shape slot antenna without varactor.

substrate with dielectric constant of 2.2 and 1.2 mm thickness. The antenna's electrical parameters were calculated and optimized by using ADS software. The dimensions of the slot antenna shown in Figure 16.26 are 66 × 60 × 1.2 mm. An enhancement-mode pseudo-morphic high-electron-mobility transistor (E-PHEMT) low noise amplifier (LNA) was connected to a slot antenna. The radiating element is connected to the LNA via an input matching network. An output matching network connects the amplifier port to the receiver. A direct current (DC) bias network supplied the required voltages to the amplifiers. The amplifier specification is listed in Table 15.1.

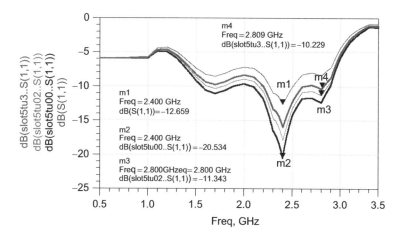

FIGURE 16.25   S11 of a wideband tunable wearable T shape slot antenna.

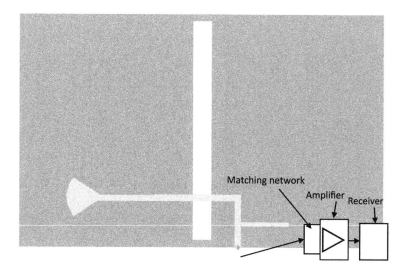

FIGURE 16.26   A wideband active receiving wearable slot antenna.

The amplifier complex S parameters are listed in Table 15.2. The amplifier noise parameters are listed in Table 15.3. Active slot antenna S11 parameter is presented in Figure 16.27. Active slot antenna S22 parameter is presented in Figure 16.28. The antenna bandwidth is around 40% for VSWR better than 3:1. The active slot antenna S21 parameter, gain, is presented in Figure 16.29. The active antenna gain is 18 ± 2.5 dB for frequencies ranging from 200 to 580 MHz. The active antenna gain is 12 ± 2 dB for frequencies ranging from 1.3 to 3.3 GHz. The active slot antenna noise figure is presented in Figure 16.30. The active slot antenna noise figure is 0.5 ± 0.3 dB for frequencies ranging from 200 MHz to 3.3 GHz.

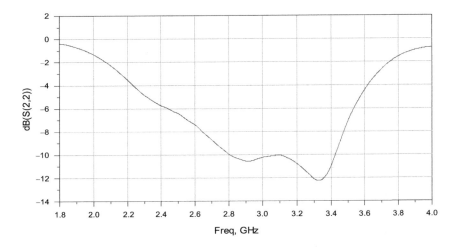

FIGURE 16.27   Active slot antenna S11 parameter.

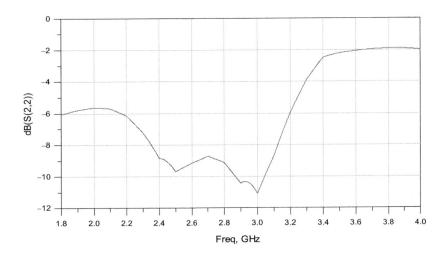

FIGURE 16.28   Active slot antenna S22 parameter.

16.10 WEARABLE ACTIVE T SHAPE SLOT ANTENNAS FOR 5G COMMUNICATION SYSTEMS

A wideband active wearable receiving T shape slot antenna is shown in Figure 16.31. The radiating element is designed to provide the optimal load to the active elements. The slot antenna is printed on RT-Duroid 5880 dielectric substrate with dielectric constant of 2.2 and 1.2 mm thickness. The antenna's electrical parameters were calculated and optimized by using ADS software. The dimensions of the slot antenna shown in Figure 16.31 are 66 × 60 × 1.2 mm. An E-PHEMT LNA was connected to a slot antenna. The radiating element is connected to the LNA via an input matching network. An output matching network connects the amplifier port to the receiver.

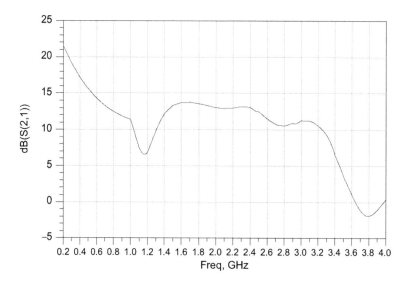

FIGURE 16.29   Active slot antenna S21 parameter, gain.

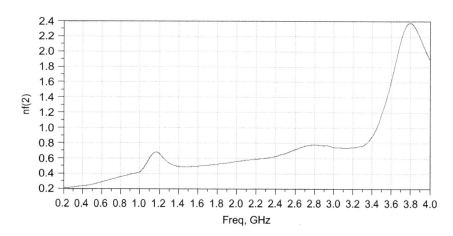

FIGURE 16.30   Active slot antenna noise figure.

A DC bias network supplies the required voltages to the amplifiers. The amplifier specification is listed in Table 15.1. The amplifier complex S parameters are listed in Table 15.2. The amplifier noise parameters are listed in Table 15.3. Active slot antenna S11 parameter is presented in Figure 16.32. The antenna bandwidth is around 40% for VSWR better than 2:1. The active slot antenna S21 parameter, gain, is presented in Figure 16.33. The active antenna gain is 18 ± 2.5 dB for frequencies ranging from 200 to 580 MHz. The active antenna gain is 12.5 ± 2.5 dB for frequencies ranging from 1 to 3 GHz. The active slot antenna noise figure is presented in Figure 16.34. The active slot antenna noise figure is 0.5 ± 0.3 dB for frequencies ranging from 300 MHz to 3.2 GHz. Active slot antenna S22 parameter is presented in Figure 16.35.

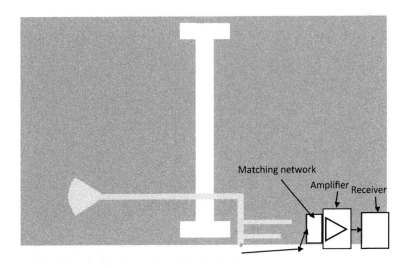

FIGURE 16.31   A wideband active receiving wearable slot antenna.

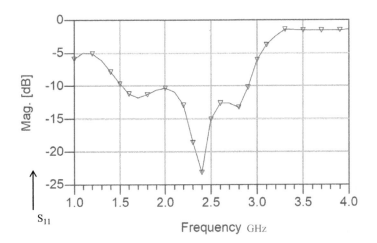

FIGURE 16.32   Active T shape slot antenna S11 parameter.

16.11 NEW FRACTAL COMPACT ULTRA-WIDEBAND, 1–6 GHZ, NOTCH ANTENNA

A wideband notch antenna with fractal structure has been designed. The antenna is printed on RT-Duroid 5880 dielectric substrate with dielectric constant of 2.2 and 1.2 mm thickness. The notch antenna is shown in Figure 16.36. The notch antenna's dimensions are 74.5 × 57.1 mm. The antenna center frequency is 2.75 GHz.

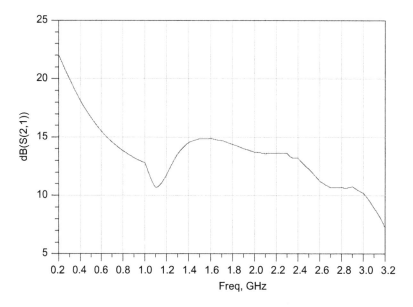

FIGURE 16.33   Active T shape slot antenna S21 parameter.

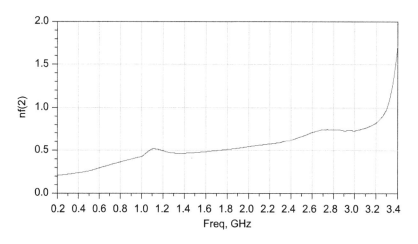

FIGURE 16.34   Active slot antenna noise figure.

The antenna bandwidth is around 200% for S11 lower than −6.5dB, as presented in Figure 16.37. The notch antenna VSWR is better than 3:1 for frequencies from 1 to 5.5 GHz. The antenna beam width is around 84°. The antenna gain is around 3.5 dBi as presented in Figure 16.38. The H plane radiation pattern of the wideband notch antenna with fractal structure is presented in Figure 16.39.

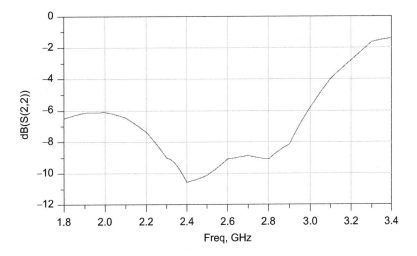

FIGURE 16.35 Active T shape slot antenna S22 parameter.

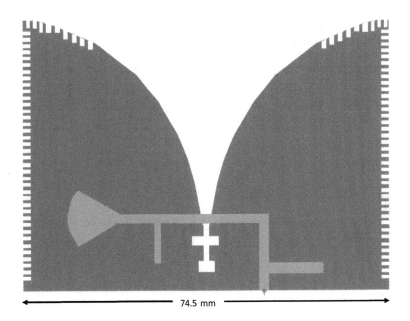

FIGURE 16.36 A wideband notch antenna with fractal structure.

16.12 NEW COMPACT ULTRA-WIDEBAND NOTCH ANTENNA 1.3–3.9 GHZ

A wideband notch antenna with fractal structure has been designed. The antenna is printed on RT-Duroid 5880 dielectric substrate with dielectric constant of 2.2 and 1.2 mm thickness. The notch antenna is shown in Figure 16.40. The notch antenna's

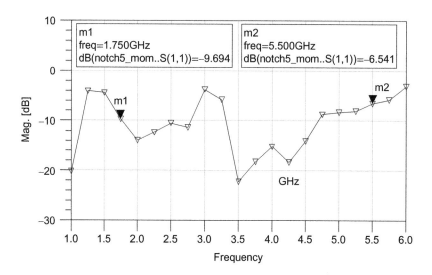

FIGURE 16.37 A wideband notch antenna with fractal structure, computed S11.

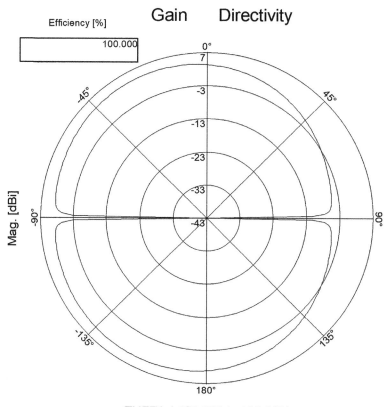

FIGURE 16.38 E plane radiation pattern of the wideband notch antenna with fractal structure.

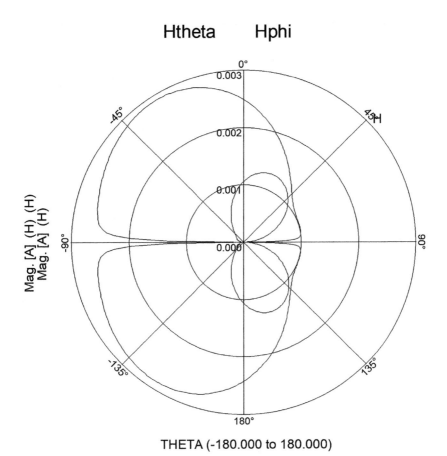

FIGURE 16.39 H plane radiation pattern of the wideband notch antenna with fractal structure.

dimensions are 52.2 × 36.8 mm. The antenna center frequency is 2.7 GHz. The antenna bandwidth is around 100% for S11 lower than −6.5 dB, as presented in Figure 16.41. The notch antenna VSWR is better than 3:1 for frequencies from 1.3 to 3.9 GHz. The antenna beam width is around 84°. The antenna gain is around 3.5 dBi.

By using fractal structure, the notch antenna's length and width were reduced by around 50%.

16.13 NEW COMPACT ULTRA-WIDEBAND NOTCH ANTENNA 5.8–18 GHZ

A wideband notch antenna with fractal structure has been designed. The antenna is printed on RT-Duroid 5880 dielectric substrate with dielectric constant of 2.2 and 1.2 mm thickness. The notch antenna is shown in Figure 16.42. The notch antenna's dimensions are 11 × 7.7 mm. The antenna center frequency is 12 GHz. The antenna bandwidth is around 100% for S11 lower than −5dB, as presented in Figure 16.43.

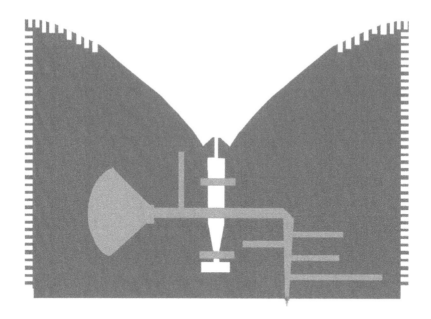

FIGURE 16.40 A wideband 1.3–3.9 GHz notch antenna with fractal structure.

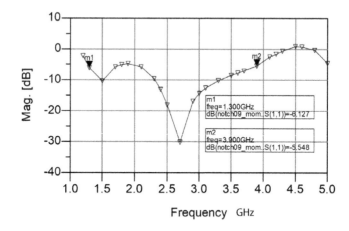

FIGURE 16.41 A wideband 1.3–3.9 GHz notch antenna with fractal structure, S11 results.

The notch antenna VSWR is better than 3:1 for more than 90% of the frequency range from 5.8 to 18 GHz. The antenna beam width is around 84°. The antenna gain is around 3.5 dBi. Figure 16.44 presents the radiation pattern of the wideband notch antenna with fractal structure at 8 GHz.

The antenna matching network was optimized to get better S11 results at 16–18 GHz. The length and width of the stubs were tuned to get better S11 results at 16–18 GHz.

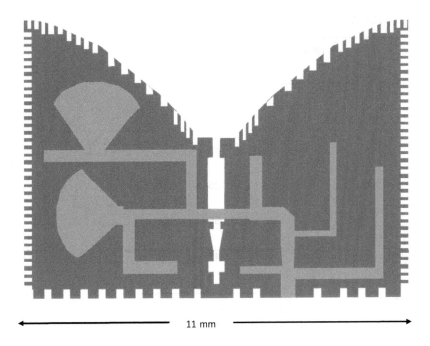

11 mm

FIGURE 16.42 A wideband 5.8–18 GHz notch antenna with fractal structure.

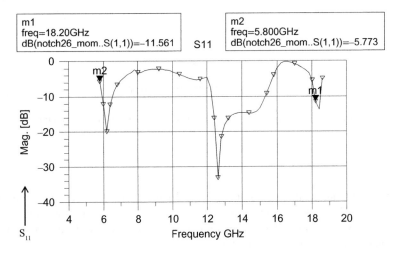

FIGURE 16.43 A wideband 5.8–18 GHz notch antenna with fractal structure, S11 results.

16.14 NEW FRACTAL ACTIVE COMPACT ULTRA-WIDEBAND, 0.5–3 GHZ, NOTCH ANTENNA

A wideband active notch antenna with fractal structure has been designed. The antenna is printed on RT-Duroid 5880 dielectric substrate with dielectric constant of 2.2 and 1.2 mm thickness. The active notch antenna is shown in Figure 16.45.

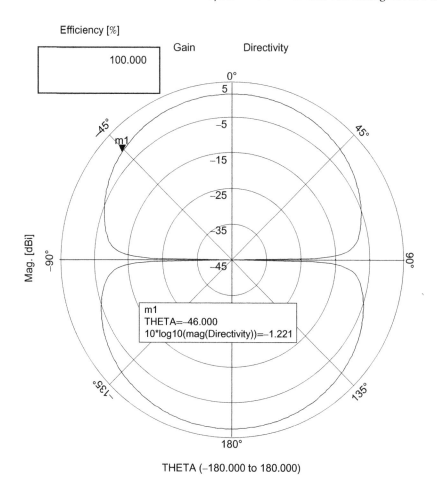

FIGURE 16.44 Radiation pattern of the wideband notch antenna with fractal structure at 8 GHz.

The notch antenna's dimensions are 74.5 × 57.1 mm. The antenna center frequency is 1.75 GHz. The active antenna bandwidth is around 200% for S11 lower than −5dB. The active notch antenna VSWR is better than 3:1 for frequencies from 0.5 to 3 GHz. The antenna beam width is around 84°.

An E-PHEMT LNA was connected to a notch antenna. The radiating element is connected to the LNA via an input matching network. An output matching network connects the amplifier port to the receiver. A DC bias network supplies the required voltages to the amplifiers. The amplifier specification is listed in Table 15.1. The amplifier complex S parameters are listed in Table 15.2. The amplifier noise parameters are listed in Table 15.3. The active notch antenna S21 parameter, gain, is presented in Figure 16.46. The active antenna gain is 22 ± 2.5 dB for frequencies ranging from 200 to 900 MHz. The active antenna gain is 12.5 ± 2.5 dB for frequencies

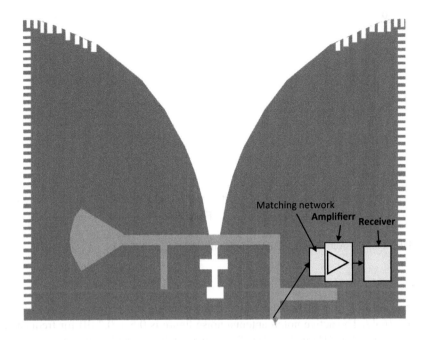

FIGURE 16.45   A wideband active notch antenna with fractal structure.

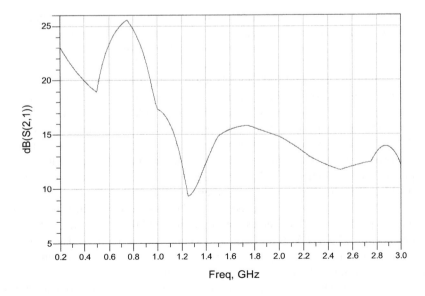

FIGURE 16.46   Active notch antenna S21 parameter.

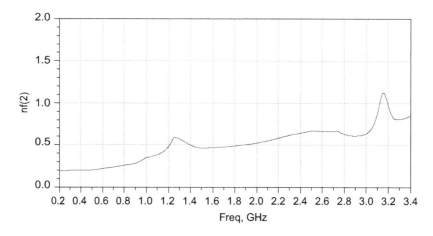

FIGURE 16.47 Active notch antenna noise figure.

ranging from 1 to 3 GHz. The active notch antenna noise figure is presented in Figure 16.47. The active notch antenna noise figure is 0.5 ± 0.3 dB for frequencies ranging from 300 MHz to 3.0 GHz. The active notch antenna S22 parameter is lower than −5dB for frequencies from 0.5 to 3 GHz.

16.15 NEW COMPACT ULTRA-WIDEBAND ACTIVE NOTCH ANTENNA 0.4–3 GHZ

A wideband notch antenna with fractal structure has been designed. The antenna is printed on RT-Duroid 5880 dielectric substrate with dielectric constant of 2.2 and 1.2 mm thickness. The notch antenna is shown in Figure 16.48. The notch antenna's dimensions are 52.2 × 36.8 mm. The antenna center frequency is 1.7 GHz. The antenna bandwidth is around 100% for S11 lower than −5dB as shown in Figure 16.49. The notch antenna VSWR is better than 3:1 for frequencies from 0.4 to 3 GHz. The antenna beam width is around 84°. An E-PHEMT LNA was connected to a notch antenna. The radiating element is connected to the LNA via an input matching network. An output matching network connects the amplifier port to the receiver. A DC bias network supplies the required voltages to the amplifiers. The amplifier specification is listed in Table 15.1. The amplifier complex S parameters are listed in Table 15.2. The amplifier noise parameters are listed in Table 15.3. The active notch antenna S21 parameter, gain, is presented in Figure 16.50. The active antenna gain is 20 ± 2.5 dB for frequencies ranging from 400 MHz to 1.3 GHz. The active antenna gain is 12.5 ± 2.5 dB for frequencies ranging from 1.3 to 3 GHz. The active notch antenna noise figure is presented in Figure 16.51. The active notch antenna noise figure is 0.5 ± 0.3 dB for frequencies ranging from 300 MHz to 3.0 GHz. The active notch antenna S22 parameter is lower than −5 dB for frequencies from 0.5 to 3 GHz.

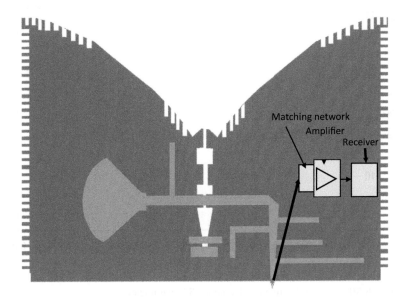

FIGURE 16.48   A wideband fractal active notch antenna with fractal structure.

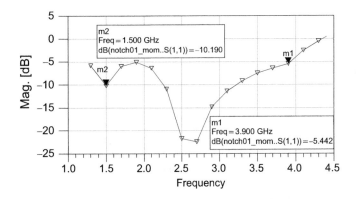

FIGURE 16.49   A fractal active notch antenna S11 parameter.

16.16 CONCLUSIONS

This chapter presents new compact ultra-wideband slot and notch antennas in frequencies ranging from 1 to 18 GHz. The slot and notch antennas were analyzed by using 3D full-wave software. The antenna bandwidth is from 50% to 100% with VSWR better than 3:1. The antenna gain is around 3 dBi with efficiency higher than 90%. The antennas' electrical parameters were computed in the vicinity of the human body. Compact new ultra-wideband notch antenna 1–6 GHz and a wideband notch antenna 5.8–18 GHz are presented. Space-filling technique and Hilbert curves were

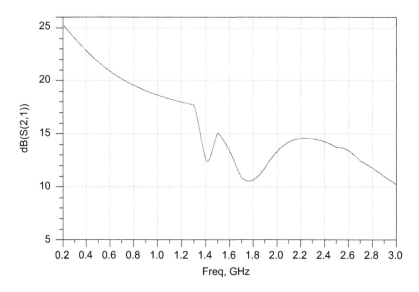

FIGURE 16.50 A fractal active notch antenna S21 parameter.

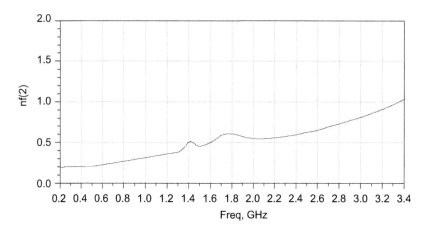

FIGURE 16.51 Active fractal notch antenna noise figure.

employed to design the fractal notch antennas. The fractal notch antennas were ana-lyzed by using 3D full-wave software. The antenna bandwidth is around 100% with VSWR better than 3:1. The antenna gain is around 3.5 dBi with efficiency higher than 90%. By using fractal structure, a notch antenna's length and width can be reduced by up to 50%.

This chapter also presents new compact ultra-wideband active slot and notch antennas in frequencies ranging from 1 to 6 GHz.

REFERENCES

[1] A. Sabban, "Active Compact Wearable Body Area Networks for Wireless Communication, Medical and IOT Applications," *Applied System Innovation Journal*, 46, 1–20, December 2018.

[2] A. Sabban, "New Wideband Passive and Active Wearable Antennas for Energy Harvesting Applications," *Journal of Sensor Technology*, 7, 53–70, December 2017.

[3] A. Sabban, "New Wideband Compact Wearable Slot Antennas for Medical and Sport Applications," *Journal of Sensor Technology*, 08, 18–34, December 2017.

[4] A. Sabban, *Novel Wearable Antennas for Communication and Medical Systems*, Taylor & Francis Group, Boca Raton, FL, October 2017.

[6] A. Sabban, *Wideband RF Technologies and Antenna in Microwave Frequencies*", John Wiley & Sons, New York, July 2016.

[7] L. C. Godara (Ed.), *Handbook of Antennas in Wireless Communications*, CRC Press LLC, Florida, USA, 2002.

[8] J. D. Kraus and R.J. Marhefka, *Antennas for All Applications*, 3rd edn., McGraw Hill, New Delhi, India, 2002.

[9] J. R. James, P. S. Hall, and C. Wood, *Microstrip Antenna Theory and Design*, P. Peregrinus, Cop., London 1981.

[10] A. Sabban and K. C. Gupta, "Characterization of Radiation Loss from Microstrip Discontinuities Using a Multiport Network Modeling Approach," *IEEE Transactions on Microwave Theory and Techniques*, 39(4), 705–712, April 1991.

[11] A. Sabban, Multiport Network Model for Evaluating Radiation Loss and Coupling Among Discontinuities in Microstrip Circuits, Ph.D. thesis, University of Colorado at Boulder, January 1991.

[12] A. Sabban, Microstrip Antenna Arrays, U.S. Patent 1986/4,623,893, 1986.

[13] A. Sabban, "A New Wideband Stacked Microstrip Antenna," in *IEEE Antenna and Propagation Symposium*, Houston, TX, June 1983.

[14] A. Sabban, *Low Visibility Antennas for Communication Systems*, Taylor & Francis Group, New York, 2015.

[15] A. Sabban, "Wideband Microstrip Antenna Arrays," in *IEEE Antenna and Propagation Symposium MELCOM*, Tel-Aviv, June 1981.

[16] A. Sabban, *RF Engineering, Microwave and Antennas*, Saar Publications, Israel 2014.

[17] K. Fujimoto and J. R. James (Eds.), *Mobile Antenna Systems Handbook*, Artech House, Boston, MA, 1994.

[18] A. Sabban, "New Wideband Notch Antennas for Communication Systems," *Wireless Engineering and Technology Journal*, 7, 75–82, April 2016.

[19] A. Sabban, Dual Polarized Dipole Wearable Antenna, U.S. Patent number: 8203497, June 19, 2012.

[20] A. Sabban, "Wideband Tunable Printed Antennas for Medical Applications," in *IEEE Antenna and Propagation Symposium*, Chicago, IL, July 2012.

[21] A. Sabban, "New Wideband Printed Antennas for Medical Applications," *IEEE Transactions on Antennas and Propagation*, 61(1), 84–91, January 2013.

[22] A. Sabban, "Comprehensive Study of Printed Antennas on Human Body for Medical Applications," *International Journal of Advance in Medical Science (AMS)*, 1, 1–10, February 2013.

[23] ADS Software, Keysight web page, https://www.keysight.com/il/en/products/software/pathwave-design-software/pathwave-advanced-design-system.html.

[24] C.A. Balanis, *Antenna Theory: Analysis and Design*, 2nd edn. Wiley, Hoboken, NJ, 1996.

17 Design and Measurements Process of Wearable Communication, Medical and IOT Systems

Albert Sabban

CONTENTS

17.1 INTRODUCTION

This chapter describes electromagnetics, microwave engineering, wearable systems, and antenna design process and measurements. More information on electromagnetics, microwave engineering, wearable systems, and antenna design process and measurements is presented in [1–27]. Computer-aided design (CAD) commercial software is presented in this chapter, see [24–27]. Measurement techniques of wearable antennas and radio frequency (RF) medical systems in the vicinity of a human body are presented in Sections 17.3–17.15. Basic RF measurement theory is presented in Sections 17.3–17.7, see [1–23]. The measurement of S parameters is the first stage in electromagnetics, microwave engineering and antenna measurements. Setups for microwave engineering measurements will be presented in this chapter. Maximum input and output measurements of communication systems are presented in this chapter. Intermodulation measurements, second intercept point (IP2) and third intercept point (IP3), are discussed in this chapter in Sections 17.12 and 17.13.

Antenna measurements will be discussed in this chapter in Sections 17.15 and 17.16. It is more convenient to measure antennas in the receiving mode. If the measured antenna is reciprocal, the antenna's radiation characteristics are identical for the receiving and transmitting modes. Active antennas are not reciprocal. The radiation characteristics of antennas are usually measured in the far field. Far-field antenna measurements suffer from some disadvantages. A long free space area is needed. Reflection from the ground from walls affect measured results and add errors to measured results. It is difficult and almost impossible to measure an antenna in the antenna's operating environment, such as an airplane or satellite. Facilities for antenna measurements are expensive. Some of these drawbacks may be solved by near-field and indoor measurements. Near-field measurements are presented in [1]. Small communication companies do not own antenna measurement facilities. However, there are several companies around the world that provide antenna measurements services, including near-field and far-field measurements. One day near-field measurements may cost around 5000 USD. One day far-field measurements may cost around 3000 USD.

17.2 CAD COMMERCIAL SOFTWARE

In the last decade, several electromagnetic (EM) commercial software packages were developed, see [24–27]. The most popular software that is used in the design and development of wearable systems and antennas will be presented in this section.

17.2.1 HIGH FREQUENCY STRUCTURE SIMULATOR (HFSS) SOFTWARE

Ansys HFSS is a three-dimensional (3D) EM simulation software for designing and simulating high-frequency devices, such as antennas, antenna arrays, RF and microwave components, high-speed interconnects, filters, connectors, integrated circuits (ICs) packages and printed circuit boards (PCBs). Ansys HFSS is employed to design high-frequency, high-speed communications systems, radar systems, RF components and modules, satellites, Internet of Things (IOT) products and other high-speed RF and digital devices. For more information see, https://www.ansys.com/products/electronics/ansys-hfss.

HFSS employs several solvers to give the RF engineer deep insight into all the 3D EM problems. Through integration with Ansys thermal, structural and fluid dynamics tools, HFSS provides a multi-physics analysis of RF products, ensuring their thermal and structural reliability. The Ansys HFSS simulation suite consists of a comprehensive set of solvers to address diverse EM problems. Its automatic adaptive mesh refinement lets the designer focus on the design instead of spending time determining and creating the best mesh.

17.2.1.1 High-Frequency EM Solvers

Ansys HFSS uses the highly accurate finite element method (FEM), the large-scale method of moments (MoM) technique and the ultra-large-scale asymptotic method of shooting and bouncing rays (SBR) with advanced diffraction and creeping wave physics for enhanced accuracy (SBR+). The Ansys HFSS simulation suite has the following solvers to tackle EM problems:

HFSS solvers

- Frequency Domain.
- Time Domain.
- Integral Equations (IE).
- Hybrid Technologies.

HFSS SBR+

- SBR.
- Physical Optics.
- Physical Theory of Diffraction (PTD).

- Uniform Theory of Diffraction (UTD).
- Creeping Wave.

HFSS contains multiple simulation engines in one package, each targeted toward a specific application or simulation output.

17.2.1.1.1 HFSS Hybrid Technologies

The FEM-IE hybrid technology is built upon HFSS FEM, IE MoM and the patented Ansys domain decomposition method (DDM) to solve electrically large and complex systems. By applying the appropriate solver technology, local regions of high geometric detail and complex materials are addressed with finite element HFSS, while regions of large objects or installed platforms are addressed with 3D MoM HFSS-IE. The solution is delivered in a single setup through a single, scalable and fully coupled system matrix.

17.2.1.1.2 FEM (Frequency Domain)

This is a 3D, full-wave, frequency domain EM solver based on the FEM. Engineers can calculate S, Y and Z parameters and resonant frequency, visualize EM fields, and can generate component models to evaluate signal quality, transmission path loss, impedance mismatch, parasitic coupling and far-field radiation. Typical applications include antennas, mobile communication devices, ICs, high-speed digital and RF interconnects, waveguides, connectors, filters and EM compatibility (EMC).

17.2.1.1.3 Finite Element Transient (Time Domain)

The Finite Element Time Domain solver is used to simulate transient EM field behavior and visualize fields and system responses in typical applications like time domain reflectometry (TDR), lightning strikes, pulsed ground-penetrating radar (GPR), electrostatic discharge (ESD) and EM interference (EMI). It leverages the same finite element meshing approach as the frequency domain solver, without the need to switch meshing technologies to switch simulation domains. The transient solver complements the frequency domain HFSS solver and enables engineers to understand the EM characteristics on the same mesh in both time and frequency domains.

17.2.1.1.4 IE

The IE solver employs the 3D MoM technique for efficiently solving open radiation and scattering problems. It is ideal for radiation studies like antenna design and antenna placement, and for scattering studies such as radar cross section (RCS). The solver can employ either multilevel fast multipole methods (MLFMM) or adaptive cross-approximation (ACA) to reduce memory requirements and solve times, allowing this tool to be applied to very large problems.

17.2.1.1.5 Ansys HFSS SBR+

SBR+ is an EM solver to empower SBR technology with simultaneous and consistent implementations of PTD, UTD and Creeping Wave for simulating installed antenna performance on electrically large platforms that are hundreds or thousands of wavelengths in size.

SBR uses a ray tracing technique to model induced surface currents on the antenna platform or scattering geometry composed of conductors and dielectrics. With the SBR+ solver, engineers can obtain fast and accurate predictions of far-field installed antenna radiation patterns, near-field distributions and antenna-to-antenna coupling (S parameters) on electrically medium, large and enormous platforms. Transmissions and reflections can be modeled in large structures like vehicles, aircraft and antenna radomes. HFSS SBR+ also provides efficient radar signature modeling, including inverse synthetic aperture radar (ISAR) images of electrically large targets.

17.2.1.1.6 PTD

The PTD wedge correction feature is used for correcting PO currents along sharp edges of installed antenna platforms to refine EM field diffraction.

17.2.1.1.7 UTD

Engineers can model UTD edge diffraction rays created by illuminated geometry edges and identified by PTD wedges. This is important for cases where the significant parts of the scattering geometry are otherwise shadowed from direct or multi-bounce illumination.

17.2.1.2 Ansys RF Option

The Ansys RF Option combined with HFSS creates an end-to-end high-performance RF simulation flow. It includes Ansys EMIT, a unique multi-fidelity approach for predicting RF system performance in complex RF environments with multiple sources of interference, and provides the diagnostic tools needed to quickly identify root-cause RF interference (RFI) issues. The RF Option also includes Ansys circuits, which include a harmonic balance (HB) circuit simulation, 2.5D planar MoM solver, filter synthesis and more.

17.2.1.3 RF Option Features

- RF link budget analysis.
- Built-in wireless propagation models.
- RF co-site and antenna coexistence analysis.
- Automated diagnostics for rapid root-cause analysis.
- Quick assessment and comparison of potential mitigation measures.
- RF radio and component libraries.
- Multi-fidelity behavioral radio models.
- Antenna-to-antenna coupling models.

17.2.1.4 Circuit Analyses

- Linear.
- Transient.
- Direct current (DC) analysis with multiple continuation options.
- Multitone HB analysis.

For more information see, https://www.ansys.com/products/electronics/ansys-hfss.

17.2.2 Advanced Design System (ADS)

ADS is an electronic design automation software system. It offers complete design integration to designers of products, such as cellular and portable phones, pagers, wireless networks, and radar and satellite communications systems.

ADS supports communication system and RF design engineers to develop all types of RF designs, from RF and microwave modules and printed antennas to integrated monolithic microwave ICs (MMICs) for communication, medical, IOT and aerospace defense applications.

With a complete set of simulation technologies ranging from frequency and time domain circuit simulation to EM field simulation, ADS lets designers fully characterize and optimize designs, such as Harmonic Balance, Circuit Envelope, Transient Convolution, Ptolemy, X-parameter, Momentum and 3D EM simulators, including both FEM and finite-difference time-domain (FDTD) solvers.

17.2.2.1 ADS Features

- Project design environment to input schematics, launch simulations and manage design projects.
- Data display to manipulate and plot data.
- Linear simulator for S parameters, DC and small signal, alternating current (AC) analysis.
- RF system simulator.
- Statistical simulator including Design-of-Experiments (DOE) for ensuring robust designs.
- Optimizers to optimize designs for performance and yield.
- Filter design guide to synthesize lumped and distributed filters.
- Passive circuit design guide to synthesize matching networks and passive functions.
- Design guide developer studio to create customized design guides.
- Additional model libraries – RF system, RF passive and multilayer interconnects.
- Dynamic link to Cadence for simulating Composer radio-frequency IC (RFIC) schematic in ADS.
- Mentor dynamic access for bidirectional RF board design transfer.
- Connection manager for bidirectional data transfer.

17.2.2.2 ADS Functionality

- **Design Environment** – Graphical user interface (UI) for schematic entry and simulation setup within the ADS.
- **Data Display** – Allows simulation results to be viewed either graphically, as they would appear on analyzers or other instruments, or as raw data.
- **Connection Manager** – Reads data from, downloads data to, and controls selected instruments. The W2200BP includes the connection manager client only. The Connection Manager Server must be downloaded and installed separately.

17.2.2.3 Simulators

- **Linear Simulator** – Frequency-domain circuit simulator that analyzes a large variety of RF and microwave circuits operating under linear conditions.
- **Statistical Design** – Optimization, Sensitivity Analysis, Yield, Yield Optimization (also known as Design Centering), DOE and Yield Sensitivity Histograms.
- **RF System Simulator** – Model a complete RF system with accurate block level models that can later be replaced with device level models for further, more detailed verification.

17.2.2.4 Model Sets

- **RF Passive Circuit Models** – Models for many common RF parts including inductors, transformers, couplers, crystals and bond-wires.
- **Multilayer Interconnect Models** – Models for multiple coupled lines used in multilayer structures such as PCBs, multichip devices and IC packages.
- **RF System Models** – Gain blocks, mixers, modulators and demodulators and, phase lock loop (PLL) components. In addition, verification model extractors and verification models for RF system-level analysis and verification are included.

17.2.2.5 Design Guides

- **Passive Circuit Design Guide** – Synthesis and analysis of commonly used microstrip components familiar to microwave designers, such as branch-line couplers, Wilkinson dividers, coupled line filters, quarter-wave matching networks and lumped-element matching.
- **Filter Design Guide** – Synthesis and analysis of commonly used passive filters familiar to microwave designers. Microwave circuits can be synthesized in seconds.
- **Design Guide Developer Studio** – Create an ADS Design Guide menu interface.

Momentum Circuit Designer – The Momentum Circuit Designer is a basic high-frequency physical design suite that integrates standard and advanced layout editing features with Momentum EM simulation technology and Linear (S parameters) circuit simulation to speed physical circuit design.

Momentum Circuit Designer is an extensible entry-level product bundle that includes ADS Schematic capture, Data Display, Momentum EM simulator, and Layout editor, as well as layout/schematic design synchronization capabilities.

17.2.2.6 FEM Simulator

The FEM simulator has full-wave three dimensions EM simulation capabilities, based on the FEM. The RF-Pro UI, which comes with this FEM element, makes setting up RF circuit co-simulation in ADS fast with no errors for the design of multi-technology RF modules that integrate RFIC, MMIC, package and PCB. It also automates the extraction of nets and components for EM simulation without modifying the layout. The FEM simulator enables simulation of 3D structures, such as connectors, wire-bonds and packaging with circuit and system components. It is important especially for RF module designs where 3D interconnects and packaging must be simulated along with the circuit. For more details see in, https://www.keysight.com/en/pc-1297113/advanced-design-system-ads?nid=-34346.0&cc=IL&lc=eng.

The frequency domain simulation can be used in the Electromagnetic Professional (EM-Pro) software, in the 3D EM platform and in ADS.

17.2.3 CST Software

CST is a high-performance 3D EM analysis software for designing, analyzing and optimizing EM components and systems.

EM field solvers for applications across the EM spectrum are contained within a single UI in CST software. The solvers can be coupled to perform hybrid simulations, giving engineers the flexibility to analyze whole systems made up of multiple components in an efficient and straightforward way. Co-design with other SIMULA software products allows EM simulation to be integrated into the design flow and drives the development process from the earliest stages to the final stages.

Common subjects of EM analysis include the performance and efficiency of antennas and filters, EMC and EMI, exposure of the human body to EM fields, electro-mechanical effects in motors and generators, and thermal effects in high-power devices. More information is presented in CST Software, https://www.3ds.com/products-services/simulia/products/cst-studio-suite/

With System Assembly and Modeling (SAM), CST provides an environment that simplifies the management of simulation projects, enabling the intuitive construction of EM systems and straightforward management of complex simulation flows using schematic modeling.

The SAM framework can be used for analyzing and optimizing an entire device that consists of multiple individual components. These are described by relevant physical quantities such as currents, fields or S parameters. SAM enables the use of the most efficient solver technology for each component.

SAM helps users to compare the results of different solvers or model configurations within one simulation project and perform post-processing automatically. SAM facilitates the set-up of a linked sequence of solver runs for hybrid and multi-physics simulations. For example, using the results of EM simulation to calculate thermal effects, then structural deformation, and then another EM simulation to analyze detuning. This combination of different levels of simulation helps to reduce the computational effort required to analyze a complex model accurately.

The CST software design environment is an intuitive UI used by all the modules. It comprises a 3D interactive modeling tool, a schematic layout tool, a pre-processor for the EM solvers and post-processing tools tailored to industry needs. The ribbon-based interface uses tabs to display all the tools and options needed to set up, carry out and analyze a simulation, grouped according to their position in the workflow. Contextual tabs mean that the most relevant options for the task are always just a click away. In addition, the Project Wizard and the QuickStart Guide provide guidance to new users and offer access to a wide range of features.

The 3D interactive modeling tool at the heart of the interface uses the ACIS 3D CAD kernel. This powerful tool enables complex models to be constructed within CST software.

17.2.3.1 CST Solvers

- Transient solver.
- Transmission line model (TLM) solver.
- Frequency domain solver.
- Eigenmode solver.
- Resonant solver.
- IE Solver.
- Asymptotic Solver.

17.2.3.2 CST Applications

- CST's products cover an extremely wide range of EM components.
- Applications include static, stationary, low and high-frequency problems, as well as devices with movement of charged particles.
- Typical applications are couplers, filters, planar structures, connectors, EMC and Specific Absorption Rate (SAR) problems, all kind of antennas, packages, low-temperature co-fired ceramic (LTCC) structures, inductors, capacitors, waveguides, plasma sources, optical devices, sensors, recording units, actuators, motors and EM brakes.

17.2.4 MICROWAVE OFFICE, AWR

Microwave Office design suite provides a flexible RF/microwave design tool.

Built on AWR high-frequency design platform with its open design environment and advanced unified data model. Microwave Office seamlessly interoperates with Visual System Simulator (VSS) system design and AXIEM and Analyst EM simulation software tools in the NI AWR Design Environment platform to deliver a complete RF and microwave circuit, system and EM co-simulation environment.

AWR's AXIEM EM software is an EM analysis.

The AXIEM product was developed specifically for 3D planar applications such as RF PCBs and modules, LTCC, MMIC and RFIC designs.

The APLAC simulator offers multi-level analyses which includes:

- DC operation point.
- Linear frequency domain.
- Time domain.
- HB.
- Phase noise.
- Linear/non-linear noise including AC noise contributors, temperature.
- Yield predictions and optimization.

More information can be found in https://www.awr.com/serve/microwave-office-brochure-1.

AWR is fast and accurate simulation technology that offers robust circuit analysis and design insight, providing the linear and nonlinear time and frequency domain measurements required to properly characterize and optimize high-frequency devices.

Simulation-Ready models, comprehensive libraries of high-frequency distributed transmission models, surface-mount vendor components and process design kits (PDKs) from leading MMIC/RFIC foundries enable accurate simulation of designs prior to manufacture, resulting in fewer and faster design iterations.

Capabilities Design Entry – The AWR UI is tailored to provide project management and design entry for high-frequency circuits, enabling designers to quickly build networks from a comprehensive library of RF-aware components. The library supports parameterization for tuning/optimization and hierarchical design with circuit, system, and EM co-simulation, simulation controls and result graphs for standard and user customized RF/microwave measurements.

Automation – Automation features expedite design tasks and manage network and measurement data, including labor-saving wizards to import PCB layout and/or Open Access schematic information from third-party tools, as well as an easy-to-use application programming interface (API) and scripting functionality to support customization and user-defined automation.

Load-Pull Analysis – Readily develop amplifier input/output matching circuits using complex swept load-pull data sets based on either measured or simulated data. Performance contours include available output power, gain, efficiency Physical Address Extension (PAE), intermodulation distortion levels or any amplifier-related performance metric.

Synthesis and Design Assistance – Accelerate design starts with synthesis modules and design assist wizards that generate topologies based on a set of user-specified RF/ microwave performance criteria. Synthesized filters, impedance matching, mixer and passive component networks are available for further refinement, optimization, EM verification and physical design.

Simulation APLAC – This robust HB simulator provides linear and nonlinear circuit analysis with powerful multi-rate HB, transient-assisted HB and time variant (circuit envelope) analysis, supporting large-scale and highly nonlinear RF/microwave circuits.

Planar EM – AXIEM provides the speed and accuracy to characterize and optimize passive structures, transmission lines, planar antennas and large (more than 100K unknowns) patch arrays.

3D EM – Analyst helps accelerate high-frequency product development from early physical design characterization through to full 3D EM verification. Its 3D finite element solver provides fast and accurate EM analysis of interconnects such as bond wires, vias/ via fencing and ball grids.

Automated Circuit Extraction (ACE) – Using layout-based models for circuit extraction, ACE dramatically reduces the time required to model complex interconnects by automatically identifying transmission lines from the layout and partitioning them into existing models.

17.2.4.1 Microwave Office for MMIC Design

- Design capture with industry-leading tuning.
- Linear and nonlinear frequency-domain and time-domain simulation.
- Layout with production ready GDSII export.

- EM analysis from layout or schematic to most commercial EM solvers.
- Design rule check (DRC) and layout versus schematic (LVS) from a simulation schematic to ICED or Mentor.

17.2.4.2 System Simulation and Frequency Planning (with VSS)
- PDKs from a wide range of foundries.
- Co-design packaging and module flow.

17.2.4.3 Microwave Office for Module Design
- Full integration of APLAC's HB and time-domain simulators.
- ACE technology delivers fast and accurate interconnect modeling.
- iNets technology provides automated interconnect construction for MMIC, module and PCB designs.
- EXTRACT flow enables EM and circuit extraction to be controlled by the schematic.
- AXIEM 3D planar EM solver technology delivers capacity, accuracy and speed.
- Multi-technology PDKs support circuit, system, schematic and layout co-design.

17.2.4.4 ACE Technology
- ACE circuit extraction technology dramatically reduces from hours to seconds the time required to do the initial modeling of complex interconnects. The new technology enables designers to perform interconnect modeling at the earliest stages of the design flow, where problems can be identified and corrected before costly and time-consuming redesigns are required.
- Microwave Office users can leverage layout-based models for circuit extraction as opposed to traditional schematic-based designs flows.

17.3 MODELING AND REPRESENTATION OF WEARABLE SYSTEMS WITH N PORTS

Antenna systems and communication systems may be represented as multiport networks with N ports as shown in Figure 17.1. We may assume that only one mode propagates in each port. The EM fields in each port represents incident and reflected waves. The EM fields may be represented by equivalent voltages and currents as given in Equations 17.1 and 17.2, see [1–5].

$$V_n^- = Z_n I_n^-$$
$$V_n^+ = Z_n I_n^+$$
(17.1)

$$I_n^- = Y_n V_n^-$$
$$I_n^+ = Y_n V_n^+$$
(17.2)

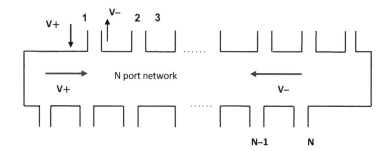

FIGURE 17.1 Multiport networks with N ports.

The voltages and currents in each port are given in Equation 17.3.

$$I_n = I_n^+ - I_n^-$$
$$V_n = V_n^+ + V_n^-$$
$$(17.3)$$

The relations between the voltages and currents may be represented by the Z matrix as given in Equation 17.4. The relations between the currents and voltages may be represented by the Y matrix as given in Equation 17.5. The Y matrix is the inverse of the Z matrix.

$$[V] = \begin{bmatrix} V_1 \\ V_2 \\ \vdots \\ V_N \end{bmatrix} = \begin{bmatrix} Z_{11} & Z_{12} & & Z_{1N} \\ Z_{21} & Z_{22} & & Z_{2N} \\ & & & \\ Z_{N1} & Z_{N2} & & Z_{NN} \end{bmatrix} \begin{bmatrix} I_1 \\ I_2 \\ \vdots \\ I_N \end{bmatrix} = [Z][I] \qquad (17.4)$$

$$[I] = \begin{bmatrix} I_1 \\ I_2 \\ \vdots \\ I_N \end{bmatrix} = \begin{bmatrix} Y_{11} & Y_{12} & & Y_{1N} \\ Y_{21} & Y_{22} & & Y_{2N} \\ & & & \\ Y_{N1} & Y_{N2} & & Y_{NN} \end{bmatrix} \begin{bmatrix} V_1 \\ V_2 \\ \vdots \\ V_N \end{bmatrix} = [Y][V] \qquad (17.5)$$

17.4 SCATTERING MATRIX

We cannot measure voltages and currents in microwave networks. However, we can measure power, voltage standing wave ratio (VSWR) and the location of the minimum field strength. We can calculate the reflection coefficient from these data. The scattering matrix is a mathematical presentation that describes how EM energy propagates through a multi-port network. The S matrix allows us to accurately describe the properties of complicated networks. S parameters are defined for a given frequency and system impedance and vary as a function of frequency for any non-ideal network. The scattering S matrix describes the relation between the forward and reflected waves as written in Equation 17.6. S parameters describe the response of an N-port network to voltage signals at each port. The first number in the subscript refers to the responding port, while the second number refers to the incident port. So S21 means the response at port 2 due to a signal at port 1.

$$[V^-] = \begin{bmatrix} V_1^- \\ V_2^- \\ V_2^+ \end{bmatrix} = \begin{bmatrix} S_{11} & S_{12} & S_{1N} \\ S_{21} & S_{22} & S_{2N} \\ S_{N1} & S_{N2} & S_{NN} \end{bmatrix} \begin{bmatrix} V_1^+ \\ V_2^+ \\ V_2^+ \end{bmatrix} = [S][V^+] \quad (17.6)$$

The Snn elements represent reflection coefficients. The Snm elements represent transmission coefficients as written in Equation 17.7, where a_i represents the forward voltage in the, i port.

$$S_{nn} = \frac{V_n^-}{V_n^+}|a_i = 0 \quad i \neq n$$

$$S_{nm} = \frac{V_n^-}{V_m^+}|a_i = 0 \quad i \neq m \quad (17.7)$$

By normalizing the S matrix we can represent the forward and reflected voltages as written in Equation 17.8. S parameters depend on the frequency and are given as a function of frequency. In a reciprocal microwave network $S_{nm} = S_{mn}$ and $[S]^t = [S]$.

$$I = I^+ - I^- = V^+ - V^-$$
$$V = V^+ + V^-$$
$$V^+ = \frac{1}{2}(V + I)$$
$$V^- = \frac{1}{2}(V - I) \quad (17.8)$$

The relation between Z and S matrix is derived by using Equations 17.8 and 17.9 and is given in Equations 17.10 and 17.11.

$$I_n = I_n^+ - I_n^- = V_n^+ - V_n^-$$
$$V_n = V_n^+ + V_n^- \quad (17.9)$$

$$[V] = [V^+] + [V^-] = [Z][I] = [Z][V^+] - [Z][V^-]$$
$$([Z] + [U])[V^-] = ([Z] - [U])[V^+]$$
$$[V^-] = ([Z] + [U])^{-1}([Z] - [U])[V^+] \quad (17.10)$$
$$[V^-] = [S][V^+]$$
$$[S] = ([Z] + [U])^{-1}([Z] - [U])$$

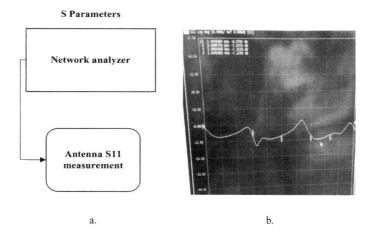

a.   b.

FIGURE 17.2   (a) Antenna S parameter measurements, (b) measured antenna S_{11}.

$$\left[V^+\right] = \frac{1}{2}\left(\left[V\right]+\left[I\right]\right) = \frac{1}{2}\left(\left[Z\right]+\left[U\right]\right)\left[I\right]$$

$$\left[V^-\right] = \frac{1}{2}\left(\left[V\right]-\left[I\right]\right) = \frac{1}{2}\left(\left[Z\right]-\left[U\right]\right)\left[I\right]$$

$$\frac{1}{2}\left[I\right] = \left(\left[Z\right]+\left[U\right]\right)^{-1}\left[V^+\right]$$ (17.11)

$$\left[V^-\right] = \left(\left[Z\right]-\left[U\right]\right)\left(\left[Z\right]+\left[U\right]\right)^{-1}\left[V^+\right]$$

$$\left[S\right] = \left(\left[Z\right]-\left[U\right]\right)\left(\left[Z\right]+\left[U\right]\right)^{-1}$$

A network analyzer is employed to measure S parameters, as shown in Figure 17.2a. A network analyzer may have 2 to 16 ports.

17.5 S PARAMETERS MEASUREMENTS

Antenna S parameters measurement is usually a one port measurement. First, we calibrate the network analyzer to the desired frequency range. A one port, S1P, calibration process consists of three steps:

- Short calibration.
- Open calibration.
- Load calibration.

Connect the antenna to the network analyzer and measure S11 parameter. Save and plot S11 results. Antenna S parameters measured result is shown in Figure 17.2. A setup for S parameters measurement is shown in Figure 17.3a. S_{11} parameters measurement results are shown in Figure 17.3b. A two port S parameter

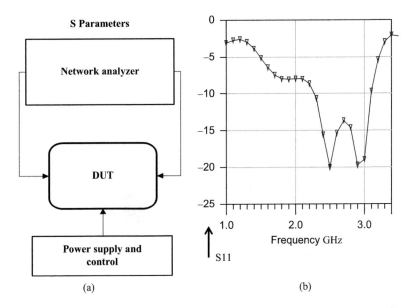

FIGURE 17.3 (a) Two port S parameter measurements, (b) S_{11} parameter results.

measurements setup is shown in Figure 17.4. A two port, S2P, calibration process consists of four steps:

– Short calibration.
– Open calibration.
– Load calibration.
– Through calibration.

Measure S parameters: S_{11}, S_{22}, S_{12}, and S_{21} for N channels. RF Head gain is given by S_{21} parameter. Gain flatness and phase balance between channels can be measured by comparing S_{21} magnitude and phase measured values. RF Head gain and flatness measurement setup is presented in Figure 17.4a. A two port network analyzer is shown in Figure 17.4b. Table 17.1 presents a typical table of measured S parameter results. S parameters in dB may be calculated by using Equation 17.12.

$$Sij(dB) = 20 * log\left[Sij(magnitude)\right] \qquad (17.12)$$

17.5.1 TYPES OF S PARAMETERS MEASUREMENTS

Small signal S parameters measurements – In small signal S parameter measurements the signals have only linear effects on the network so that gain compression does not take place. Passive networks are linear at any power level.

Large signal S parameters measurements – S parameters are measured for different power level. The S matrix will vary with input signal strength.

Gain and flatness measurements

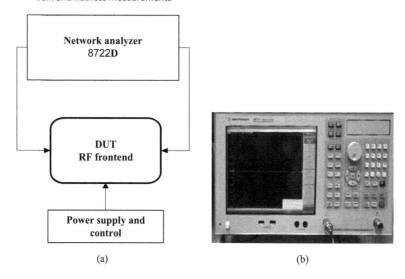

(a)   (b)

FIGURE 17.4   (a) Radio frequency Head gain and flatness measurements, (b) network analyzer.

TABLE 17.1
S parameter results

Channel	S11(E/T) dB	S22(E/T) dB	S12(E/T) dB	S21(E/T) dB
1	−12	−11	−22	31
2	−10.5	−11	−23	29
3	−11	−10	−20	29
n−1	−10	−9	−20	29
n	−9	−10.5	−19	28

17.6 TRANSMISSION MEASUREMENTS

A block diagram of transmission measurements setup is shown in Figure 17.5a. The transmission measurements setup consists of a sweep generator, device under test (DUT), transmitting and receiving antennas, and spectrum analyzer. Measured transmission results by using spectrum analyzer are shown in Figure 17.5b.

The received power may be calculated by using Friis equation as given in Equations 17.13 and 17.14. The receiving antenna may be a standard gain antenna with a known gain. Where r represents the distance between the antennas.

$$P_R = P_T G_T G_R \left(\frac{\lambda}{4\pi r} \right)^2$$

$$For - G_T = G_R = G \qquad (17.13)$$

$$G = \sqrt{\frac{P_R}{P_T}} \left(\frac{4\pi r}{\lambda} \right)$$

Transmitting channel meassurements

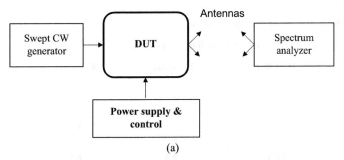

(a)

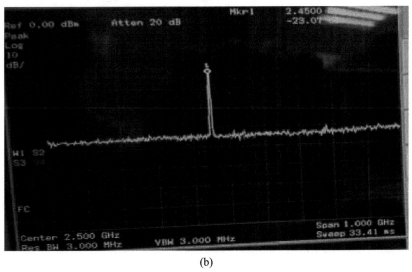

(b)

FIGURE 17.5 (a) Transmission measurements setup, (b) measured transmission results.

$$P_R = P_T G_T G_R \left(\frac{\lambda}{4\pi r}\right)^2$$

$$For - G_T \neq G_R \tag{17.14}$$

$$G_T = \frac{1}{G_R} \frac{P_R}{P_T} \left(\frac{4\pi r}{\lambda}\right)^2$$

Transmission measurements results may be summarized as in Table 17.2.

17.7 OUTPUT POWER AND LINEARITY MEASUREMENTS

A block diagram of the setup of output power and linearity measurements is shown in Figure 17.6. The output power and linearity measurements setup consists of a sweep generator, DUT and a spectrum analyzer, or power meter. In output power and

TABLE 17.2

Transmission measurements results

Transmission results for Antennas Under Test (AUT) dBm

Antenna	F1(MHz)	F2(MHz)	F3(MHz)	Remarks
1	10	9	8	
2	9	8	7	
3	9.5	8.5	7.5	
4	10	9	8	
5	9	8	7	
6	10.5	9.5	8.5	
7	9	8	7	
8	11	10	9	

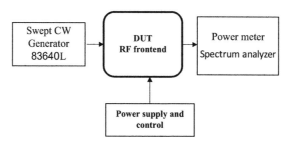

Output powerrange and linearity

FIGURE 17.6   Output power and linearity measurements setup.

linearity measurements we increase the synthesizer power in 1 dB steps and measure the output power level and linearity.

17.8 POWER INPUT PROTECTION MEASUREMENT

A block diagram of power input protection measurements' setup is shown in Figure 17.7. The output power and linearity measurements setup consists of a sweep generator, power amplifier, DUT, attenuator and a spectrum analyzer, or power meter. In power input protection measurements, we increase the synthesizer power in 1 dB steps from 0 dBm and measure the output power level and observe that the DUT functions with no damage.

17.9 NON-HARMONIC SPURIOUS MEASUREMENTS

A block diagram of non-harmonic spurious measurements' setup is shown in Figure 17.8a. A spectrum analyzer is shown in Figure 17.8b.

The non-harmonic spurious measurements setup consists of a sweep generator, DUT and a spectrum analyzer. In non-harmonic spurious measurements, we

Power input protection

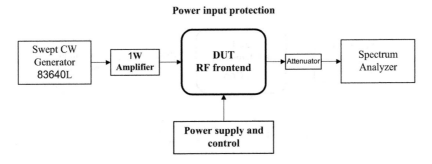

FIGURE 17.7 Power input protection measurement.

Non harmonic spurious measurements

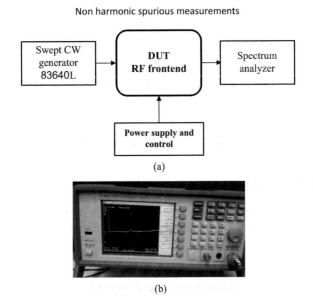

(a)

(b)

FIGURE 17.8 (a) Non-harmonic spurious measurements, (b) spectrum analyzer.

increase the synthesizer power in 1 dB steps, up to 1 dBc point, and measure the spurious level.

17.10 SWITCHING TIME MEASUREMENTS

A block diagram of switching time measurements setup is shown in Figure 17.9. The switching time measurements setup consists of a sweep generator, DUT, detector, pulse generator and oscilloscope. In switching time measurements, we transmit an RF signal through the DUT. We inject a pulse via the switch control port. The pulse envelope may be observed on an oscilloscope. Switching time may be measured by using the oscilloscope.

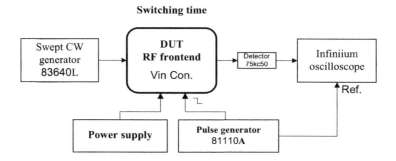

FIGURE 17.9   Switching time measurement setup.

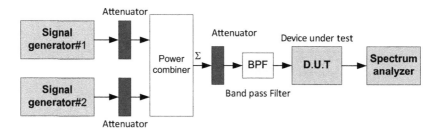

FIGURE 17.10   Setup for second intercept point and third intercept point measurements.

17.11 IP2 MEASUREMENTS

The setup for IP2 and IP3 measurements is shown in Figure 17.10. Second order inter-modulation results are shown in Figure 17.10. IP2 may be computed by Equation 17.15.

$$IP_{2\,[dBm]} = P_{out\,[dBm]} + \Delta_{[dB]} \tag{17.15}$$

Second order intermodulation results are shown in Figure 17.11.

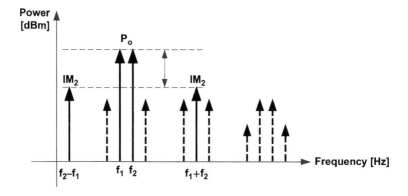

FIGURE 17.11   Second order intermodulation results.

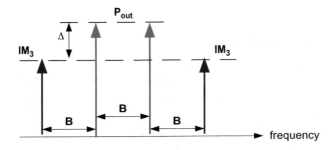

FIGURE 17.12 Third order intermodulation results.

Third order intermodulation results are shown in Figure 17.12. IP3 may be computed by Equation 17.16.

$$IP_{3\,[dBm]} = P_{out\,[dBm]} + \frac{\Delta_{[dB]}}{2}$$ (17.16)

17.12 IP3 MEASUREMENTS

A block diagram of output IP3 measurements setup is shown in Figure 17.13. An IP3 setup consists of two sweep generators, DUT and a spectrum analyzer. In an IP3 test we inject two signals to DUT and measure the intermodulation signals. Test results listed in Table 17.3 present IP3 measurements of a receiving channel. Input and output measured IP3 may be calculated by using Equations 17.16 to 17.18.

$$IP3_{out} = P_{F1} + \left(\left(P_{F1} + P_{F2} \right)/2 \right) - P_{IM1})/2$$ (17.17)

$$IP3_{IN} = \frac{\left(P_{F1} + P_{F2} \right)}{2} + \frac{\left(\dfrac{\left(P_{F1} + P_{F2} \right)}{2} - P_{IM2} \right)}{2} - \left(\frac{\left(P_{F1} + P_{F2} \right)}{2} - P_{in} \right)$$ (17.18)

IP3

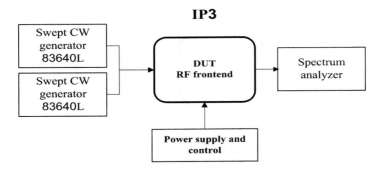

FIGURE 17.13 Two tone measurements.

TABLE 17.3
Third intercept point measurements

Pin (dBm)	19.999 (GHz) IM1	F1 = 20 (GHz)	F2 = 20.001 (GHz)	20.002 (GHz) IM2	IP3 OUT	IP3 INPUT
B	C (dBm)	D (dBm)	E (dBm)	F (dBm)	(dBm)	(dBm)
−17	−10.8	10	10	−13.8	20.4	−5.1
Pin (dBm)	29.999 (GHz) IM1	F1 = 30 (GHz)	F2 = 30.001 (GHz)	30.002 (GHz) IM2	IP3 OUT	IP3 INPUT
−14.50	−17	10	10	−18.8	23.5	−0.1
Pin (dBm)	39.998 (GHz) IM1	F1 = 39.999 (GHz)	F2 = 40 (GHz)	40.001 (GHz) IM2	IP3 OUT	IP3 INPUT
−14.5	−12	10	10	−11.2	21	−3.9

As an example, the first signal is at 20 GHz with power level of 10 dBm. The second signal is at 20.001 GHz with power level of 10 dBm. The first intermodulation signal is at 19.999 GHz with power level of −10.8 dBm. The second intermodulation signal is at 20.001 GHz with power level of −13.8 dBm. The power level of IP3 output is 20.4 dBm. The power level of IP3 at the input is −5.1 dBm.

17.13 NOISE FIGURE MEASUREMENTS

A block diagram of noise figure measurements setup is shown in Figure 17.14. The noise figure measurements setup consists of a noise source, DUT, amplifier and a spectrum analyzer. The noise level is measured without the DUT as a calibration level. We measure the difference, Delta (Δ) value, in the noise figure when the noise source is on to the measured noise figure when the noise source is off.

$$NF = 10LOG\left(10^{0.1*EN}\right)\Big/\left(\left(10^{0.1*\Delta}\right)-1\right) \qquad (17.19)$$

Where ENR is listed on the noise source for a given frequency. Delta, Δ, is the difference in the noise figure measurement when the noise source is on to the noise

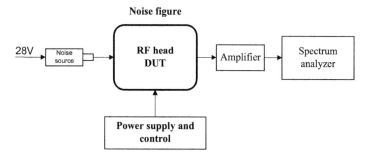

FIGURE 17.14 Noise figure measurements setup.

TABLE 17.4
Noise figure measurements

Parameter	Measurement 1	Measurement 2	Measurement 3
ENR	23.62	23.62	24
Delta [dB]	14	15	14
NF [dB]	9.777	8.740	10.159

figure measurement when the noise source is off. Measured NF is calculated by using Equation 17.19. Noise figure measurements are listed in Table 17.4.

17.14 ANTENNA MEASUREMENTS

Typical parameters of antennas are radiation pattern, gain, directivity, beam width, polarization and impedance. During antenna measurements we ensure that the antenna meets the required specifications and we can characterize the antenna parameters.

17.14.1 Radiation Pattern Measurements

A radiation pattern is the antenna's radiated field as a function of the direction in space. The radiated field is measured at various angles at a constant distance from the antenna. The radiation pattern of an antenna can be defined as the locus of all points where the emitted power per unit surface is the same. The radiated power per unit surface is proportional to the square of the electric field of the EM wave. The radiation pattern is the locus of points with the same electrical field strength. Usually the antenna's radiation pattern is measured in a far-field antenna range. The antenna under test is placed at a far-field distance from the transmitting antenna. Due to the size required to create a far-field range for large antennas, near-field techniques are employed. Near-field techniques allow measurement of the fields on a surface close to the antenna (usually 3–10 wavelength). Near field is transferred to far field by using Fourier transform.

The far-field distance or Fraunhofer distance, R, is given in Equation 17.20.

$$R = 2D^2 \big/ \lambda \qquad (17.20)$$

Where D is the maximum antenna dimension and λ is the antenna wavelength.

The radiation pattern graphs can be drawn using Cartesian (rectangular) coordinates as shown in Figure 17.15, see [1–5]. A polar radiation pattern plot is shown in Figure 17.16. The polar plot is useful to measure the beam width, which is the angle at the −3 dB points around the maximum gain. A 3D radiation pattern of a loop antenna is shown in Figure 17.17.

Main beam – Main beam is the region around the direction of maximum radiation, usually the region that is within 3 dB of the peak of the main lobe.

Beam width – Beam width is the angular range of the antenna pattern in which at least half of the maximum power is emitted. This angular range, of the major lobe, is

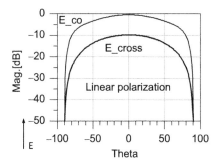

FIGURE 17.15   Radiation pattern of folded dipole dual polarized antenna.

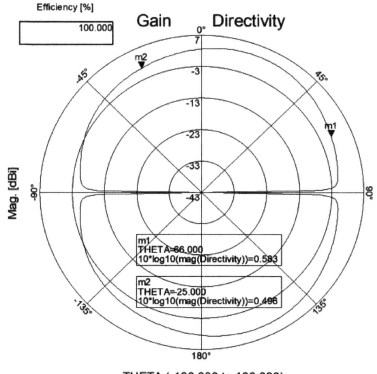

FIGURE 17.16   Radiation pattern of a wideband wearable printed slot antenna at 2 GHz.

defined as the points at which the field strength falls around 3 dB regarding to the maximum field strength.

Side lobes level – side lobes are smaller beams that are away from the main beam. Side lobes present radiation in undesired directions. The side lobe level is a parameter used to characterize the antenna's radiation pattern. It is the maximum value of the side lobes away from the main beam and is expressed usually in decibels.

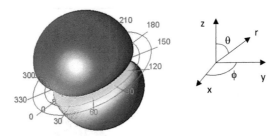

FIGURE 17.17 Loop antenna three-dimensional radiation pattern.

Radiated power – Total radiated power when the antenna is excited by a current or voltage of known intensity.

17.14.2 Directivity and Antenna Effective Area (Aeff)

Antenna directivity is the ratio between the amounts of energy propagating in a certain direction compared to the average energy radiated to all directions over a sphere as given in Equation 17.21, see [1–4].

$$D = \frac{P(\theta, \phi)\,\text{maximal}}{P(\theta, \phi)\,\text{average}} = 4\pi \frac{P(\theta, \phi)\,\text{maximal}}{P\,\text{rad}} \tag{17.21}$$

Where, $P(\theta, \phi)$average = $1/(4\pi) \iint P(\theta, \phi) \sin \theta \, d\theta \, d\phi$ = (P rad)/(4π)
An approximation used to calculate antenna directivity is given in Equation 17.22.

$$D \sim \frac{4\pi}{\theta E \times \theta H} \tag{17.22}$$

θE – Measured beam width *in radian in elevation* (EL) *plane*

θH – Measured beam width *in azimuth* (AZ) *plane*

Measured beam width *in radian in AZ plane and in EL plane allow us to calculate antenna directivity*.

Aeff – The antenna area which contributes to the antenna directivity is given in Equation 17.23.

$$\text{Aeff} = \frac{D\lambda^2}{4\pi} \sim \frac{\lambda^2}{\theta E \times \theta H} \tag{17.23}$$

17.14.3 Radiation Efficiency (α)

Radiation efficiency is the ratio of power radiated to the total input power, $\alpha = G/D$. The efficiency of an antenna takes into account losses and is equal to the total

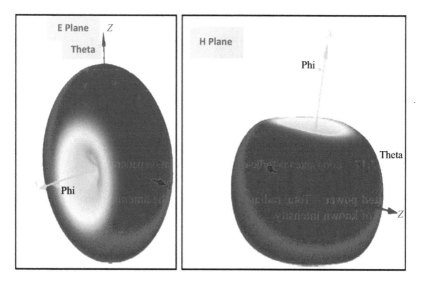

FIGURE 17.18 E and H plane radiation pattern of loop antenna in free space.

radiated power divided by the radiated power of an ideal lossless antenna. Efficiency is equal to the radiation resistance divided by total resistance (real part) of the feed-point impedance. Efficiency is defined as the ratio of the power that is radiated to the total power used by the antenna as given in Equation 17.24. Total power is equal to power radiated plus power loss.

$$\alpha = \frac{P_r}{P_r + P_l} \tag{17.24}$$

E and H plane radiation pattern of a wire loop antenna in free space is shown in Figure 17.18.

17.14.4 TYPICAL ANTENNA RADIATION PATTERN

A typical antenna radiation pattern is shown in Figure 17.19. The antenna main beam width is measured between the points that the maximum relative field intensity

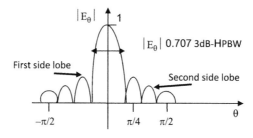

FIGURE 17.19 Antenna typical radiation pattern.

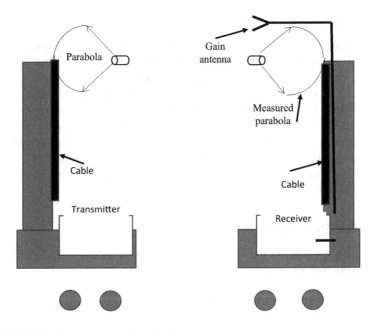

FIGURE 17.20 Antenna range for radiation pattern measurements.

E decays to, 0.707E. Half of the radiated power is concentrated in the antenna main beam. The antenna main beam is called 3 dB beam width. Radiation to undesired direction is concentrated in the antenna side lobes.

An antenna's radiation pattern is usually measured in free space ranges. An elevated free space range is shown in Figure 17.20. An anechoic chamber is shown in Figure 17.21.

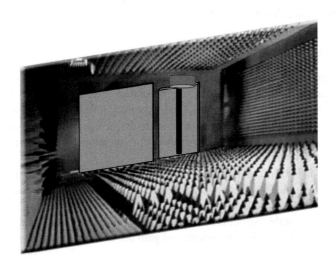

FIGURE 17.21 Anechoic chamber.

17.14.5 GAIN MEASUREMENTS

Antenna gain (G) – The ratio between the amounts of energy propagating in a certain direction compared to the energy that would be propagating in the same direction if the antenna were not directional, isotropic radiator, is known as its gain.

Figure 17.20 presents antenna far field range for radiation pattern measurements. Antenna gain is measured by comparing the field strength measured by the antenna under test to the field strength measured by a standard gain horn as shown in Figure 17.20. The gain as a function of frequency of the standard gain horn is supplied by the standard gain horn manufacturer. Figure 17.21 presents an anechoic chamber used in indoor antenna measurements. The chamber's metallic walls are covered with absorbing materials.

17.15 ANTENNA RANGE SETUP

An antenna range setup is shown in Figure 17.20. An antenna range setup consists of the following instruments:

- Transmitting system that consists of a wideband signal generator and transmitting antenna.
- Measured receiving antenna.
- Receiver.
- Positioning system.
- Recorder and plotter.
- Computer and data processing system.

The signal generator should be stable with controlled frequency value, good spectral purity and controlled power level. A low cost receiving system consists of a detector and amplifiers. Several companies sell antenna measurement setups, such as Agilent, Tektronix, Anritsu and others.

17.16 CONCLUSIONS

In this chapter, electromagnetics, microwave engineering, wearable systems and antennas design and measurements are presented. Electromagnetics commercial CAD software is presented in this chapter. Electromagnetics CAD software improves the design and development of wearable systems for communication and medical applications. The S parameters of wearable communication systems are measured by using a network analyzer setup.

Wearable antennas' radiation characteristics on a human body may be measured by using a phantom. The effect of the antenna's location on a human body should be considered in the antenna design process.

Setups for wearable systems and antennas measurement were presented in this chapter. An antenna's radiation pattern and gain are measured in a far–field antenna range.

REFERENCES

[1] Balanis, C.A., *Antenna Theory: Analysis and Design*, 2nd edition, Wiley, NJ, USA, 1996.

[2] Godara, L.C. Ed., *Handbook of Antennas in Wireless Communications*, CRC Press LLC, FL, USA, 2002.

[3] Kraus, J.D. and Marhefka, R.J., *Antennas for All Applications*, 3rd edition, McGraw Hill, New York, USA, 2002.

[4] Sabban, A., *RF Engineering, Microwave and Antennas*, Saar Publication, Tel Aviv, Israel, 2014.

[5] Sabban, A., *Low Visibility Antennas for Communication Systems*, Taylor & Francis Group, New York, USA, 2015.

[6] Sabban, A., *Wideband RF Technologies and Antenna in Microwave Frequencies*, Taylor & Francis Group, New York, USA, 2016.

[7] Sabban, A., Small Wearable Meta Materials Antennas for Medical Systems, *The Applied Computational Electromagnetics Society Journal*, 31(4), 434–443, April 2016.

[8] Sabban, A., New Compact Wearable Meta-material Antennas, *Global Journal for Research and Analysis*, IV, 268–271, August 2015.

[9] Sabban, A., New Wideband Meta Materials Printed Antennas for Medical Applications, *Journal of Advance in Medical Science (AMS)*, 3, 1–10, April 2015. Invited paper.

[10] Sabban, A., New wideband printed antennas for medical applications, *IEEE Journal, Transactions on Antennas and Propagation*, 61(1), 84–91, January 2013.

[11] Sabban, A., Comprehensive Study of Printed Antennas on Human Body for Medical Applications, *Journal of Advance in Medical Science (AMS)* 1, 1–10, February 2013.

[12] Sabban, A., Wearable Antennas, in *Advancements in Microstrip and Printed Antennas*, Ahmed Kishk (Ed.)., Intech, Croatia, 2013. ISBN 980-953-307-543-8, http://www.intechopen.com/books/show/title/advancement-in-microstrip-antennas-with-recent-applications.

[13] Sabban, A., Microstrip antenna arrays, in *Microstrip Antennas*, Nasimuddin Nasimuddin (Ed.), InTech, Croatia, 361–384, 2011, ISBN: 978-953-307-247-0, Available from: http://www.intechopen.com/articles/show/title/microstrip-antenna-arrays.

[14] Sabban, A., Dually Polarized Tunable Printed Antennas for Medical Applications, IEEE European Antennas and Propagation Conference, EUCAP 2015, Lisbon, Portugal, April 2015.

[15] Sabban, A., New Microstrip Meta Materials Antennas, IEEE Antennas and Propagation, Memphis, USA, July 2014.

[16] Sabban, A., Wearable Antennas for Medical Applications, IEEE BodyNet 2013, Boston, USA, October 2013, 1–7.

[17] Sabban, A., New Meta Materials Antennas, IEEE Antennas and Propagation, Orlando, USA, July 2013.

[18] Sabban, A., Meta Materials Antennas, *New Tech Magazine*, Tel Aviv, Israel, June 2013, 16–19.

[19] A. Sabban, Wideband Tunable Printed Antennas for Medical Applications, IEEE APS/URSI Conference, Chicago, IL, USA, July 2012, 1–2.

[20] Sabban, A., MM Wave Microstrip Antenna Arrays, *New Tech Magazine*, Tel Aviv, Israel, June 2012, 16–21.

[21] Sabban, A., New Compact Wideband Printed Antennas for Medical Applications, IEEE APS/URSI Conference, Spokane Washington USA, July 2011, 251–254.

[22] Sabban, A., Interaction between New Printed Antennas and Human Body in Medical applications, Asia Pacific Symp., Japan, December 2010, 187–190.

[23] Sabban, A., Wideband Printed Antennas for Medical Applications, APMC 2009 Conference, Singapore, 12/2009, 393–396

[24] Ansys HFSS, https://www.ansys.com/products/electronics/ansys-hfss.

[25] Keysight ADS, https://www.keysight.com/en/pc-1297113/advanced-design-system-ads?nid=-34346.0&cc=IL&lc=eng.

[26] CST Software, https://www.3ds.com/products-services/simulia/products/cst-studio-suite/.

[27] Microwave Office AWR, https://www.awr.com/serve/microwave-office-brochure-1.

18 Wearable Antennas in Vicinity of Human Body for 5G, IOT and Medical Applications

Albert Sabban

CONTENTS

18.1 INTRODUCTION

This chapter presents the analysis and measurements of wearable antennas in the vicinity of a human body. The electrical properties of human body tissues have a significant effect on the electrical characteristics of wearable antennas, see [1–15]. Printed antennas and compact printed wearable antennas were presented in books and papers, see [16–21]. The antenna's input impedance variation as a function of distance from a body was computed by employing electromagnetic software [22]. The antenna's radiation characteristics on human body have been measured by using a phantom. The phantom's electrical characteristics represent the human body's electrical characteristics. The phantom contains a mix of 55% water, 44% sugar and 1% salt.

The antenna under test was placed on the phantom during the measurement of the antenna's radiation characteristics. S_{11} and S_{12} parameters were measured directly on a human body by using a network analyzer. The measured results were compared to a known reference antenna.

18.2 ANALYSIS OF WEARABLE ANTENNAS IN VICINITY OF HUMAN BODY

The major issue in the design of wearable antennas is the interaction between radio frequency (RF) transmission and the human body. Electrical properties of human body tissues have been used in the design of wearable antennas presented in this book. The dielectric constant and conductivity of human body tissues may be used to calculate the attenuation α of RF transmission through the human body, as given in Equations 18.1 and 18.2. Properties of human body tissues are listed in Table 18.1 see [8].

$$\gamma = \sqrt{j\omega\mu(\sigma + j\omega\varepsilon)} = \alpha + j\beta \tag{18.1}$$

$$\alpha = \mathrm{Re}(\gamma) \tag{18.2}$$

The attenuation α of RF transmission through stomach, skin and pancreas tissues in dB/cm is listed in Table 18.2. The attenuation α of RF transmission through stomach and pancreas tissues is around 2.2 dB/cm at 1 GHz.

Figure 18.1 presents attenuation via several human tissues. Stomach tissue attenuation at 500 MHz is around 1.6 dB/cm. There is a good agreement between measured and computed results. The attenuation α of RF transmission through blood, fat and

TABLE 18.1
Properties of human body tissues

Tissue	Property	434 MHz	800 MHz	1000 MHz
Prostate	σ	0.75	0.90	1.02
	ε	50.53	47.4	46.65
Stomach	σ	0.67	0.79	0.97
	ε	42.9	40.40	39.06
Colon, heart	σ	0.98	1.15	1.28
	ε	63.6	60.74	59.96
Kidney	σ	0.88	0.88	0.88
	ε	117.43	117.43	117.43
Nerve	σ	0.49	0.58	0.63
	ε	35.71	33.68	33.15
Fat	σ	0.045	0.056	0.06
	ε	5.02	4.58	4.52
Lung	σ	0.27	0.27	0.27
	ε	38.4	38.4	38.4

TABLE 18.2

The attenuation, α, of radio frequency transmission through stomach, skin and pancreas tissues

Frequency (MHz)	α Stomach (dB/cm)	α Skin (dB/cm)	α Pancreas (/cm)
150	1.09643156	0.99356488	1.15470908
200	1.20939308	1.08475696	1.2702312
250	1.31292812	1.16787664	1.37479916
300	1.41391992	1.24912144	1.4755566
350	1.48670172	1.30428284	1.54858144
400	1.55709652	1.35728292	1.61897624
450	1.62247428	1.40642908	1.68601188
500	1.67844292	1.44867464	1.74763988
550	1.73350016	1.49011296	1.80851272
600	1.78806264	1.53108256	1.86901232
650	1.83840664	1.57730356	1.9296508
700	1.88836872	1.62329888	1.99015908
750	1.93809644	1.669164	2.05065868
800	1.98771132	1.71496836	2.11124508
850	2.0487404	1.7427704	2.15668488
900	2.1098476	1.77030336	2.20203788
950	2.17106764	1.79764536	2.2473822
1000	2.23242656	1.82481376	2.29276992

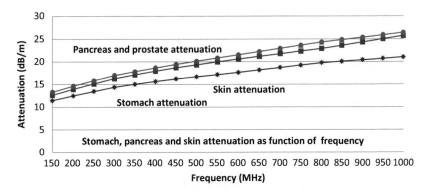

FIGURE 18.1 The attenuation, α, of radio frequency transmission through stomach, skin and pancreas tissues.

small intestine tissues in dB/m is listed in Table 18.3. The attenuation α of RF transmission through blood tissues is around 4.6 dB/cm at 1 GHz. However, the attenuation α of RF transmission through fat tissues is around 0.45 dB/cm at 1 GHz. Figure 18.2 presents attenuation of human blood tissues. Blood tissue attenuation at 500 MHz is around 4 dB/cm.

TABLE 18.3

The attenuation α of radio frequency transmission through blood, fat and small intestine tissues

Frequency (MHz)	α Blood (dB/m)	α Fat (dB/m)	α Small intestine (dB/m)
150	26.5193	1.8752	16.7351
200	29.4816	2.30063	21.4619
250	31.9065	2.59202	23.9958
300	34.0036	2.85079	25.4946
350	35.5104	3.10219	26.4352
400	36.7922	3.35519	27.0538
450	37.9234	3.49384	27.4771
500	38.9634	3.62996	27.7771
550	39.9081	3.77162	27.996
600	40.7788	3.92648	28.16
650	41.5575	4.08133	28.2857
700	42.2879	4.23651	28.384
750	42.9794	4.28321	28.462
800	43.6395	4.32899	28.5251
850	44.3448	4.3741	28.5766
900	45.0293	4.41874	28.6193
950	45.6965	4.4789	28.655
1000	46.3492	4.53912	28.6851

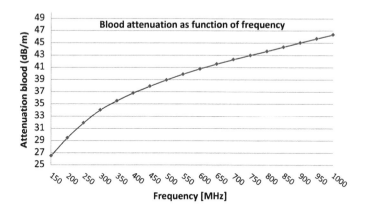

FIGURE 18.2   The attenuation α of radio frequency transmission through blood.

18.3 DESIGN OF WEARABLE ANTENNAS IN PRESENCE OF HUMAN BODY

An antenna's input impedance variation as a function of distance from a body had been computed by employing Momentum software. The analyzed structure is presented in Figure 18.3a. These properties were employed in the antenna design. The antenna was placed inside a belt with thickness between 1 and 4 mm as shown in

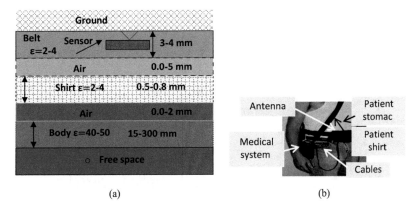

(a)   (b)

FIGURE 18.3 (a) Analyzed structure model, (b) medical system on patient.

Figure 18.3b. The patient's body thickness was varied from 15 to 300 mm. The dielectric constant of the body was varied from 40 to 50. The antenna was placed inside a belt with thickness between 2 and 4 mm with dielectric constant from 2 to 4. The air layer between the belt and the patient's shirt may vary from 0 to 8 mm. The shirt thickness was varied from 0.5 to 1 mm. The dielectric constant of the shirt was varied from 2 to 4. Figure 18.4 presents S_{11} results (of the antenna shown in Figure 4.5) for different belt thicknesses, shirt thicknesses and air spacing between the antennas and a human body. One may conclude from the results shown in Figure 18.4 that the antenna has voltage standing wave ratio (VSWR) better than 2.5:1 for air spacing up to 8 mm between the antennas and the patient's body. For frequencies ranging from 415 to 445 MHz the antenna has VSWR better than 2:1 when there is no air spacing between the antenna and the patient's body. Results shown in Figure 18.5 indicate that the folded antenna (the antenna shown in Figure 4.9) has VSWR better than 2.0:1 for air spacing up to 5 mm between the antennas and the patient's body. Figure 18.6 presents S_{11} results of the folded antenna's results for different positions relative to the human body. An explanation of Figure 18.5 is given in Table 18.4. If the air

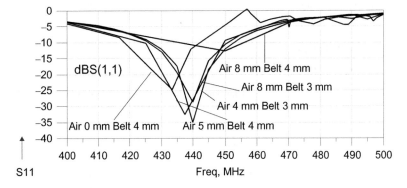

FIGURE 18.4 S_{11} results for different antenna positions relative to the human body.

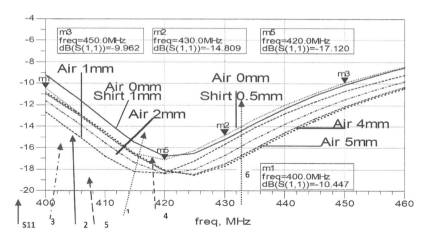

FIGURE 18.5 Folded antenna S_{11} results for different antenna positions relative to the human body.

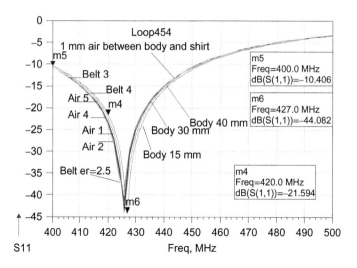

FIGURE 18.6 Loop antenna S_{11} results for different antenna positions relative to the body.

TABLE 18.4
Explanation of Figure 18.5

Picture #	Line type	Sensor position
1	Dot ······	Shirt thickness 0.5 mm
2	Line ——	Shirt thickness 1 mm
3	Dash dot -·-·	Air spacing 2 mm
4	Dash ---	Air spacing 4 mm
5	Long dash – –	Air spacing 1 mm
6	Big dots ••••	Air spacing 5 mm

TABLE 18.5
Explanation of Figure 18.6

Plot color	Sensor position
Red	Body 15 mm air spacing 0 mm
Blue	Air spacing 5 mm Body 15 mm
Pink	Body 40 mm air spacing 0 mm
Green	Body 30 mm air spacing 0 mm
Sky	Body 15 mm Air spacing 2 mm
Purple	Body 15 mm Air spacing 4 mm

spacing between the sensors and the human body is increased from 0 to 5 mm the antenna's resonant frequency is shifted by 5%. The loop antenna, see Figure 4.20, has VSWR better than 2.0:1 for air spacing up to 5 mm between the antennas and the patient's body, as presented in Figure 18.6. If the air spacing between the sensors and the human body is increased from 0 to 5 mm the computed antenna's resonant frequency is shifted by 2%. However, if the air spacing between the sensors and the human body is increased up to 5 mm, the measured loop antenna's resonant frequency is shifted by 5%. An explanation of Figure 18.6 is given in Table 18.5.

18.4 WEARABLE ANTENNA ARRAYS FOR MEDICAL AND 5G APPLICATIONS

An application of the proposed antenna is shown in Figure 18.7a. Three to four folded dipole or loop antennas may be assembled in a belt and attached to the patient's stomach, see Figure 18.7b. The cable from each antenna is connected to a recorder. The received signal is routed to a switching matrix. The signal with the highest level is selected during the medical test. The antennas receive a signal that is transmitted from various positions in the human body. Folded antennas may also be attached on the patient's back in order to improve the level of the received signal from different

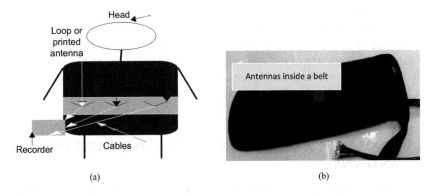

FIGURE 18.7 (a) Printed wearable antenna, (b) wearable antennas inside a belt.

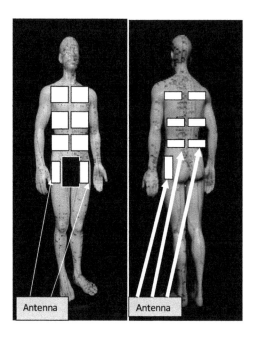

FIGURE 18.8   Printed patch antenna locations for various medical applications.

locations in the human body. Figure 18.8 shows various antenna locations on the back and front of the human body for different medical applications. In several applications the distance separating the transmitting and receiving antennas is less than $2D^2/\lambda$. D is the largest dimension of the radiator. In these applications the amplitude of the electromagnetic field close to the antenna may be quite powerful, but because of rapid fall-off with distance, the antenna does not radiate energy to infinite distances, but instead the radiated power remains trapped in the region near to the antenna. Thus, the near fields only transfer energy to close distances from the receivers. The receiving and transmitting antennas are magnetically coupled. A change in current flow through one wire induces a voltage across the ends of the other wire through electromagnetic induction. The amount of inductive coupling between two conductors is measured by their mutual inductance. In these applications we have to refer to the near field and not to the far field radiation.

In Figure 18.9a several microstrip antennas for medical applications at 434 MHz are shown. The backside of the antennas is presented in Figure 18.9b. The diameter of the loop antenna presented in Figure 18.10 is 50 mm. The dimensions of the folded dipole antenna are 7 × 6 × 0.16 cm. The dimensions of the compact folded dipole presented in Figure 18.10 are 5 × 5 × 0.05 cm.

18.5 SMALL WIDE BAND DUAL POLARIZED WEARABLE PRINTED ANTENNA

A small microstrip loaded dipole antenna has been designed. The antenna consists of two layers. The first layer consists of flame retardant-4 (FR-4) 0.4 mm dielectric

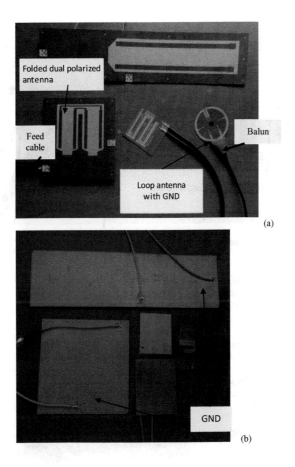

FIGURE 18.9   (a) Microstrip antennas for medical applications, (b) backside of the antennas.

substrate. The second layer consists of Kapton 0.4 mm dielectric substrate. The substrate thickness determines the antenna bandwidth. However, with thinner substrate we may achieve better flexibility. The proposed antenna is dual polarized. The printed dipole and the slot antenna provide dual orthogonal polarizations. The dual polarized antenna dimensions are shown in Figure 18.11a. The antenna dimensions are 4 × 4 × 0.08 cm. The dual polarized antenna layout is shown in Figure 18.11b.

The antenna may be attached to the patient's shirt on the patient's stomach or back zone. The antenna has been analyzed by using Keysight Momentum software. There is a good agreement between measured and computed results. The antenna bandwidth is around 15% for VSWR better than 2:1. The antenna beam width is around 100°. The antenna gain is around 0 dBi. The computed S_{11} parameters are presented in Figure 18.12. Figure 18.13 presents the antenna measured S_{11} parameters on human body. The antenna cross-polarized field strength may be adjusted by varying the slot feed location. The computed 3D radiation pattern of the antenna is shown in Figure 18.14. The computed radiation pattern is shown in Figure 18.15.

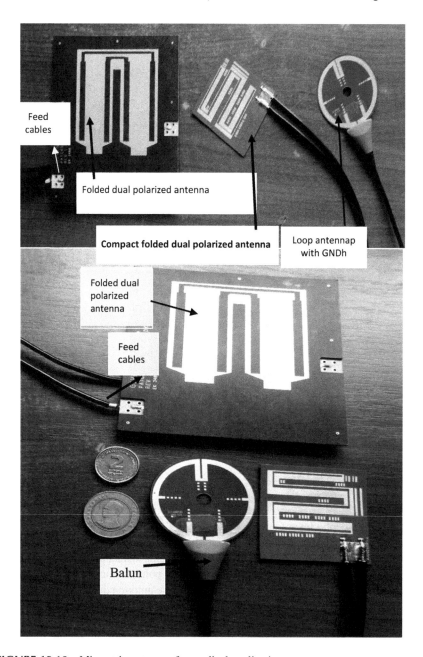

FIGURE 18.10 Microstrip antennas for medical applications.

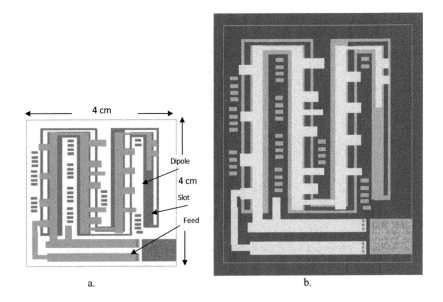

FIGURE 18.11   Printed compact dual polarized antenna. (a) Antenna dimensions, (b) layout.

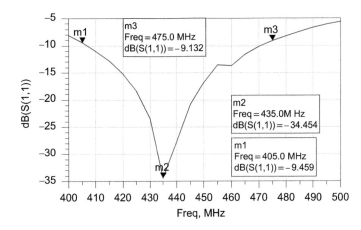

FIGURE 18.12   Computed S_{11} results of the wide band antenna.

18.6 WEARABLE HELIX ANTENNA PERFORMANCE ON HUMAN BODY

In order to compare the variation of the antenna's input impedance as a function of distance from the body to other antennas, a helix antenna has been designed. A helix antenna with nine turns is shown in Figure 18.16. A photo of the helix antenna with nine turns is shown in Figure 18.17. The wearable helix antenna matching network was printed on a dielectric substrate with dielectric constant of 3.55. The backside of the circuit is copper under the microstrip matching stubs. However, in the helix

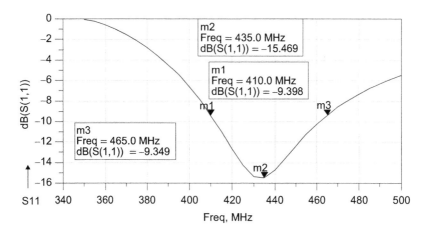

FIGURE 18.13   Measured S_{11} of the small wide band antenna on human body.

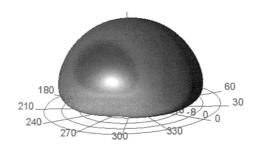

FIGURE 18.14   Small wide band antenna three-dimensional radiation pattern.

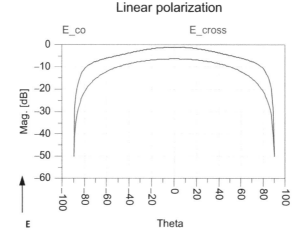

FIGURE 18.15   Small dual polarized antenna radiation pattern.

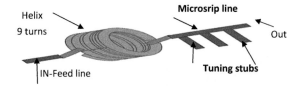

FIGURE 18.16 Wearable helix antenna layout

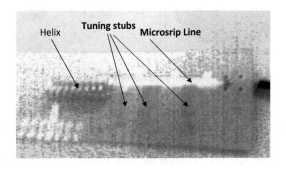

FIGURE 18.17 Helix antenna for medical applications.

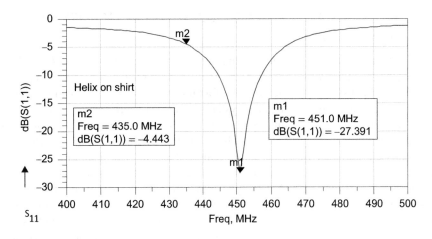

FIGURE 18.18 Measured S_{11} of the helix antenna on human body.

antenna area there is no ground plane. The antenna has been designed to operate on a human body. A matching microstrip line network has been designed on RO4003 substrate with 0.8 thickness. The helix antenna has VSWR better than 3:1 at the frequency range from 440 to 460 MHz. The antenna dimensions are 4 × 4 × 0.6 cm. Figure 18.18 presents the measured S_{11} parameters on a human body. The computed E and H radiation planes of the helix antenna are shown in Figure 18.19. The helix

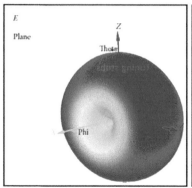
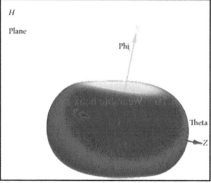

FIGURE 18.19   E and H plane radiation patterns of the helix antenna.

antenna's input impedance variation as a function of distance from the body is very sensitive. If the air spacing between the helix antenna and the human body is increased from 0 to 2 mm the antenna's resonant frequency is shifted by 5%.

However, if the air spacing between the dual polarized antenna and the human body is increased from 0 to 5 mm the antenna's resonant frequency is shifted only by 5%.

18.7 WEARABLE ANTENNA MEASUREMENTS IN VICINITY OF HUMAN BODY

This section presents measurement techniques of wearable antennas and RF medical systems in the vicinity of a human body. Measurement results of wearable antennas and RF medical systems in the vicinity of a human body were presented in [1–15]. Wearable antennas and RF medical systems' radiation characteristics on a human body may be measured by using a phantom. The phantom's electrical characteristics represent the human body's electrical characteristics. The phantom has a cylindrical shape with a diameter of 40 cm and a length of 1.5 m. The phantom's electrical characteristics are similar to a human body's electrical characteristics. The wearable antenna under test was placed on the phantom during the measurement of the antenna's radiation characteristics. The phantom was employed to compare the electrical performance of several new wearable antennas. The phantom was also employed to measure the electrical performance of several antenna belts in the vicinity of a human body.

18.8 PHANTOM CONFIGURATION

The phantom represents human body tissues. The phantom contains a mix of water, sugar and salt. The relative concentration of water, sugar and salt determines the electrical characteristics of the phantom environment. A mixture of 55% water, 44% sugar and 1% salt presents the electrical characteristics of stomach tissues. The phantom may be used to measure electromagnetic radiation from inside or outside the

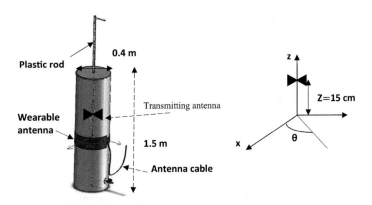

FIGURE 18.20 Phantom configuration.

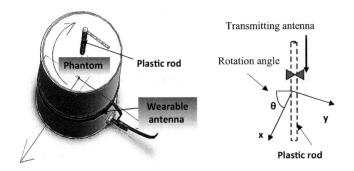

FIGURE 18.21 Transmitting antenna rotation.

phantom. The phantom is a fiberglass cylinder with 1.5 m height and 0.4 m diameter as shown in Figure 18.20. The thickness of the cylinder surface is around 2.5 mm. The phantom contains a plastic rod with 5 mm thickness. The position of the plastic rod inside the phantom may be adjusted. The plastic rod may be rotated as shown in Figure 18.21. A small transmitting antenna may be attached to the plastic rod at different height positions. The antenna may be rotated in the x-y plane.

18.9 MEASUREMENT OF WEARABLE ANTENNAS BY USING A PHANTOM

The electrical characteristics of several wearable antennas were measured by using the phantom. The position of the transmitting antennas along the z axis and x axis was varied as listed in Tables 18.6 to 18.10, ($Z = 0$, $Z = -15$ cm and $Z = 15$ cm). The angular angle of the transmitting antenna was varied from $0°$ to $270°$. Signal reception levels and immunity to noise were measured for several types of wearable antenna. The antennas' electrical performances were compared and the best antenna was chosen according to the system's electrical requirements.

Test procedure and process
The test procedure and process is described below.

Test procedure
The test checks two parameters of the antennas array:

- Signal reception levels.
- Immunity to noise.

Measured antennas
The test checks the following antenna arrays:

- Four-sensor antenna array in a belt, antennas in orientations of +45°, +45°, +45°, −45°.
- Sensor belt with four loop antennas in orientations of +45°, +45°, +45°, −45°.
- Sensor belt with four antennas in orientations of +45°, −45°, +45°, −45°.
- Thin belt with four-sensor array in orientations of +45°, −45°, +45°, −45°.
- Four-loop antenna array in a sleeve.

Test process

- Place the antenna array on the phantom as shown in Figure 18.21 and connect to the recorder.
- Signal reception levels.
- Place the transmitter in the phantom in the following coordinates, 5 minutes per each location.
- Z values: from −15 cm to +15 cm in 15 cm increments, 0 being the level of the antennas' center.
- X values: from −5 cm to −20 cm in 15 cm increments, 0 being the container's wall where the antennas are attached.
- θ values: from 0° to 270° in 90° increments, 0° being perpendicular to the middle of the antennas set.

Immunity to noise
Place the transmitter outside the phantom in the following coordinates, 5 minutes per each location:

Z values: from −40 cm to +40 cm in 10 cm increments, 0 being the level of the antennas' center.
X values: 100 cm from container's wall θ values: from 0° to 270° in 90° increments, 0° being perpendicular to the middle of the antennas set.

18.10 MEASUREMENT RESULTS OF WEARABLE ANTENNAS

The measurements of a five-antenna array configuration were measured. For all the configurations the lowest measured signal level was when the transmitting antenna was located at z = 15 cm and x = −5 cm. For all the configurations, the highest measured signal level was when the transmitting antenna was located at z = 0 cm and x = −5 cm.

18.10.1 Measurements of Antenna Array 1

The measured antenna consists of four loop antennas with a tuning capacitor as shown in Figure 18.22. The loop antennas are printed on FR-4 substrate with dielectric constant of 4.8 and 0.25 mm thickness. The loop radiators' orientations are: +45°, −45°, −45°, −45°. The antennas were inserted in a thin belt. Measurement results of antenna number 1 are listed in Table 18.6.

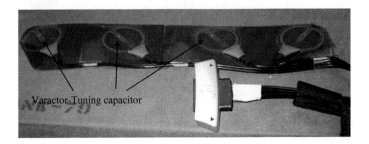

FIGURE 18.22   Sensor belt with four antennas in orientations of +45°, +45°, +45°, −45°.

TABLE 18.6
Measurements of antenna 1

	Angle θ			
Antenna 1 (X cm, Z cm)	**0°**	**90°**	**180°**	**270°**
Z = 0 cm				
Signal level (dB) X = −5, Z = 0	−60	−63	−81	−65
Signal level (dB) X = −20, Z = 0	−78	−70	−65	−74
Z = 15 cm				
Signal level (dB) X = −5, Z = 15	−81	−89	−83	−82
Signal level (dB) X = −20, Z = 15	−86	−89	−81	−85
Z = −15 cm				
Signal level (dB) X = −5, Z = −15	−70	−76	−81	−68
Signal level (dB) X = −20, Z = −15	−79	−64	−74	−81
	Angle θ			
Noise test	**0°**	**90°**	**180°**	**270°**
Z = 0 cm				
Signal level (dB)	−72	−74	−74	−74
Noise level (dB)	−95	−95	−95	−95
Z = 40 cm				
Signal level (dB)	−72	−74	−74	−74
Noise level (dB)	−88	−88	−88	−88
Z = −40 cm				
Signal level (dB)	−72	−74	−74	−74
Noise level (dB)	−94	−93	−94	−95

The highest signal level is at z = 0 X = −5 cm θ = 0°. At θ = 180° the signal level is lower by 21 dB. The lowest signal level is at z = ±15 cm X = −5 cm θ = 90°. The noise level is lower by 14 to 21 dB than the signal level.

18.10.2 MEASUREMENTS OF ANTENNA ARRAY 2

The measured antenna consists of four loop antennas without a tuning capacitor as shown in Figure 18.23. The loop radiators' orientations are: +45°, −45°, −45°, −45°. The antennas were inserted in a thin belt. Measurement results of antenna number 2 are listed in Table 18.7.

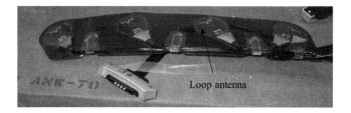

FIGURE 18.23 Sensor belt with four antennas in orientations of +45°, +45°, +45°, −45° without tuning capacitor.

TABLE 18.7
Measurements of antenna 2

Antenna 1 (X cm, Z cm)	Angle θ			
	0°	**90°**	**180°**	**270°**
Z = 0 cm				
Signal level (dB) X = −5, Z = 0	−63	−70	−82	−72
Signal level (dB) X = −20, Z = 0	−84	−66	−66	−79
Z = 15 cm				
Signal level (dB) X = −5, Z = 15	−82	−93	−90	−84
Signal level (dB) X = −20, Z = 15	−88	−83	−79	−90
Z = −15 cm				
Signal level (dB) X = −5, Z = −15	−74	−82	−85	−92
Signal level (dB) X = −20, Z = −15	−86	−70	−77	−81
Noise test	Angle θ			
	0°	**90°**	**180°**	**270°**
Z = 0 cm				
Signal level (dB)	−67	−69	−72	−68
Noise level (dB)	−85	−85	−86	−85
Z = 40 cm				
Signal level (dB)	−66	−69	−72	−68
Noise level (dB)	−80	−83	−86	−85
Z = −40 cm				
Signal level (dB)	−68	−69	−72	−69
Noise level (dB)	−84	−82	−83	−83

The highest signal level is at z = 0 cm X = −5 cm θ = 0°. At θ = 180° the signal level is lower by 19 dB. The lowest signal level is at z = 15 cm X = −5 cm θ = 90°. The noise level is lower by 14 to 18 dB than the signal level.

18.10.3 Measurements of Antenna Array 3

The measured antenna consists of four tuned loop antennas with a tuning capacitor as shown in Figure 18.24. The loop radiators' orientations are: +45°, −45°, +45°, −45°. The antennas were inserted in a belt.

Measurements results of antenna number 3 are listed in Table 18.8.

FIGURE 18.24 Sensor belt with four antennas in orientations of +45°, −45°, +45°, −45°.

TABLE 18.8
Measurements of antenna 3

	Angle θ			
Antenna 1 (X cm, Z cm)	**0°**	**90°**	**180°**	**270°**
Z = 0 cm				
Signal level (dB) X = −5, Z = 0	−63	−63	−82	−69
Signal level (dB) X = −20, Z = 0	−83	−70	−68	−76
Z = 15 cm				
Signal level (dB) X = −5, Z = 15	−85	−86	−85	−86
Signal level (dB) X = −20, Z = 15	−89	−88	−86	−85
Z = −15 cm				
Signal level (dB) X = −5, Z = −15	−72	−79	−83	−74
Signal level (dB) X = −20, Z = −15	−85	−69	−77	−82
	Angle θ			
Noise test	**0°**	**90°**	**180°**	**270°**
Z = 0 cm				
Signal level (dB)	−68	−70	−76	−70
Noise level (dB)	−91	−91	−92	−92
Z = 40 cm				
Signal level (dB)	−68	−70	−76	−70
Noise level (dB)	−90	−88	−90	−88
Z = −40 cm				
Signal level (dB)	−68	−70	−76	−70
Noise level (dB)	−89	−87	−90	−87

The highest signal level is at z = 0 cm X = −5 cm θ = 0°. At θ = 180° the signal level is lower by 19 dB. The lowest signal level is at z = 15 cm X = −20 cm θ = 0°. The noise level is lower by 18 to 23 dB than the signal level.

18.10.4 Measurements of Antenna Array 4 in a Thinner Belt

The measured antenna consists of four loop antennas with a tuning capacitor as shown in Figure 18.25. The loop radiators orientations are: +45°, −45°, +45°, −45°. The antennas were inserted in a thinner belt. Measurements results of antenna number 4 are listed in Table 18.9.

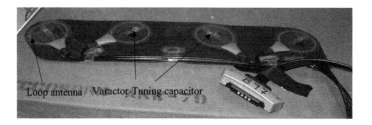

FIGURE 18.25   Four antennas in orientations of +45°, −45°, +45°, −45°, in a thinner belt.

TABLE 18.9
Measurements of antenna 4

	Angle θ			
Antenna 1 (X cm, Z cm)	**0°**	**90°**	**180°**	**270°**
Z = 0 cm				
Signal level (dB) X = −5, Z = 0	−61	−62	−81	−63
Signal level (dB) X = −20, Z = 0	−79	−69	−67	−73
Z = 15 cm				
Signal level (dB) X = −5, Z = 15	−86	−88	−88	−83
Signal level (dB) X = −20, Z = 15	−90	−84	−82	−86
Z = −15 cm				
Signal level (dB) X = −5, Z = −15	−70	−81	−82	−67
Signal level (dB) X = −20, Z = −15	−80	−67	−74	−81
	Angle θ			
Noise test	**0°**	**90°**	**180°**	**270°**
Z = 0 cm				
Signal level (dB)	−70	−70	−76	−70
Noise level (dB)	−95	−95	−95	−95
Z = 40 cm				
Signal level (dB)	−70	−70	−76	−70
Noise level (dB)	−92	−92	−91	−92
Z = −40 cm				
Signal level (dB)	−70	−70	−76	−70
Noise level (dB)	−92	−92	−91	−92

The highest signal level is at z = 0 cm X = −5 cm θ = 0°. At θ = 180° the signal level is lower by 20 dB. The lowest signal level is at z = 15 cm X = −20 cm θ = 0°. The noise level is lower by 18 to 25 dB than the signal level.

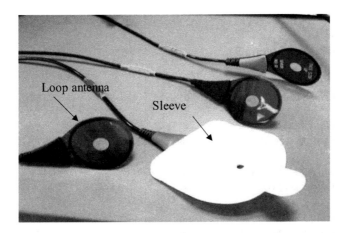

FIGURE 18.26 Four loop antennas in a sleeve, radiators' orientations are: (90°, 90°, 90°, 90°).

18.10.5 MEASUREMENTS OF ANTENNA ARRAY 5

The measured antenna consists of four loop antennas without a tuning capacitor inserted in a sleeve, as shown in Figure 18.26. The sleeve improves the antenna VSWR from 4:1 to 2:1. The loop radiators' orientations are: (90°, 90°, 90°, 90°).

The antennas were inserted in a thin sleeve. Measurements results of antenna number 5 are listed in Table 18.10.

Four compact antennas in a thinner belt are shown in Figure 18.27.

18.11 FABRICATION OF THE SENSOR BELT ARRAY

Fabricated wearable antenna belt arrays' and RF medical systems' electrical and radiation characteristics on a human body may be measured by using a compact phantom. The phantom's electrical characteristics represent the human body's electrical characteristics. The phantom has a bottle cylindrical shape with a 20 cm diameter and a length around 50 cm. The phantom's electrical characteristics are similar to the human body's electrical characteristics. The manufactured wearable antennas under test were placed on the phantom during the measurement of the antenna S_{11} parameter. The phantom's electrical characteristics represent the human body's electrical characteristics. The phantom contains a mix of 55% water, 44% sugar and 1% salt. A photo of the fabricated sensor belt array on the phantom is shown in Figure 18.28. A photo of the network analyzer used to measure the sensor belt array S_{11} parameter is shown in Figure 18.29. The S_{11} parameter of the fabricated sensor is compared to the S_{11} parameter of a calibration sensor as shown in Figure 18.30. The measured S_{11} of the calibration sensor is shown in Figure 18.31. A photo of the measured fabricated sensor belt array is shown in Figure 18.32. S_{11} test results of the fabricated sensor belt array are listed in Table 18.11. The S_{11} parameter of the fabricated sensor is shown in Figure 18.33.

TABLE 18.10

Measurements of antenna 5

Antenna 1 (X cm, Z cm)	Angle θ			
	0°	**90°**	**180°**	**270°**
Z = 0 cm				
Signal level (dB) X = −5, Z = 0	−67	−52	−58	−60
Signal level (dB) X = −20, Z = 0	−68	−70	−77	−78
Z = 15 cm				
Signal level (dB) X = −5, Z = 15	−90	−86	−92	−90
Signal level (dB) X = −20, Z = 15	−84	−86	−90	−86
Z = −15 cm				
Signal level (dB) X = −5, Z = −15	−85	−90	−85	−92
Signal level (dB) X = −20, Z = −15	−90	−85	−75	−78

Noise test	Angle θ			
	0°	**90°**	**180°**	**270°**
Z = 0 cm				
Signal level (dB)	−70	70	−70	−70
Noise level (dB)	−90	−90	−90	−90
Z = 40 cm				
Signal level (dB)	−72	−72	−72	−72
Noise level (dB)	−90	−90	−90	−90
Z = −40 cm				
Signal level (dB)	−73	−73	−73	−73
Noise level (dB)	−90	−90	−90	−90

The highest signal level is at z = 0 cm X = −5 cm θ = 90°. At θ = 180° the signal level is lower by 6 dB. At θ = 270° the signal level is lower by 8 dB. The lowest signal level is at z = 15 cm X = −20 cm θ = 180°. The noise level is lower by 18 to 22 dB than the signal level.

FIGURE 18.27   Four compact antennas in orientations of 45°, −45°, 0°, 90°, in a thinner belt.

FIGURE 18.28 A photo of the fabricated sensor belt array on the phantom.

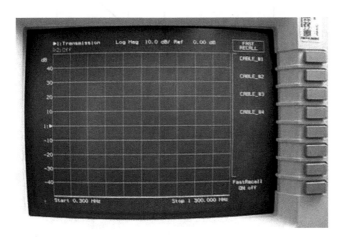

FIGURE 18.29 A photo of the network analyzer used to measure the sensor belt array S_{11} parameter.

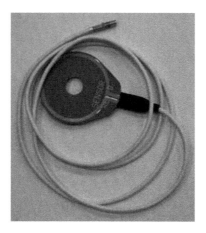

FIGURE 18.30   A photo of the calibration sensor.

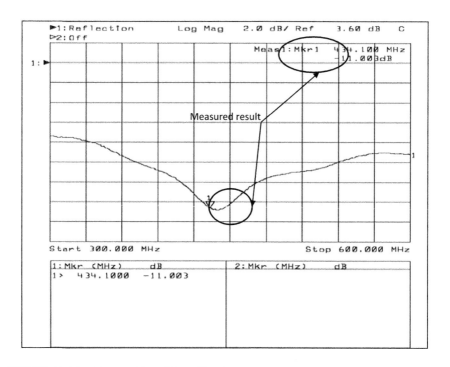

FIGURE 18.31   Measured S_{11} of the calibration sensor.

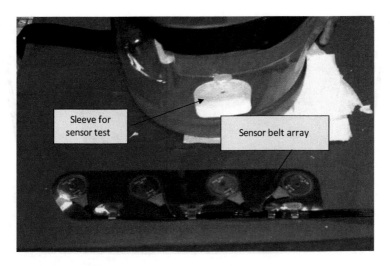

FIGURE 18.32   A photo of the fabricated sensor belt array.

TABLE 18.11
S_{11} test results of the fabricated sensor belt array

Sensor belt array – belt identification

Sensor belt array	Measured return loss (S_{11}) (dB)	Return loss (S_{11}) (dB) required
1	−4.5	<−4.5
2	−5	<−4
3	−4	<−4
4	−5	<−4.5

18.12 CONCLUSIONS

This chapter presents the analysis and measurements of wearable antennas in the vicinity of a human body. The antenna S_{11} results for different belt thicknesses, shirt thicknesses and air spacing between the antennas and a human body are presented in this chapter. If the air spacing between the dual polarized antenna and the human body is increased from 0 to 5 mm the antenna's resonant frequency is shifted by 5%. However, if the air spacing between the helix antenna and the human body is increased only from 0 to 2 mm the antenna's resonant frequency is shifted by 5%. The effect of the antenna's location on the human body should be considered in the antenna design process. The proposed antenna may be used in medical RF systems.

An antenna's radiation characteristics on a human body have been measured by using a phantom. The phantom's electrical characteristics represent the human body's

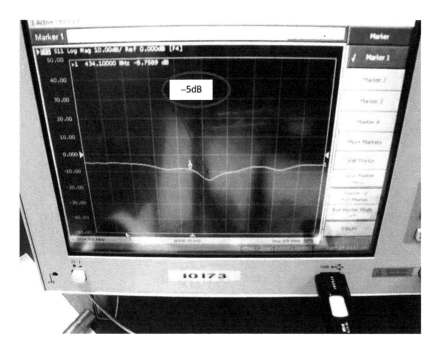

FIGURE 18.33   Measured S_{11} of the measured sensor.

electrical characteristics. The phantom contains a mix of 55% water, 44% sugar and 1% salt. The antenna under test was placed on the phantom during the measurement of the antenna's radiation characteristics. S_{11} and S_{12} parameters were measured directly on a human body by using a network analyzer.

REFERENCES

[1] A. Sabban, *Novel Wearable Antennas for Communication and Medical Systems*, Taylor & Francis Group, Boca Raton, FL, October 2017.

[2] A. Sabban, *Wideband RF Technologies and Antenna in Microwave Frequencies*, John Wiley & Sons, New York, July 2016.

[3] A. Sabban, *Low Visibility Antennas for Communication Systems*, Taylor & Francis Group, New York, 2015.

[4] A. Sabban, "Small Wearable Meta Materials Antennas for Medical Systems," *The Applied Computational Electromagnetics Society Journal*, 31, 4, 434–443, April 2016.

[5] A. Sabban, Dual Polarized Dipole Wearable Antenna, U.S. Patent number: 8203497, June 19, 2012.

[6] A. Sabban, "New Wideband Printed Antennas for Medical Applications," *IEEE Transactions on Antennas and Propagation*, 61(1), 84–91, January 2013.

[7] A. Sabban, *Microstrip Antenna Arrays*; N. Nasimuddin (Ed.), *Microstrip Antennas*, InTech, New York, pp. 361–384, 2011, ISBN: 978-953-307-247-0, http://www.intechopen.com/articles/show/title/microstrip-antenna-arrays.

[8] L. C. Chirwa, P. A. Hammond, S. Roy, and D. R. S. Cumming, "Electromagnetic Radiation from Ingested Sources in the Human Intestine between 150 MHz and 1.2 GHz," *IEEE Transaction on Biomedical Engineering*, 50(4), 484–492, April 2003.

[9] D. Werber, A. Schwentner, and E. M. Biebl, "Investigation of RF Transmission Properties of Human Tissues," *Advances in Radio Science*, 4, 357–360, 2006.

[10] K. Fujimoto and J.R. James (Eds.), *Mobile Antenna Systems Handbook*, Artech House, Boston, MA, 1994.

[11] B. Gupta, S. Sankaralingam, and S. Dhar, "Development of Wearable and Implantable Antennas in the Last Decade," in *Microwave Mediterranean Symposium (MMS)*, Guzelyurt, Turkey, August 2010, pp. 251–267.

[12] T. Thalmann, Z. Popovic, B. M. Notaros, and J. R. Mosig, "Investigation and Design of a Multi-Band Wearable Antenna," in *3rd European Conference on Antennas and Propagation, EuCAP 2009*, Berlin, Germany, 2009, pp. 462–465.

[13] P. Salonen, Y. Rahmat-Samii, and M. Kivikoski, "Wearable Antennas in the Vicinity of Human Body," in *IEEE Antennas and Propagation Society Symposium*, Monterey, CA, 2004, Vol.1, pp. 467–470.

[14] T. Kellomaki, J. Heikkinen, and M. Kivikoski, "Wearable Antennas for FM Reception," in *First European Conference on Antennas and Propagation, EuCAP 2006*, the Hague, the Netherlands, 2006, pp. 1–6.

[15] A. Sabban, "Wideband Printed Antennas for Medical Applications," in *APMC 2009 Conference*, Singapore, Decmeber 2009.

[16] Y. Lee, "Antenna Circuit Design for RFID Applications," Microchip Technology Inc., Microchip AN 710c, Arizona USA.

[17] A. Sabban, "A New Wideband Stacked Microstrip Antenna," in *IEEE Antenna and Propagation Symposium*, Houston, TX, June 1983.

[18] A. Sabban, "Wideband Microstrip Antenna Arrays," in *IEEE Antenna and Propagation Symposium MELCOM*, Tel-Aviv, June 1981.

[19] J. R. James, P. S. Hall, and C. Wood, *Microstrip Antenna Theory and Design*, P. Peregrinus, cop., London, UK, 1981.

[20] A. Sabban and K. C. Gupta, "Characterization of Radiation Loss from Microstrip Discontinuities Using a Multiport Network Modeling Approach," *IEEE Transactions on Microwave Theory and Techniques*, 39(4), 705–712, April 1991.

[21] A. Sabban, Multiport Network Model for Evaluating Radiation Loss and Coupling among Discontinuities in Microstrip Circuits, Ph.D. thesis, University of Colorado at Boulder, January 1991.

[22] Software, Keysightt, http://www.keysight.com/en/pc-1297113/advanced-design-system-ads?cc=IL&lc=eng.

Index

519

9780367622169